SOLIDWORKS 2023 Tutorial

A Step-by-Step Project Based Approach
Utilizing 3D Solid Modeling

David C. Planchard
CSWP & SOLIDWORKS Accredited Educator

PUBLICATIONS

SDC Publications
P.O. Box 1334
Mission, KS 66222
913-262-2664
www.SDCpublications.com

Examination Copies
Books received as examination copies are for review purposes only and may not be made available for student use. Resale of examination copies is prohibited.

Electronic Files
Any electronic files associated with this book are licensed to the original user only. These files may not be transferred to any other party.

Trademarks
SOLIDWORKS®, eDrawings®, SOLIDWORKS Simulation®, SOLIDWORKS Flow Simulation, and SOLIDWORKS Sustainability are a registered trademark of Dassault Systèmes SOLIDWORKS Corporation in the United States and other countries; certain images of the models in this publication courtesy of Dassault Systèmes SOLIDWORKS Corporation.

The Y14 ASME Engineering Drawing and Related Documentation Publications utilized in this text are as follows: ASME Y14.1 1995, ASME Y14.2M-1992 (R1998), ASME Y14.3M-1994 (R1999), ASME Y14.41-2003, ASME Y14.5-1982, ASME Y14.5-1999, and ASME B4.2. Note: By permission of The American Society of Mechanical Engineers, Codes and Standards, New York, NY, USA. All rights reserved.

Download all needed model files from the SDC Publication website (www.SDCpublications.com/downloads/978-1-63057-556-4)

ISBN-13: 978-1-63057-556-4
ISBN-10: 1-63057-556-9

Printed and bound in the United States of America.

INTRODUCTION

The **SOLIDWORKS® 2023 Tutorial** is written to assist students, designers, engineers and professionals who are new to SOLIDWORKS. The text provides a step-by-step project-based learning approach featuring machined components with additional design projects. Over 100 models are provided as classroom examples using various SOLIDWORKS features, sketches and mate types.

It is a great starting point for those new to SOLIDWORKS or as a teaching aid in classroom training to become familiar with the software interface, basic commands and strategies as users complete a series of models while learning different ways to accomplish a particular task. The book is divided into four sections with 11 chapters.

Chapter 1 - Chapter 5: Explore the SOLIDWORKS User Interface and CommandManager, Document and System properties, simple and complex parts and assemblies, proper design intent, design tables, configurations, multi-sheet, multi-view drawings, BOMs, and Revision tables using basic and advanced features.

Follow the step-by-step instructions and develop multiple assemblies that combine over 100 extruded machined parts and components. Formulate the skills to create, modify and edit sketches and solid features.

Learn the techniques to reuse features, parts and assemblies through symmetry, patterns, copied components, apply proper design intent, design tables, and configurations.

Chapter 6: Create and learn about large assemblies using Standard, Advanced and Mechanical mates. All assemblies and components for the final ROBOT assembly are provided.

Chapter 7 - Chapter 10: Certified SOLIDWORKS Associate - Mechanical Design (CSWA) exam chapters provide detailed information and resources about the exam.

Information is provided on the login procedure at https://3dexperience.virtualtester.com/#home for students to register, examine their certifications, manage their account and take Segment 1 & 2 of the certification.

Take the free online sample CSWA exam. There is a 30-day waiting period before retaking the sample online CSWA exam.

Note: The sample online CSWA exam is provided in the book (.pdf) format along with the SOLIDWORKS models, solutions, and additional practice exam problems.

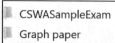

Each chapter addresses one of the various categories in the CSWA exam: Drafting Competencies, Basic Part Creation and Modification, Intermediate Part Creation and Modification, Advanced Part Creation and Modification, Assembly Creation and Modification.

Chapter 11: Additive Manufacturing - 3D printing fundamentals chapter provides a basic understanding between Additive vs. Subtractive manufacturing. Discuss Fused Filament Fabrication (FFF), STereoLithography (SLA), and Selective Laser Sintering (SLS) printer technology. Select suitable filament material. Comprehend 3D printer terminology. Knowledge of preparing, saving, and printing a model on a Fused Filament Fabrication 3D printer.

If your school is an academic certification provider, your instructor can allocate free exam credits for the CSWA (Segment 1 & 2), CSWA-SD, CSWA-S, and CSWA-AM certifications.

About the Author

David Planchard is the founder of D&M Education LLC. Before starting D&M Education, he spent over 35 years in industry and academia holding various engineering, marketing, and teaching positions. He holds five U.S. patents. He has published and authored numerous papers on Machine Design, Product Design, Mechanics of Materials, and Solid Modeling. He is an active member of the SOLIDWORKS Users Group and the American Society of Engineering Education (ASEE). David holds a BSME, MSM with the following professional certifications: CCAI, CCNP, CSWA-SD, CSWA-S, CSWA-AM, CSWP, CSWP-DRWT and SOLIDWORKS Accredited Educator. David is a SOLIDWORKS Solution Partner, a faculty member and the SAE advisor at Worcester Polytechnic Institute in the Mechanical Engineering department.

In 2012, David's senior Major Qualifying Project team (senior capstone) won first place in the Mechanical Engineering department at WPI.

In 2014, 2015 and 2016, David's senior Major Qualifying Project teams won the Provost award in the Mechanical Engineering department for design excellence.

In 2018, David's senior Major Qualifying Project team (Co-advisor) won the Provost award in the Electrical and Computer Engineering department. Subject area: Electrical System Implementation of Formula SAE Racing Platform.

In 2020, he was awarded Emeritus status at Worcester Polytechnic Institute (WPI). His ME design class achieved world recognition, featured in *Compass* published by Dassault Systèmes, in the technical article, "Open Innovation in a Pandemic."

In 2022, David's senior Major Qualifying Project team won second place in the Provost award in the Electrical and Computer Engineering department (ECE) at WPI. His FSAE Electric team won the IEEE Excellence in Electric Vehicle Engineering Award and achieved 4th place at the Formula Hybrid + Electric competition at the New Hampshire Motor Speedway in Loudon, NH. The competition is part of the Society of Automotive Engineers (SAE) Collegiate Design Series and is regarded as the most complex and dynamic of the series.

David Planchard is the author of the following books:

- **SOLIDWORKS® 2021 Reference Guide** 2020, 2019, 2018, and 2016

- **Engineering Design with SOLIDWORKS® 2023**, 2022, 2021, 2020, 2019, 2018, 2017, and 2016

- **Engineering Graphics with SOLIDWORKS® 2023,** 2022, 2021, 2020, 2019, 2018, 2017, and 2016

- **SOLIDWORKS® 2023 Quick Start**, 2022, 2021, 2020, 2019, and 2018

- **SOLIDWORKS® 2023 Tutorial**, 2021, 2020, 2019, 2018, 2017, and 2016

- **Drawing and Detailing with SOLIDWORKS® 2022**, 2014, and 2012

- **Official Certified SOLIDWORKS® Professional (CSWP) Certification Guide 2020 - 2023**, 2019 - 2020, and 2015 - 2017

- **Official Guide to Certified SOLIDWORKS® Associate Exams: CSWA, CSWA-SD, CSWA-S, CSWA-AM 2020 - 2023**, 2019 - 2021, 2017 - 2019, and 2015 - 2017

Acknowledgements

Writing this book was a substantial effort that would not have been possible without the help and support of my loving family and of my professional colleagues. I would like to thank Professor John M. Sullivan Jr., Professor Jack Hall, and the community of scholars at Worcester Polytechnic Institute who have enhanced my life, my knowledge and helped to shape the approach and content to this text.

The author is greatly indebted to my colleagues from Dassault Systèmes SOLIDWORKS Corporation for their help and continuous support: Mike Puckett, Avelino Rochino, Yannick Chaigneau, Terry McCabe and the SOLIDWORKS Partner team.

Thanks also to Professor Richard L. Roberts of Wentworth Institute of Technology, Professor Dennis Hance of Wright State University, Professor Jason Durfess of Eastern Washington University and Professor Aaron Schellenberg of Brigham Young University - Idaho who provided vision and invaluable suggestions.

Contact the Author

We realize that keeping software application books current is imperative to our customers. We value the hundreds of professors, students, designers, and engineers that have provided us input to enhance the book. Please contact me directly with any comments, questions or suggestions on this book or any of our other SOLIDWORKS books at dplanchard@msn.com.

Note to Instructors

Download all needed model files (initial and final) from the SDC Publications website www.SDCpublications.com/downloads/978-1-63057-556-4. Instructors are provided PowerPoint lectures, labs, and student and instructor chapter quizzes. Initial and final SOLIDWORKS models are included for each chapter and homework section. Over 100 models are provided as classroom examples using various SOLIDWORKS features, sketches and mate types. Additional design projects are included.

The sample online CSWA exam is provided in the book (.pdf) format along with the SOLIDWORKS models, solutions, and additional practice exam problems. An instructor's folder is provided with log in and allocating student exam credit information.

Trademarks, Disclaimer and Copyrighted Material

SOLIDWORKS®, eDrawings®, SOLIDWORKS Simulation®, SOLIDWORKS Flow Simulation, and SOLIDWORKS Sustainability are a registered trademark of Dassault Systèmes SOLIDWORKS Corporation in the United States and other countries; certain images of the models in this publication courtesy of Dassault Systèmes SOLIDWORKS Corporation.

The publisher and the author make no representations or warranties with respect to the accuracy or completeness of the contents of this work and specifically disclaim all warranties, including without limitation warranties of fitness for a particular purpose. No warranty may be created or extended by sales or promotional materials. Dimensions of parts are modified for illustration purposes. Every effort is made to provide an accurate text. The authors and the manufacturers shall not be held liable for any parts, components, assemblies or drawings developed or designed with this book or any responsibility for inaccuracies that appear in the book. Web and company information was valid at the time of this printing.

References

- SOLIDWORKS Help Topics and What's New, SOLIDWORKS Corporation, 2023.

- 80/20 Product Manual, 80/20, Inc., Columbia City, IN, 2012.

- Ticona Designing with Plastics - The Fundamentals, Summit, NJ, 2009.

- Emerson-EPT Bearing Product Manuals and Gear Product Manuals, Emerson Power Transmission Corporation, Ithaca, NY, 2009.

- Emhart - A Black and Decker Company, On-line catalog, Hartford, CT, 2012.

During the initial SOLIDWORKS installation, you are requested to select either the ISO or ANSI drafting standard. ISO is typically a European drafting standard and uses First Angle Projection. The book is written using the ANSI (US) overall drafting standard and Third Angle Projection for drawings.

TABLE OF CONTENTS

Introduction   **I-1**
About the Author   I-3
Acknowledgements   I-4
Contact the Author   I-4
Note to Instructors   I-4
Trademarks, Disclaimer, and Copyrighted Material   I-5
References   I-5
Table of Contents   I-6
What is SOLIDWORKS?   I-15
Overview of Chapters   I-17
About the Book   I-24
Windows Terminology in SOLIDWORKS   I-25

Chapter 1 - Overview of SOLIDWORKS and the User Interface   **1-1**
Chapter Objective   1-3
What is SOLIDWORKS?   1-3
Basic concepts in SOLIDWORKS   1-3
Start a SOLIDWORKS Session   1-4
  Tutorial: Start a SOLIDWORKS Session   1-4
  Welcome dialog box   1-4
  Home Tab   1-5
  Recent Tab   1-5
  Learn Tab   1-5
  Alerts Tab   1-6
SOLIDWORKS User Interface (UI) and CommandManager   1-7
  Menu Bar toolbar   1-7
  Menu Bar menu (No model open)   1-8
Drop-down menu (Model open)   1-8
Drop-down menu (Open part document)   1-8
Create a New Part Document   1-9
  Novice Mode   1-10
  Advanced Mode   1-10
Graphic Window (Default)   1-11
  View Default Sketch Planes   1-12
Download the SOLIDWORKS folder   1-12
Open the Bracket Part   1-12
  Part FeatureManager   1-13
  FeatureManager Rollback Bar   1-13
Heads-up View toolbar   1-15
  Dynamic Annotation Views   1-15
  Zoom to Fit   1-15
  Zoom to Area   1-15
  Window-Select   1-15

Rotate   1-15
Front View   1-16
Right View   1-16
Top View   1-16
Trimetric view   1-16
SOLIDWORKS Help   1-16
SOLIDWORKS Tutorials   1-17
User Interface Tools   1-17
Right-click   1-18
Consolidated toolbar   1-18
System feedback icons   1-18
Confirmation Corner   1-19
Heads-up View toolbar   1-19
CommandManager (Default Part tab)   1-22
CommandManager (Default Drawing tab)   1-23
CommandManager (Default Assembly tab)   1-24
CommandManager (Float/Fit)   1-25
Collapse the CommandManager   1-25
FeatureManager Design Tree   1-26
FeatureManager design tree tab   1-26
PropertyManager tab   1-26
Configuration Manager tab   1-26
DimXpertManager tab   1-26
DisplayManager tab   1-26
CAM tab   1-26
Hide/Show tab   1-26
Sensors tool   1-26
Tags   1-27
Split   1-27
Fly-out FeatureManager   1-28
Task Pane   1-29
3DEXPERIENCE MARKETPLACE   1-29
SOLIDWORKS Resources   1-30
Design Library   1-30
File Explorer   1-31
View Palette   1-31
Appearances, Scenes and Decals   1-31
Custom Properties   1-32
Motion Study tab   1-32
3D Views tab   1-34
Dynamic Reference Visualization   1-34
Mouse Movements   1-35
Single-Click   1-35
Double-Click   1-35
Right-Click   1-35
Scroll Wheel   1-35

Summary   1-36

Chapter 2 - Parts and Assembly Creation   **2-1**
Chapter Objective   2-3
Chapter Overview   2-4
Start a SOLIDWORKS Session   2-6
AXLE Part   2-9
  AXLE Part-Extruded Boss/Base Feature   2-10
  AXLE Part-Save   2-13
  AXLE Part-Edit Appearance   2-14
  AXLE Part-View Modes   2-15
SHAFT-COLLAR Part   2-18
  SHAFT-COLLAR Part-Extruded Boss/Base Feature   2-19
  SHAFT-COLLAR Part-Extruded Cut Feature   2-21
  SHAFT-COLLAR-Modify Dimensions and Edit Color   2-23
FLATBAR Part   2-24
  FLATBAR Part-Extruded Boss/Base Feature   2-25
  FLATBAR Part-Extruded Cut Feature   2-28
  FLATBAR Part-Linear Pattern Feature   2-30
LINKAGE Assembly   2-31
Mate Types   2-32
  Standard Mates   2-32
  Advanced Mates   2-33
  Mechanical Mates   2-34
AirCylinder Assembly-Open and Save As option   2-35
LINKAGE Assembly-Insert FLATBAR Part   2-39
LINKAGE Assembly-Insert SHAFT-COLLAR Part   2-42
Motion Study - Basic Motion tool   2-45
LINKAGE Assembly-Basic Motion   2-45
Summary   2-48
Questions   2-49
Exercises   2-50

Chapter 3 - Front Support Assembly   **3-1**
Chapter Objective   3-3
Chapter Overview   3-4
Reference Planes and Orthographic Projection   3-5
HEX-STANDOFF Part   3-9
  HEX-STANDOFF Part-Extruded Boss/Base Feature   3-10
  HEX-STANDOFF Part-HOLE Wizard Feature   3-14
ANGLE-13HOLE Part   3-15
  ANGLE-13HOLE Part-Documents Properties   3-17
  ANGLE-13HOLE Part-Extruded Thin Feature   3-18
  ANGLE-13HOLE Part-Extruded Cut Feature   3-20
  ANGLE-13HOLE Part-Linear Pattern Feature   3-22
  ANGLE-13HOLE Part-Fillet Feature   3-23
  ANGLE-13HOLE Part-Second Extruded Cut and Linear Pattern   3-24

ANGLE-13HOLE Part-Third Extruded Cut Feature   3-26
TRIANGLE Part   3-31
  TRIANGLE Part-Mirror, Offset and Fillet Sketch Tools   3-33
  TRIANGLE Part-Extruded Boss/Base Feature   3-36
  TRIANGLE Part-First Extruded Cut Feature   3-37
  TRIANGLE Part-Second Extruded Cut Feature   3-39
  TRIANGLE Part-Mirror Feature   3-41
  TRIANGLE Part-Third Extruded Cut Feature   3-42
  TRIANGLE Part-Circular Pattern Feature   3-44
SCREW Part   3-45
  SCREW Part-Documents Properties   3-47
  SCREW Part-Revolved Feature   3-47
  SCREW Part-Extruded Cut Feature   3-51
  SCREW Part-Circular Pattern Feature   3-53
  SCREW Part-Fillet Feature   3-53
  SCREW Part-Chamfer Feature   3-54
FRONT-SUPPORT Assembly   3-56
  FRONT-SUPPORT Assembly-Insert ANGLE-13HOLE   3-56
  FRONT-SUPPORT Assembly-Insert HEX-STANDOFF   3-58
  FRONT-SUPPORT Assembly-Insert TRIANGLE   3-61
  FRONT-SUPPORT Assembly-Insert SCREW   3-64
Chapter Summary   3-66
Questions   3-68
Exercises   3-69

Chapter 4 - Fundamentals of Drawing   **4-1**
Chapter Objective   4-3
Chapter Overview   4-4
Drawing Template and Sheet Format   4-5
Create a new Drawing   4-7
Drawing-Document Properties   4-9
Title Block   4-10
Create a Title Block   4-11
Company Logo   4-15
Insert a Company Logo   4-15
Save Sheet Format and Save As Drawing Template   4-17
FLATBAR Drawing   4-21
  FLATBAR Drawing-Open the FLATBAR Part   4-21
  Move views and Properties of the Sheet   4-25
  FLATBAR Drawing-Position views   4-27
  Detail Drawing   4-28
  FLATBAR Drawing-Dimensions and Annotations   4-30
  FLATBAR Drawing-Part Number and Document Properties   4-36
  FLATBAR Drawing-Linked Note   4-38
LINKAGE Assembly Drawing-Sheet1   4-41
  LINKAGE Assembly Drawing-Exploded view   4-45
  LINKAGE Assembly Drawing-Animation   4-47

LINKAGE Assembly Drawing-Bill of Materials   4-48
LINKAGE Assembly Drawing-Automatic Balloons   4-50
LINKAGE Assembly Drawing-Sheet2   4-51
LINKAGE Assembly Drawing-Sheet2 Section view   4-53
LINKAGE Assembly Drawing-Sheet2 Detail view   4-53
FLATBAR Part-Design Table   4-55
FLATBAR Drawing-Sheet2   4-59
FLATBAR-SHAFTCOLLAR Assembly   4-61
Insert a Center of Mass Point   4-66
Chapter Summary   4-68
Questions   4-68
Exercises   4-70

Chapter 5 - Advanced Features   **5-1**
Chapter Objective   5-3
Chapter Overview   5-4
WEIGHT Part   5-6
WEIGHT Part-Lofted Feature   5-12
WEIGHT Part-Instant 3D Extruded Cut Feature   5-13
HOOK Part   5-14
HOOK Part-Swept Base Feature   5-20
HOOK Part-Swept Profile   5-20
HOOK Part-Dome Feature   5-20
HOOK Part-Thread Feature   5-21
WHEEL Part   5-24
WHEEL Part-Extruded Boss/Base Feature   5-27
WHEEL Part-Revolved Cut Feature   5-28
WHEEL Part-First Extruded Cut Feature   5-31
WHEEL Part-Second Extruded Cut Feature   5-33
WHEEL Part-Circular Pattern Feature   5-36
Modify a Part   5-39
HEX-ADAPTER Part   5-39
HEX-ADAPTER Part-Extruded Boss/Base Feature   5-42
HEX-ADAPTER Part-Extruded Cut Feature   5-42
AXLE-3000 Part   5-45
SHAFTCOLLAR-500 Part   5-46
Chapter Summary   5-49
Questions   5-50
Exercises   5-51

Chapter 6 - PNEUMATIC-TEST-MODULE and ROBOT Assembly   **6-1**
Chapter Objective   6-3
Chapter Overview   6-4
Assembly Techniques   6-6
PNEUMATIC-TEST-MODULE Layout   6-7
FLATBAR Sub-assembly   6-9
3HOLE-SHAFTCOLLAR Assembly   6-9

WHEEL-FLATBAR Assembly	6-14
WHEEL-FLATBAR Assembly-Insert 3HOLE-SHAFT-COLLAR	6-17
WHEEL-FLATBAR Assembly-Insert 5HOLE-SHAFT-COLLAR	6-19
WHEEL-AND-AXLE Assembly	6-23
WHEEL-AND-AXLE Assembly-Insert HEX-ADAPTER	6-26
WHEEL-AND-AXLE Assembly-Insert SHAFTCOLLAR-500	6-28
PNEUMATIC-TEST-MODULE Assembly	6-30
Modify the LINKAGE Assembly	6-31
PNEUMATIC-TEST-MODULE-Insert LINKAGE Assembly	6-40
PNEUMATIC-TEST-MODULE-Insert AIR-RESERVOIR-SUPPORT	6-42
PNEUMATIC-TEST-MODULE-Component Pattern	6-45
PNEUMATIC-TEST-MODULE-Linear Component Pattern	6-46
PNEUMATIC-TEST-MODULE-Insert FRONT-SUPPORT	6-48
PNEUMATIC-TEST-MODULE-Mirrored Component	6-51
PNEUMATIC-TEST-MODULE-MIRRORFRONT-SUPPORT	6-53
Component Properties	6-54
PNEUMATIC-TEST-MODULE-Insert WHEEL-AND-AXLE	6-54
PNEUMATIC-TEST-MODULE-Remove Rigid State	6-56
PNEUMATIC-TEST-MODULE-Review AirCylinder Configurations	6-57
Final ROBOT Assembly	6-62
Create the Robot Assembly	6-63
Insert the PNEUMATIC-TEST-MODULE Assembly	6-63
Insert the basic_integration Assembly	6-65
Chapter Summary	6-66
Questions	6-67
Exercises	6-69
Chapter 7 - CSWA Introduction and Drafting Competencies	**7-1**
Introduction	7-1
Taking the Exam (Segment 1 & 2)	7-5
Objective	7-6
Procedure to Create a Named Drawing view	7-7
Tutorial: Drawing Named Procedure 7-1	7-8
Tutorial: Drawing Named Procedure 7-2	7-8
Tutorial: Drawing Named Procedure 7-3	7-8
Tutorial: Drawing Named Procedure 7-4	7-9
Tutorial: Drawing Named Procedure 7-5	7-9
Tutorial: Drawing Named Procedure 7-6	7-10
Tutorial: Drawing Named Procedure 7-7	7-10
Tutorial: Drawing Named Procedure 7-8	7-10
Intended Audience	7-11
Summary	7-12
Questions	7-13

Chapter 8 - CSWA Basic and Intermediate Part Creation and Modification   **8-1**

Objectives   8-1
Introduction   8-1
Read and Understand an Engineering Document   8-2
Build a Basic Part from a Detailed Illustration   8-5
  Tutorial: Volume/Center of Mass 8-1   8-5
  Tutorial: Volume/Center of Mass 8-2   8-6
  Tutorial: Mass-Volume 8-3   8-9
  Tutorial: Mass-Volume 8-4   8-10
  Tutorial: Mass-Volume 8-5   8-12
Build Additional Basic Parts   8-16
  Tutorial: Mass-Volume 8-6   8-16
  Tutorial: Mass-Volume 8-7   8-18
  Tutorial: Basic/Intermediate-Part 8-1   8-20
  Tutorial: Basic/Intermediate-Part 8-2   8-23
Summary   8-26
Questions   8-27

Chapter 9 - CSWA Advanced Part Creation and Modification   **9-1**

Objectives   9-1
Introduction   9-1
Build an Advanced Part from a Detailed Illustration   9-2
  Tutorial: Advanced Part 9-1   9-2
  Tutorial: Advanced Part 9-2   9-7
Calculate Center of Mass Relative to a Created Coordinate System Location   9-10
  Tutorial: Coordinate Location 9-1   9-10
  Tutorial: Coordinate Location 9-2   9-12
  Tutorial: Advanced Part 9-3   9-13
  Tutorial: Advanced Part 9-3A   9-17
  Tutorial: Advanced Part 9-3B   9-18
  Tutorial: Advanced Part 9-4   9-20
  Tutorial: Advanced Part 9-4A   9-26
Summary   9-27
Questions   9-28

Chapter 10 - CSWA - Assembly Creation and Modification   **10-1**

Objectives   10-1
Assembly Modeling   10-2
Build an Assembly from a Detailed Dimensioned Illustration   10-3
  Tutorial: Assembly Modeling 10-1   10-5
  Tutorial: Assembly Modeling 10-2   10-11
  Tutorial: Assembly Modeling 10-3   10-16
Summary   10-21
Questions   10-22

Chapter 11 - Additive Manufacturing - 3D Printing Fundamentals   **11-1**
Chapter Objective   11-3
Additive vs. Subtractive Manufacturing   11-4
3D Printer Technology   11-5
   Stages of 3D Printing   11-5
  Fused Filament Fabrication (FFF)   11-6
  StereoLithography (SLA)   11-9
  Selective Laser Sintering (SLS)   11-11
Select the Correct Filament Material for FFF   11-12
  PLA (Polylactic Acid)   11-13
  Flex/Soft PLA   11-13
  PLA Storage   11-14
  PLA Part Accuracy   11-14
  ABS (Acrylonitrile-Butadiene-Styrene)   11-14
  ABS Storage   11-15
  ABS Part Accuracy   11-15
Nylon   11-16
  Nylon 618   11-16
  Nylon 645   11-16
  Nylon Storage   11-17
  Nylon Accuracy   11-17
PVA (Polyvinyl Alcohol)   11-17
STereoLithography (*.stl) file   11-18
  Save an STL (*.stl) file   11-18
Additive Manufacturing (*.amf) file   11-19
  Save an Additive Manufacturing (*.amf) file   11-19
3D Manufacturing Format (*.3mf) file   11-20
  Save a 3D Manufacturing Format (*.3mf) file   11-20
What is a Slicer?   11-21
How does a Slicer Work?   11-21
Slicer Parameters   11-21
  Layer Height   11-21
  Shell (Wall) Thickness   11-22
  Infill Density/Overlap   11-22
  Infill Patterns   11-22
  Print Speed   11-23
  Support Types   11-23
   Touching Buildplate   11-23
   Everywhere   11-24
  Bed Platform Adhesion   11-24
   Raft   11-24
   Skirt   11-24
   Brim   11-24
Part Orientation   11-25
  Example 1   11-25
  Example 2   11-26

Optimize Print Direction   11-26
  Thin Region   11-26
  Area of Overhang   11-26
  Amount of needed Support   11-26
Remove Model from the Build Plate   11-28
  Non-heated Build Plate   11-28
  Heated Build Plate   11-28
Know the Printer's Limitations   11-29
Tolerance for Interlocking Parts   11-29
General Printing Tips   11-29
  Reduce Infill/Overlap   11-29
  Control Build Area Temperature   11-30
  Add Pads   11-31
  Safe Zone Rule   11-31
  First Layer Not Sticking   11-31
  Level Build Platform   11-32
  Minimize Internal Support   11-32
  Design a Water Tight Mesh   11-32
  Clearance   11-32
In General   11-33
Print Directly from SOLIDWORKS   11-34
  Add-in   11-34
SOLIDWORKS Additive Manufacturing Certification (CSWA-AM) Exam   11-35
Summary   11-36

Appendix   **A-1**
SOLIDWORKS Keyboard Shortcuts   A-1
Modeling - Best Practices   A-3
Helpful On-Line Information   A-5
SOLIDWORKS Document Types   A-6
CSWA Homework Answers   A-7

Glossary   **G-1**
Index   **I-1**

What is SOLIDWORKS?

SOLIDWORKS® is a mechanical design automation software package used to build parts, assemblies and drawings that takes advantage of the familiar Microsoft® Windows graphical user interface.

SOLIDWORKS is an easy to learn design and analysis tool (SOLIDWORKS Simulations, SOLIDWORKS Motion, SOLIDWORKS Flow Simulation etc.), which makes it possible for designers to quickly sketch 2D and 3D concepts, create 3D parts and assemblies and detail 2D drawings.

In SOLIDWORKS, you create 2D and 3D sketches, 3D parts, 3D assemblies and 2D drawings. The part, assembly and drawing documents are related. Additional information on SOLIDWORKS and its family of products can be obtained at their URL, www.SOLIDWORKS.com.

Features are the building blocks of parts. Use feature tools such as Extruded Boss/Base, Extruded Cut, Fillet, etc. from the Features tab in the CommandManager to create 3D parts.

Extruded features begin with a 2D sketch created on a Sketch plane.

The 2D sketch is a profile or cross section. Use sketch tools such as Line, Center Rectangle, Slot, Circle Centerline, Mirror, etc. from the Sketch tab in the CommandManager to create a 2D sketch. Sketch the general shape of the profile. Add geometric relationships and dimensions to control the exact size of the geometry and your Design Intent. Design for change.

Create features by selecting edges or faces of existing features, such as a Fillet. The Fillet feature rounds sharp corners.

Dimensions drive features. Change a dimension, and you change the size of the part.

Use Geometric relationships: Vertical, Horizontal, Parallel, etc. and various End Conditions to maintain the Design Intent.

Create a hole that penetrates through a part (Through All). SOLIDWORKS maintains relationships through the change.

The step-by-step approach used in this text allows you to create, edit and modify parts, assemblies and drawings.

Change is an integral part of design.

Overview of Chapters

Chapter 1: Overview of SOLIDWORKS and the User Interface.

SOLIDWORKS is a design software application used to create 2D and 3D sketches, 3D parts and assemblies and 2D drawings.

Chapter 1 introduces the user to the SOLIDWORKS Welcome dialog box, User Interface (UI) and the CommandManager: Menu bar toolbar, Menu bar menu, Drop-down menus, Context toolbars, Consolidated drop-down toolbars, System feedback icons, Confirmation Corner, Heads-up View toolbar, Document Properties and more.

Start a new SOLIDWORKS session. Create a new part. Open an existing part and view the created features and sketches using the Rollback bar. Design the part using proper design intent.

Chapter 2: Parts and Assembly Creation.

Create three parts: AXLE, SHAFT-COLLAR and FLATBAR

Understand and apply the following Sketch tools: Circle, Centerpoint Straight Slot, and Smart dimension.

Insert geometric relations and dimensions: Equal, Vertical, Horizontal, Parallel, Perpendicular, Coincident and MidPoint.

Utilize the following features: Extruded Boss/Base, Extruded Cut and Linear Pattern.

Create the LINKAGE assembly. The LINKAGE assembly utilizes the provided AirCylinder assembly. Insert the following Standard mates: Coincident, Concentric and Parallel.

Chapter 3: **Front Support Assembly**.

Chapter 3 introduces various Sketch planes to create parts. The Front, Top and Right Planes each contain the Extruded Boss/Base feature for the TRIANGLE, HEX-STANDOFF and ANGLE-13HOLE parts.

Utilize Geometric relationships in your sketches.

Create the SCREW part using the following features: Revolved Base, Extruded Cut, Circular Pattern, Fillet and Chamfer.

Create the FRONT-SUPPORT assembly.

Apply the following mate types: Concentric, Coincident, Parallel, Distance, Gear and Cam.

Chapter 4: **Fundamentals of Drawing**.

Chapter 4 covers the development of a customized Sheet format and Drawing templates.

Review the differences between the Edit Sheet and the Edit Sheet Format modes.

Develop a company logo from a bitmap or picture file.

Create a FLATBAR drawing. Insert dimensions created from the part features.

Create a LINKAGE assembly drawing with multiple views. Develop and incorporate a Bill of Materials into the drawing Custom Properties in the parts and assemblies.

Add information to the Bill of Materials in the assembly drawing.

Insert a Design Table to create multiple configurations of parts and assemblies.

Chapter 5: **Advanced Features**.

Chapter 5 focuses on creating six parts for the
PNEUMATIC-TEST-MODULE Assembly: WHEEL,
HOOK, WHEEL, HEX-ADAPTER, AXLE-3000 and
SHAFTCOLLAR-500.

Obtain the ability to reuse geometry by
modifying existing parts and to create new
parts.

Knowledge of the following
SOLIDWORKS features: Plane, Lofted
Base, Extruded Cut, Swept Base, Dome,
Thread, Extruded Boss/Base, Revolved
Cut, Extruded Cut, Circular Pattern, Axis,
Instant3D, Hole Wizard, Advanced Hole,
Split Line and Thread Wizard.

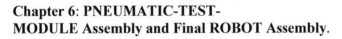

Chapter 6: **PNEUMATIC-TEST-MODULE Assembly and Final ROBOT Assembly**.

Chapter 6 focuses on the PNEUMATIC-TEST-MODULE
Assembly and the final ROBOT Assembly.

Create the WHEEL-AND-AXLE assembly. First, create the
3HOLE-SHAFTCOLLAR assembly and the 5HOLE-
SHAFTCOLLAR assembly.

Insert the WHEEL part, AXLE 3000 part, HEX-
ADAPTER part and SHAFTCOLLAR-500 part.

Insert the FLAT-PLATE part that was created in the
Chapter 3 exercises. Insert the LINKAGE assembly
and add components: HEX-STANDOFF, AXLE and
SHAFT-COLLAR.

Insert the AIR-RESERVOIR-SUPPORT
assembly. Insert the SCREW part. Utilize the
Pattern Driven Component Pattern tool and the
Linear Component Pattern tool.

Insert the FRONT-SUPPORT assembly and
apply the Mirror Components tool to complete
the Pneumatic Test Module Assembly.

Create the final ROBOT Assembly
as illustrated with the Robot-
platform sub-assembly,
PNEUMATIC-TEST-MODULE
sub-assembly, basic_integration
sub-assembly and the HEX-
ADAPTER component. Add
additional components in the
chapter exercises.

Learn the process to work with
multiple documents between parts
and assemblies and to apply the
following Assembly tools: Insert Component, Standard Mates: Concentric, Coincident,
and Parallel, Linear Component Pattern, Pattern Driven Component Pattern, Circular
Component Pattern, Mirror Components and Replace Components.

Download all needed model files from the SDC Publication website
(www.SDCpublications.com/downloads/978-1-63057-556-4).

Additional projects are included in the exercise section of chapter 6. Use the web to locate a similar assembly and to understand the parts needed in the assembly. View the provided sample parts. Create your own parts and final assembly. View the provided motion file to create the proper movement in SOLIDWORKS.

- Arbor Press Project
- Bench Vice Project
- Butterfly Valve Project
- Drill Guide Project
- Kant Twist Clamp Project
- Pipe Vice Project
- Pulley Project
- Quick Acting Clamp Project
- Radial Engine Project
- Shaper Tool Head Project
- Welder Arm Project

Chapter 7 - 10: **Certified SOLIDWORKS Associate - Mechanical Design (CSWA) exam chapters**.

The **Certified SOLIDWORKS Associate (CSWA)** Academic exam is provided either in a single 3-hour segment, or 2 - 90 minute segments.

The CSWA certification indicates a foundation in and apprentice knowledge of 3D CAD design and engineering practices and principles.

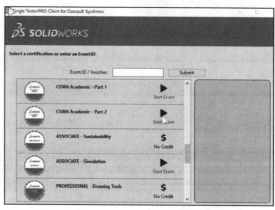

Part 1 of the CSWA Academic exam is 90 minutes, minimum passing score is 80, with 6 questions. There are two questions in the **Basic Part Creation and Modification** category, two questions in the **Intermediate Part Creation and Modification** category and two questions in the **Assembly Creation and Modification** category.

Part 2 of the CSWA Academic exam is 90 minutes, minimum passing score is 80 with 8 questions. There are three questions on the CSWA Academic exam in the **Drafting Competencies** category, three questions in the **Advanced Part Creation and Modification** category and two questions in the **Assembly Creation and Modification** category.

If your school is an academic certification provider, your instructor can allocate free exam credits for the CSWA (Segment 1 & 2), CSWA-SD, CSWA-S, and CSWA-AM certifications. The instructor will require your .edu email address.

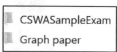

You need to pass both segments to obtain your CSWA Certification. Once you pass a segment, you will not have to take it again. You can take the segments in any order. If you fail a segment, there is a 14-day waiting period before retaking the failed segment.

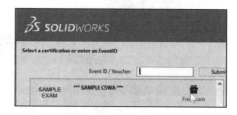

An instructor's folder is provided with log in and allocating student exam credit information.

Chapter 11: Additive Manufacturing - 3D Printing Fundamentals.

Provide a basic understanding between Additive vs. Subtractive manufacturing. Discuss Fused Filament Fabrication (FFF), STereoLithography (SLA), and Selective Laser Sintering (SLS) printer technology. Select suitable filament material. Comprehend 3D printer terminology. Knowledge of preparing, saving, and printing a model on a Fused Filament Fabrication 3D printer. Information on the Certified SOLIDWORKS Associate Additive Manufacturing (CSWA-AM) exam.

The Certified SOLIDWORKS Associate Additive Manufacturing (CSWA-AM) exam indicates a foundation in and apprentice knowledge of today's 3D printing technology and market.

On the completion of this chapter, you will be able to:

- Discuss Additive vs Subtractive manufacturing.

- Review 3D printer technology: Fused Filament Fabrication (FFF), STereoLithography (SLA), and Selective Laser Sintering (SLS).

- Select the correct filament material:
  - PLA (Polylactic acid)
  - FPLA (Flexible Polylactic acid)
  - ABS (Acrylonitrile butadiene styrene)
  - PVA (Polyvinyl alcohol)
  - Nylon 618
  - Nylon 645

- Create an STL (*.stl) file, Additive Manufacturing (*.amf) file and a 3D Manufacturing format (*.3mf) file.

- Prepare Geometric Code, "G-Code".

- Comprehend general 3D printer terminology.

- Understand optimum build orientation.

- Enter slicer parameters:
  - Raft, brim, skirt, layer height, percent infill, infill pattern, wall thickness, fan speed, print speed, bed temperature, and extruder (hot end) temperature.

- Address tolerance for interlocking parts.

- Knowledge of the Certified SOLIDWORKS Associate Additive Manufacturing exam.

About the Book

You will find a wealth of information in this book. The book is a project based step-by-step text written for new and intermediate users. The following conventions are used throughout this book:

- The term document refers to a SOLIDWORKS part, drawing or assembly file.

- The list of items across the top of the SOLIDWORKS interface is the Menu bar menu or the Menu bar toolbar. Each item in the Menu bar has a pull-down menu. When you need to select a series of commands from these menus, the following format is used: Click **Insert**, **Reference Geometry**, **Plane** from the Menu bar. The Plane PropertyManager is displayed.

- The book is organized into chapters. Each chapter is focused on a specific subject or feature.

- The ANSI overall drafting standard and Third Angle projection is used as the default setting in this text. IPS (inch, pound, second) and MMGS (millimeter, gram, second) unit systems are used.

- The following command syntax is used throughout the text. Commands that require you to perform an action are displayed in **Bold** text.

Format:	Convention:	Example:
Bold	- All commands actions. - Selected icon button. - Selected icon button. - Selected geometry: line, circle. - Value entries.	- Click **Options** from the Menu bar toolbar. - Click **Corner Rectangle** from the Sketch toolbar. - Click **Sketch** from the Context toolbar. - Select the **centerpoint**. - Enter **3.0** for Radius.
Capitalized	- Filenames. - First letter in a feature name.	- Save the **FLATBAR** assembly. - Click the **Fillet** feature.

Windows Terminology in SOLIDWORKS

The mouse buttons provide an integral role in executing SOLIDWORKS commands. The mouse buttons execute commands, select geometry, display Shortcut menus and provide information feedback.

A summary of mouse button terminology is displayed below:

Item:	Description:
Click	Press and release the left mouse button.
Double-click	Double press and release the left mouse button.
Click inside	Press the left mouse button. Wait a second, and then press the left mouse button inside the text box. Use this technique to modify Feature names in the FeatureManager design tree.
Drag	Point to an object, press and hold the left mouse button down. Move the mouse pointer to a new location. Release the left mouse button.
Right-click	Press and release the right mouse button. A Shortcut menu is displayed. Use the left mouse button to select a menu command.
Tool Tip	Position the mouse pointer over an Icon (button). The tool name is displayed below the mouse pointer.
Large Tool Tip	Position the mouse pointer over an Icon (button). The tool name and a description of its functionality are displayed below the mouse pointer.
Mouse pointer feedback	Position the mouse pointer over various areas of the sketch, part, assembly or drawing. The cursor provides feedback depending on the geometry.

Notes:

Chapter 1

Overview of SOLIDWORKS® 2023 and the User Interface

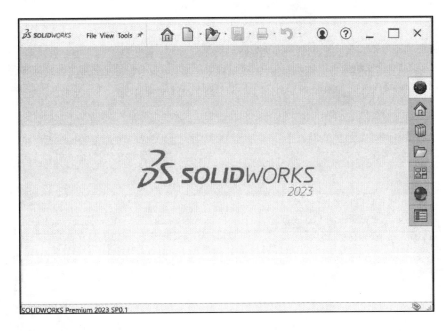

Below are the desired outcomes and usage competencies based on the completion of Chapter 1.

Desired Outcomes:	Usage Competencies:
• A comprehensive understanding of the SOLIDWORKS® User Interface (UI) and CommandManager.	• Ability to establish a SOLIDWORKS session. • Aptitude to utilize the following items: Menu bar toolbar, Menu bar menu, Drop-down menus, Context toolbars, Consolidated drop-down toolbars, System feedback icons, Task Pane, Confirmation Corner, Heads-up View toolbar, Document Properties and more. • Open a new and existing SOLIDWORKS part. • Knowledge to zoom, rotate and maneuver a three-button mouse in the SOLIDWORKS Graphics window.

Notes:

Chapter 1 - Overview of SOLIDWORKS® 2023 and the User Interface

Chapter Objective

Provide a comprehensive understanding of the SOLIDWORKS® default User Interface and CommandManager: Menu bar toolbar, Menu bar menu, Drop-down menu, Right-click Pop-up menus, Context toolbars/menus, Fly-out tool button, System feedback icons, Confirmation Corner, Heads-up View toolbar and more.

On the completion of this chapter, you will be able to:

- Utilize the Welcome - SOLIDWORKS dialog box.

- Establish a SOLIDWORKS session.

- Comprehend the SOLIDWORKS User Interface.

- Recognize the default Reference Planes in the FeatureManager.

- Open a new and existing SOLIDWORKS part.

- Utilize Help and SOLIDWORKS Tutorials.

- Zoom, rotate and maneuver a three-button mouse in the SOLIDWORKS Graphics window.

What is SOLIDWORKS®?

- SOLIDWORKS® is a mechanical design automation software package used to build parts, assemblies and drawings that takes advantage of the familiar Microsoft® Windows graphical user interface.

- SOLIDWORKS is an easy to learn design and analysis tool (SOLIDWORKS Simulation, SOLIDWORKS Motion, SOLIDWORKS Flow Simulation, Sustainability, etc.), which makes it possible for designers to quickly sketch 2D and 3D concepts, create 3D parts and assemblies and detail 2D drawings.

- Model dimensions in SOLIDWORKS are associative between parts, assemblies and drawings. Reference dimensions are one-way associative from the part to the drawing or from the part to the assembly.

- This book is written for the beginner to intermediate user.

Start a SOLIDWORKS Session

Start a SOLIDWORKS session and familiarize yourself with the SOLIDWORKS User Interface. As you read and perform the tasks in this chapter, you will obtain a sense of how to use the book and the structure. Actual input commands or required actions in the chapter are displayed in bold.

The book does not cover starting a SOLIDWORKS session in detail for the first time. A default SOLIDWORKS installation presents you with several options. For additional information, visit http://www.SOLIDWORKS.com.

Activity: Start a SOLIDWORKS Session.

Start a SOLIDWORKS session.

1)   Type **SOLIDWORKS 2023** in the Search window.

2)   Click the **SOLIDWORKS 2023** application (or if available, **double-click** the SOLIDWORKS icon on the desktop). The Welcome - SOLIDWORKS dialog box is displayed by default.

The Welcome - SOLIDWORKS box provides a convenient way to open recent documents (Parts, Assemblies and Drawings), view recent folders, access SOLIDWORKS resources, and stay updated on SOLIDWORKS news.

3)   **View** your options. Do not open a document at this time.

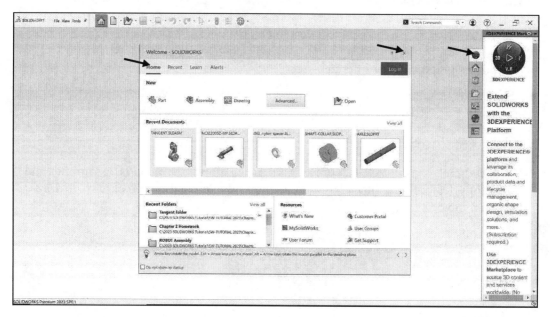

Home Tab

The Home tab lets you open new and existing documents, view recent documents and folders, and access SOLIDWORKS resources (*Part, Assembly, Drawing, Advanced mode, Open*).

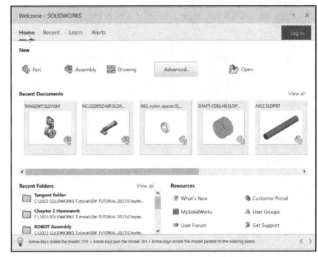

Recent Tab

The Recent tab lets you view a longer list of recent documents and folders. Sections in the Recent tab include *Documents* and *Folders*.

The Documents section includes thumbnails of documents that you have opened recently.

Click a thumbnail to open the document, or hover over a thumbnail to see the document location and access additional information about the document. When you hover over a thumbnail, the full path and last saved date of the document appears.

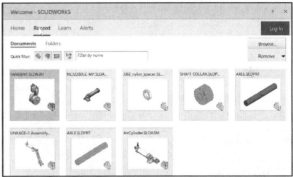

Learn Tab

The Learn tab lets you access instructional resources to help you learn more about the SOLIDWORKS software.

Sections in the Learn tab include:

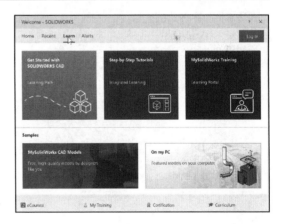

- **Get Started with SOLIDWORKS CAD**. Opens the Introducing SOLIDWORKS book.

- **Step-by-Step Tutorials**. Opens the step-by-step tutorials in the SOLIDWORKS software.

- **MySolidWorks CAD Models**. Opens the Training section at MySolidWorks.com.

- **On my PC**. Opens local folders containing sample models.

- **eCourses.** Opens the eCourses Training Catalog.

- **My Training**. Open the My Training section at MySolidWorks.com.

- **Certification**. Open the SOLIDWORKS Certification Program section at solidworks.com. You will need to create an account with a password.

- **Curriculum**. Open the Curriculum section at solidworks.com. You will need to create an account with a password.

When you install the software, if you do not install the Help Files or Example Files, the Tutorials and Samples links are unavailable.

Alerts Tab

The Alerts tab keeps you updated with SOLIDWORKS news.

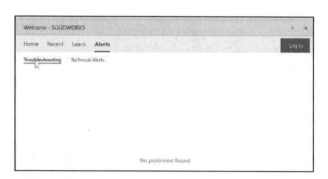

Sections in the Alerts tab include Critical, Troubleshooting, and Technical.

The Critical section does not appear if there are no critical alerts to display.

- **Troubleshooting**. Includes troubleshooting messages and recovered documents that used to be on the SOLIDWORKS Recovery tab in the Task Pane.

If the software has a technical problem and an associated troubleshooting message exists, the Welcome dialog box opens to the Troubleshooting section automatically on startup, even if you selected **Do not show at startup** in the dialog box.

- **Technical Alerts**. Open the contents of the SOLIDWORKS Support Bulletins RSS feed (Hotfixes, release news) at solidworks.com.

Close the Welcome - SOLIDWORKS dialog box.

4) Click **Close** ✕ from the Welcome - SOLIDWORKS dialog box. The SOLIDWORKS Graphics window is displayed. You can also click outside the Welcome - SOLIDWORKS dialog box, in the Graphics window.

View the SOLIDWORKS Graphics window.

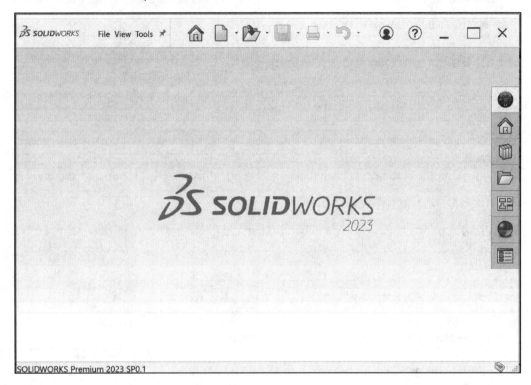

Menu Bar toolbar

The SOLIDWORKS (UI) is designed to make maximum use of the Graphics window. The Menu Bar toolbar contains a set of the most frequently used tool buttons from the Standard toolbar.

The following default tools are available:

- **Welcome to SOLIDWORKS** ⌂ - Open the Welcome dialog box, **New** ☐ - Create a new document; **Open** 📂 - Open an existing document; **Save** 💾 - Save an active document; **Print** 🖨 - Print an active document; **Undo** ↩ - Reverse the last action; **Redo** ↪ - Redoes the last action that you reverse; **Select** ▷ - Select Sketch entities, components and more; **Rebuild** - Rebuild the active part, assembly or drawing;

File Properties - Summary information on the active document;

Options - Change system options and Add-Ins for SOLIDWORKS; **Login** -

Login to SOLIDWORKS. You will need to create an account with a password; **Help**

- access to help, tutorials, updates and more.

Menu Bar menu (No model open)

SOLIDWORKS provides a context-sensitive menu
structure. The menu titles remain the same for all
three types of
documents, (Parts,
Assemblies and

Drawings) but the menu items change depending on which type of document is active.

Menu Bar menu (Model open)

The Pin option displays the Menu bar toolbar and the Menu bar menu as illustrated.
Throughout the book, the Menu bar menu and the Menu bar toolbar are referred to as the
Menu bar.

Drop-down menu (Open part document)

SOLIDWORKS takes advantage of the familiar Microsoft®
Windows user interface. Communicate with SOLIDWORKS
through drop-down menus, Context sensitive toolbars,
Consolidated toolbars or the CommandManager tabs.

To close a SOLIDWORKS drop-down menu, press the Esc key.
You can also click any other part of the SOLIDWORKS Graphics
window or click another drop-down menu.

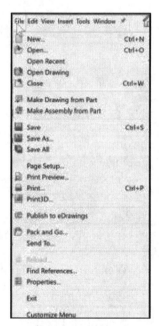

Create a New Part Document

Activity: Create a new Part Document.

A part is a 3D model, which consists of features. What are features? Features are geometry building blocks. Most features either add or remove material. Some features do not affect material (Cosmetic Thread).

Features are created either from 2D or 3D sketched profiles or from edges and faces of existing geometry.

Features are individual shapes that combined with other features make up a part or assembly. Some features, such as bosses and cuts, originate as sketches. Other features, such as shells and fillets, modify a feature's geometry.

Features are displayed in the FeatureManager.

The first sketch of a part is called the Base Sketch. The Base sketch is the foundation for the 3D model. The book focuses on 2D sketches and 3D features.

FeatureManager, CommandManager, and tree folders will vary depending on system setup, version, and Add-ins.

There are two modes in the New SOLIDWORKS Document dialog box: Novice and Advanced. The Novice option is the default option with three templates. The Advanced mode contains access to additional templates and tabs that you create in system options. Use the Advanced mode in this book.

Create a new part.

5) Click **New** 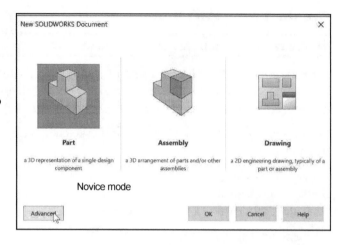 from the Menu bar. The New SOLIDWORKS Document dialog box is displayed.

Select the Advanced mode.

6) If needed, click the **Advanced** tab. The below New SOLIDWORKS Document box is displayed.

7) Click the **Templates** tab.

8) Click **Part**. Part is the default template from the New SOLIDWORKS Document dialog box.

9) Click **OK** from the New SOLIDWORKS Document dialog box.

The Advanced mode remains selected for all new documents in the current SOLIDWORKS session. When you exit SOLIDWORKS, the Advanced mode setting is saved.

The default SOLIDWORKS installation contains three tabs in the New SOLIDWORKS Document dialog box: *Templates, MBD, and Tutorial*. The *Templates* tab corresponds to the default SOLIDWORKS templates. The *MBD* tab corresponds to the templates utilized in the SOLIDWORKS (Model Based Definition). The *Tutorial* tab corresponds to the templates utilized in the SOLIDWORKS Tutorials.

Part1 is displayed in the FeatureManager and is the name of the document. Part1 is the default part window name.

The Part Origin ⊥ is displayed in blue in the center of the Graphics window. The Origin represents the intersection of the three default reference planes: *Front Plane*, *Top Plane* and *Right Plane*. The positive X-axis is horizontal and points to the right of the Origin in the Front view. The positive Y-axis is vertical and points upward in the Front view. The FeatureManager contains a list of features, reference geometry, and settings utilized in the part.

Edit the document units directly from the Graphics window as illustrated.

CommandManager tabs, FeatureManager tabs, Task Pane tabs will vary depending on system setup, version, and Add-ins.

PAGE 1 - 11

View the Default Sketch Planes.

10) Click the **Front Plane** from the FeatureManager.

11) Click the **Top Plane** from the FeatureManager.

12) Click the **Right Plane** from the FeatureManager.

13) Click the **Origin** from the FeatureManager. The Origin is the intersection of the Front, Top, and Right Planes. The Origin point is displayed.

14) Click **inside** the Graphics window.

Download all model files from the SDC Publication website (www.SDCpublications.com/downloads/978-1-63057-556-4). Open the Bracket part. Review the features and sketches in the Bracket FeatureManager. Work directly from a local hard drive.

Activity: Download the SOLIDWORKS folder. Open the Bracket Part.

Download the SW-TUTORIAL 2023 folder. Open an existing SOLIDWORKS part.

15) **Download** the SW-TUTORIAL 2023 folder.

16) **Unzip** the folder. **Work** from the unzip folder.

17) Click **Open** from the Menu bar menu.

18) Browse to the **SW-TUTORIAL 2023\Bracket** folder.

19) Double-click the **Bracket** part. The Bracket part is displayed in the Graphics window.

The FeatureManager design tree is located on the left side of the SOLIDWORKS Graphics window. The FeatureManager provides a summarized view of the active part, assembly, or drawing document. The tree displays the details on how the part, assembly or drawing document was created.

Use the FeatureManager rollback bar to temporarily roll back to an earlier state, to absorbed features, roll forward, roll to previous, or roll to the end of the FeatureManager design tree. You can add new features or edit existing features while the model is in the rolled-back state. You can save models with the rollback bar placed anywhere.

In the next section, review the features in the Bracket FeatureManager using the Rollback bar.

Activity: Use the FeatureManager Rollback Bar option.

Apply the FeatureManager Rollback Bar. Revert to an earlier state in the model.

20) Place the **mouse pointer** over the rollback bar in the FeatureManager design tree as illustrated. The pointer changes to a hand 🖐. Note the provided information on the feature. This is called Dynamic Reference Visualization.

21) Drag the **rollback bar** up the FeatureManager design tree until it is above the features you want rolled back, in this case 10.0 (10) Diameter Hole1.

22) **Release** the mouse button.

View the first feature in the Bracket Part.

23) Drag the **rollback bar** up the FeatureManager above Fillet1. View the results in the Graphics window. This is a great feature to re-engineer a part.

Return to the original Bracket Part FeatureManager.

24) Right-click **Extrude-Thin1** in the FeatureManager. The Pop-up Context toolbar is displayed.

25) Click **Roll to End**. View the results in the Graphics window.

Heads-up View toolbar

SOLIDWORKS provides the user with numerous view options. One of the most useful tools is the Heads-up View toolbar displayed in the Graphics window when a document is active.

Dynamic Annotation Views : Only available with SOLIDWORKS MBD (Model Based Definition). Provides the ability to control how annotations are displayed when you rotate models.

In the next section, apply the following tools: Zoom to Fit, Zoom to Area, Zoom out, Rotate and select various view orientations from the Heads-up View toolbar.

Activity: Utilize the Heads-up View toolbar.

Zoom to Fit the model in the Graphics window.

26) Click the **Zoom to Fit** icon. The tool fits the model to the Graphics window.

Zoom to Area on the model in the Graphics window.

27) Click the **Zoom to Area** icon. The Zoom to Area icon is displayed.

Zoom in on the top left hole.

28) **Window-select** the top left corner as illustrated. View the results.

De-select the Zoom to Area tool.

29) Click the **Zoom to Area** icon.

Fit the model to the Graphics window.

30) Press the **f** key.

Rotate the model.

31) Hold the **middle mouse button** down. Drag **upward** , **downward** , to the **left** and to the **right** to rotate the model in the Graphics window.

Display a few Standard Views.

32) Click **inside** the Graphics window.

33) Click **Front** from the drop-down Heads-up view toolbar. The model is displayed in the Front view.

34) Click **Right** from the drop-down Heads-up view toolbar. The model is displayed in the Right view.

35) Click **Top** from the drop-down Heads-up view toolbar. The model is displayed in the Top view.

Display a Trimetric view of the Bracket model.

36) Click **Trimetric** from the drop-down Heads-up view toolbar as illustrated. Note your options. View the results in the Graphics window.

SOLIDWORKS Help

Help in SOLIDWORKS is context-sensitive and in HTML format. Help is accessed in many ways, including Help buttons in all dialog boxes and

PropertyManager and Help ⓘ on the Standard toolbar for SOLIDWORKS Help.

37) Click ⑦ from the Standard toolbar.

38) Click ⑦ **Help** from the drop-down menu. The SOLIDWORKS Home Page is displayed by default. View your options.

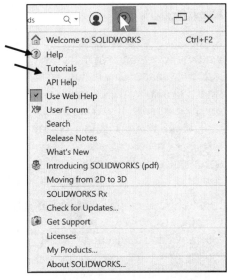

💡 SOLIDWORKS Web Help is active by default under Help in the Main menu.

Close Help. Return to the SOLIDWORKS Graphics window.

39) **Close** ❌ SOLIDWORKS Home.

SOLIDWORKS Tutorials

Display and explore the SOLIDWORKS tutorials.

40) Click ⑦ from the Standard toolbar.

41) Click **Tutorials**. The SOLIDWORKS Tutorials are displayed. The SOLIDWORKS Tutorials are presented by category.

42) Click the **Getting Started** category. The Getting Started category provides lessons on parts, assemblies, and drawings.

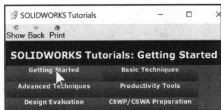

In the next section, close all models, tutorials and view the additional User Interface tools.

Activity: Close all Tutorials and Models.

Close SOLIDWORKS Tutorials and models.

43) **Close** ❌ SOLIDWORKS Tutorials.

44) Click **Window**, **Close All** from the Menu bar menu.

User Interface Tools

The book utilizes additional areas of the SOLIDWORKS User Interface. Explore an overview of these tools in the next section.

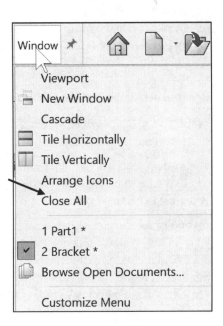

Right-click

Right-click in the Graphics window on a model, or in the FeatureManager on a feature or sketch to display the Context-sensitive toolbar. If you are in the middle of a command, this toolbar displays a list of options specifically related to that command.

Right-click an empty space in the Graphics window of a part or assembly, and a selection context toolbar above the shortcut menu is displayed. This provides easy access to the most commonly used selection tools.

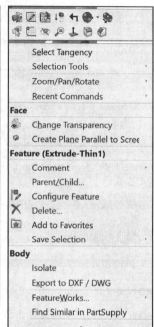

Consolidated toolbar

Similar commands are grouped together in the CommandManager. For example, variations of the Rectangle sketch tool are grouped in a single fly-out button as illustrated.

If you select the Consolidated toolbar button without expanding:

For some commands such as Sketch, the most commonly used command is performed. This command is the first listed and the command shown on the button.

For commands such as rectangle, where you may want to repeatedly create the same variant of the rectangle, the last used command is performed. This is the highlighted command when the Consolidated toolbar is expanded.

System feedback icon

SOLIDWORKS provides system feedback by attaching a symbol to the mouse pointer cursor.

The system feedback symbol indicates what you are selecting or what the system is expecting you to select.

As you move the mouse pointer across your model, system feedback is displayed in the form of a symbol, riding next to the cursor as illustrated. This is a valuable feature in SOLIDWORKS.

Confirmation Corner

When numerous SOLIDWORKS commands are active, a symbol or a set of symbols is displayed in the upper right hand corner of the Graphics window. This area is called the Confirmation Corner.

When a sketch is active, the confirmation corner box displays two symbols. The first symbol is the sketch tool icon. The second symbol is a large red X. These two symbols supply a visual reminder that you are in an active sketch. Click the sketch symbol icon to exit the sketch and to save any changes that you made.

When other commands are active, the confirmation corner box provides a green check mark and a large red X. Use the green check mark to execute the current command. Use the large red X to cancel the command.

Confirm changes you make in sketches and tools by using the D keyboard shortcut to move the OK and Cancel buttons to the pointer location in the Graphics window.

Heads-up View toolbar

SOLIDWORKS provides the user with numerous view options from the Standard Views, View and Heads-up View toolbar.

The Heads-up View toolbar is a transparent toolbar that is displayed in the Graphics window when a document is active.

For a drawing document

You can hide, move or modify the Heads-up View toolbar. To modify the Heads-up View toolbar, right-click on a tool and select or deselect the tools that you want to display.

The following views are available. Note: available views are document dependent.

- *Zoom to Fit* : Fit the model to the Graphics window.

- *Zoom to Area* : Zoom to the areas you select with a bounding box.

- *Previous View* : Display the previous view.

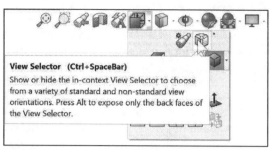

- *Section View* : Display a cutaway of a part or assembly, using one or more cross section planes.

- *Dynamic Annotation Views* : Only available with SOLIDWORKS MBD. Control how annotations are displayed when you rotate a model.

View Selector (Ctrl+SpaceBar)

Show or hide the in-context View Selector to choose from a variety of standard and non-standard view orientations. Press Alt to expose only the back faces of the View Selector.

The Orientation dialog has an option to display a view cube (in-context View Selector) with a live model preview. This helps the user to understand how each standard view orientates the model. With the view cube, you can access additional standard views. The views are easy to understand and they can be accessed simply by selecting a face on the cube.

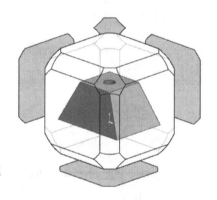

To activate the Orientation dialog box, press (Ctrl + spacebar) or click the View Orientation icon from the Heads-up View toolbar. The active model is displayed in the View Selector in an Isometric orientation (default view).

Click the View Selector icon in the Orientation dialog box to show or hide the in-context View Selector.

Press **Ctrl + spacebar** to activate the View Selector.

Press the **spacebar** to activate the Orientation dialog box.

- *View Orientation box* : Select a view orientation or the number of viewports. The options are: *Top, Left, Front, Right, Back, Bottom, Single view, Two view - Horizontal, Two view - Vertical, Four view.* Click the drop-down arrow to access Axonometric views: *Isometric, Dimetric* and *Trimetric.*

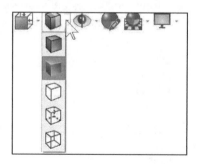

- *Display Style* : Display the style for the active view. The options are: *Wireframe, Hidden Lines Visible, Hidden Lines Removed, Shaded, Shaded With Edges.*

- *Hide/Show Items* : Select items to hide or show in the Graphics window. The available items are document dependent. Note the View Center of Mass icon.

- *Edit Appearance* : Edit the appearance of entities of the model.

- *Apply Scene* : Apply a scene to an active part or assembly document. View the available options.

- *View Setting* : Select the following settings: *RealView Graphics, Shadows In Shaded Mode, Ambient Occlusion, Perspective* and *Cartoon.*

- *Rotate view* : Rotate a drawing view. Input Drawing view angle and select the ability to update and rotate center marks with view.

- *3D Drawing View* : Dynamically manipulate the drawing view in 3D to make a selection.

To display a grid for a part, click Options , Document Properties tab. Click Grid/Snaps, check the Display grid box.

Add a custom view to the Heads-up View toolbar. Press the space key. The Orientation dialog box is displayed. Click the New View tool. The Name View dialog box is displayed. Enter a new named view. Click OK.

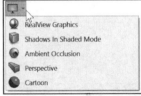

Use commands to display information about the triad or to change the position and orientation of the triad. Available commands depend on the triad's context.

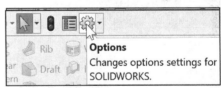

Save space in the CommandManager, limit your CommandManager tabs. **Right-click** on a CommandManager tab. Click **Tabs**. View your options to display CommandManager tabs.

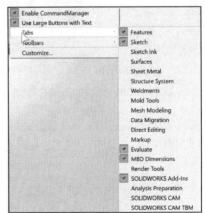

SOLIDWORKS CommandManager

The SOLIDWORKS CommandManager is a Context-sensitive toolbar. By default, it has toolbars embedded in it based on your active document type. When you click a tab below the CommandManager, it updates to display that toolbar. For example, if you click the Sketch tab, the Sketch toolbar is displayed.

For commercial users, SOLIDWORKS Model Based Definition (MBD) and SOLIDWORKS CAM is a separate application. For education users, SOLIDWORKS MBD and SOLIDWORKS CAM is included in the SOLIDWORKS Education Edition.

Below is an illustrated CommandManager for a **Part** document. Tabs will vary depending on system setup and Add-ins.

Set button size from the Toolbars tab of the Customize dialog box. To facilitate element selection on touch interfaces such as tablets, you can set up the larger Size buttons and text from the Options menu (Standard toolbar).

The SOLIDWORKS CommandManager is a Context-sensitive toolbar that automatically updates based on the toolbar you want to access. By default, it has toolbars embedded in it based on your active document type. The available tools are feature and document dependent.

Below is an illustrated CommandManager for a **Drawing** document. Tabs will vary depending on system setup and Add-ins.

To add a custom tab, right-click on a tab and click Customize. You can also select to add a blank tab and populate it with custom tools from the Customize dialog box.

The SOLIDWORKS CommandManager is a Context-sensitive toolbar that automatically updates based on the toolbar you want to access. By default, it has toolbars embedded in it based on your active document type. The available tools are feature and document dependent.

Below is an illustrated CommandManager for an **Assembly** document. Tabs will vary depending on system setup and Add-ins.

:bulb: The Markup tab is displayed by default in the CommandManager on some systems. You can draw markups with a mouse on non-touch devices, display bounding boxes for markups, create markups in drawings, and use the context toolbar to access markup options.

The Markup toolbar displays different options depending on the device. **Draw** Draw and **Touch** Touch are **not** available for non-touch devices.

Float the CommandManager. Drag the Features, Sketch or any CommandManager tab. Drag the CommandManager anywhere on or outside the SOLIDWORKS window.

To dock the CommandManager, perform one of the following:

While dragging the CommandManager in the SOLIDWORKS window, move the pointer over a docking icon - Dock above, Dock left, Dock right and click the needed command.

Double-click the floating CommandManager to revert the CommandManager to the last docking position.

Collapse the CommandManager

Collapse the CommandManager to only display the available tabs. No tools are shown.

Click the Collapsed CommandManager arrow to the right of the active CommandManager tab as illustrated. Only the tabs are displayed.

To display the CommandManager tools back, click on a CommandManager tab. Click the Pin CommandManager as illustrated.

Collapse CommandManager

Pin CommandManager

Part FeatureManager Design Tree

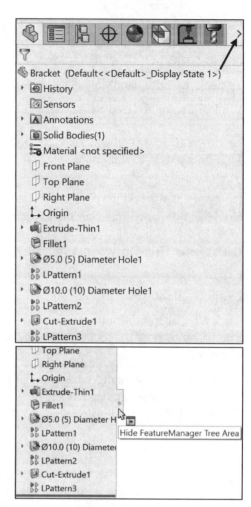

The Part FeatureManager consists of various tabs:

- *FeatureManager design tree* tab.

- *PropertyManager* tab.

- *ConfigurationManager* tab.

- *DimXpertManager* tab.

- *DisplayManager* tab.

- *CAM FeatureManager tree* tab.

- *CAM Operation tree* tab.

- *CAM Tools tree* tab.

Click the direction arrows to expand or collapse the FeatureManager design tree.

CommandManager and FeatureManager tabs and folder files will vary depending on your SOLIDWORKS applications and Add-ins.

Select the Hide/Show FeatureManager Area tab

as illustrated to enlarge the Graphics window for modeling.

The Sensors tool located in the FeatureManager monitors selected properties in a part or assembly and alerts you when values deviate from the specified limits. There are five sensor types: Simulation Data, Mass properties, Dimensions, Measurement and Costing Data.

Various commands provide the ability to control what is displayed in the FeatureManager design tree.

1. Show or Hide FeatureManager items.

Click **Options** ⚙ ˙ from the Menu bar. Click **FeatureManager** from the System Options tab. Customize your FeatureManager from the Hide/Show tree Items dialog box.

2. Filter the FeatureManager design tree. Enter information in the filter field. You can filter by *Type of features, Feature names, Sketches, Folders, Mates, User-defined tags* and *Custom properties*.

Tags are keywords you can add to a SOLIDWORKS document to make them easier to filter and to search. The Tags ✐ icon is located in the bottom right corner of the Graphics window.

Collapse all items in the FeatureManager, **right-click** and select **Collapse items**, or press the **Shift** + **C** keys.

The FeatureManager design tree and the Graphics window are dynamically linked. Select sketches, features, drawing views, and construction geometry in either pane.

Split the FeatureManager design tree and either display two FeatureManager instances, or combine the FeatureManager design tree with the ConfigurationManager or PropertyManager.

Move between the FeatureManager design tree, PropertyManager, ConfigurationManager, DimXpertManager, DisplayManager and others by selecting the tab at the top of the menu.

Split

The ConfigurationManager 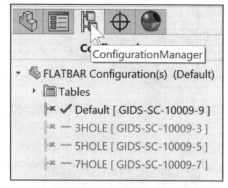 tab is located to the right of the PropertyManager tab. Use the ConfigurationManager to create, select and view multiple configurations of parts and assemblies.

The icons in the ConfigurationManager denote whether the configuration was created manually or with a design table.

The DimXpertManager ⊕ tab provides the ability to insert dimensions and tolerances manually or automatically. The options are: **Auto Dimension Scheme** ⊕, **Auto Pair Tolerance** , **Basic, Location Dimension** , **Basic Size Dimension** , **General Profile Tolerance** , **Show Tolerance Status** , **Copy Scheme** , **Import Scheme** ⊕, **TolAnalyst Study** and **Datum Target** .

TolAnalyst is available in SOLIDWORKS Premium.

Fly-out FeatureManager

The fly-out FeatureManager design tree provides the ability to view and select items in the PropertyManager and the FeatureManager design tree at the same time.

Throughout the book, you will select commands and command options from the drop-down menu, fly-out FeatureManager, Context toolbar, or from a SOLIDWORKS toolbar.

Another method for accessing a command is to use the accelerator key. Accelerator keys are special key strokes, which activate the drop-down menu options. Some commands in the menu bar and items in the drop-down menus have an underlined character.

Task Pane

The Task Pane is displayed when a SOLIDWORKS session starts. You can show, hide, and reorder tabs in the Task Pane. You can also set a tab as the default so it appears when you open the Task Pane, pin or unpin to the default location.

The Task Pane contains the following default tabs:

- *3DEXPERIENCE Marketplace* ⊙ (Add-in).

- *SOLIDWORKS Resources* ⌂.

- *Design Library* ▥ .

- *File Explorer* ▱ .

- *View Palette* ▦ .

- *Appearances, Scenes and Decals* ● .

- *Custom Properties* ▤ .

💡 Additional tabs are displayed with Add-Ins.

Use the **Back** and **Forward** buttons in the Design Library tab and the Appearances, Scenes, and Decals tab of the Task Pane to navigate in folders.

3DEXPERIENCE MARKETPLACE

Click the 3DEXPERIENCE MARKETPLACE ⊙ icon to connect to the 3DEXPERIENCE platform and leverage its collaboration, product data and lifecycle management, organic shape design, simulation solutions, and more. (Subscription required.)

Use 3DEXPERIENCE MARKETPLACE to source 3D content and services worldwide. This is an Add-in.

SOLIDWORKS Resources

The SOLIDWORKS Resources 🏠 icon displays the following default selections:

- *Welcome to SOLIDWORKS.*

- *SOLIDWORKS Tools.*

- *Online Resources.*

- *Subscription Services.*

Other user interfaces are available during the initial software installation selection: *Machine Design, Mold Design, Consumer Products Design, etc.*

Design Library

The Design Library 📚 contains reusable parts, assemblies, and other elements including library features.

The Design Library tab contains default selections. Each default selection contains additional sub categories.

The default selections are:

- *Design Library.*

- *SOLIDWORKS Content (Internet access required).*

- *Toolbox.*

- *3D Components - PartSupply.*

💡 Activate the SOLIDWORKS Toolbox. Click Tools, Add-Ins.., from the Main menu. Check the SOLIDWORKS Toolbox Library and SOLIDWORKS Toolbox Utilities box from the Add-ins dialog box or click SOLIDWORKS Toolbox from the SOLIDWORKS Add-Ins tab.

To access the Design Library folders in a non-network environment, click Add File Location and browse to the needed path. Paths may vary depending on your SOLIDWORKS version and window setup. In a network environment, contact your IT department for system details.

File Explorer

File Explorer ▱ duplicates Windows Explorer from your local computer and displays:

- *Recent Documents.*
- *Samples.*
- *Open in SOLIDWORKS*
- *Desktop.*

View Palette

The View Palette ▤ tool located in the Task Pane provides the ability to insert drawing views of an active document, or click the Browse button to locate the desired document.

Click and drag the view from the View Palette into an active drawing sheet to create a drawing view.

The selected model is Bracket.

Appearances, Scenes, and Decals

Appearances, Scenes, and Decals ⬤ provide a simplified way to display models in a photo-realistic setting using a library of Appearances, Scenes, and Decals.

An appearance defines the visual properties of a model, including color and texture. Appearances do not affect physical properties, which are defined by materials.

Scenes provide a visual backdrop behind a model. In SOLIDWORKS they provide reflections on the model. PhotoView 360 is an Add-in. Drag and drop a selected appearance, scene or decal on a feature, surface, part or assembly.

Custom Properties

The Custom Properties 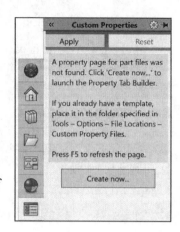 tool provides the ability to enter custom and configuration specific properties directly into SOLIDWORKS files.

Motion Study tab

Motion Studies are graphical simulations of motion for an assembly. Access the MotionManager from the Motion Study tab. The Motion Study tab is located in the bottom left corner of the Graphics window.

Incorporate visual properties such as lighting and camera perspective. Click the Motion Study tab to view the MotionManager. Click the Model tab to return to the FeatureManager design tree.

The MotionManager displays a timeline-based interface and provides the following selections from the drop-down menu as illustrated:

- *Animation:* Apply Animation to animate the motion of an assembly. Add a motor and insert positions of assembly components at various times using set key points. Use the Animation option to create animations for motion that do **not** require accounting for mass or gravity.

- *Basic Motion:* Apply Basic Motion for approximating the effects of motors, springs, collisions and gravity on assemblies. Basic Motion takes mass into account in calculating motion. Basic Motion computation is relatively fast, so you can use this for creating presentation animations using physics-based simulations. Use the Basic Motion option to create simulations of motion that account for mass, collisions or gravity.

If the Motion Study tab is not displayed in the Graphics window, click **View ➤ Toolbars ➤ MotionManager** from the Menu bar.

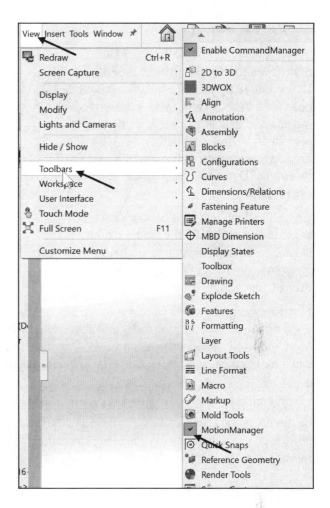

3D Views tab

Only available in SOLIDWORKS MBD.
SOLIDWORKS MBD (Model Based
Definition) lets you create models without the
need for drawings giving you an integrated
manufacturing solution. MBD helps
companies define, organize, and publish 3D
product and manufacturing information (PMI),
including 3D model data in industry standard
file formats.

Create 3D drawing views of your parts and
assemblies that contain the model settings
needed for review and manufacturing. This lets
users navigate back to those settings as they
evaluate the design.

Use the tools in the MBD Dimensions CommandManager to set up your model with selected
configurations, including explodes and abbreviated views, annotations, display states, zoom
level, view orientation and section views. Capture those settings so that you and other users
can return to them at any time using the 3D view palette.

To access the 3D View palette, click the 3D Views tab at the bottom of the SOLIDWORKS
window or the SOLIDWORKS MBD tab in the CommandManager. The Capture 3D View
button opens the Capture 3D View PropertyManager, where you specify the 3D view name,
and the configuration, display state and annotation view to capture. See help for additional
information.

Dynamic Reference Visualization (Parent/Child)

Dynamic Reference Visualization provides the ability
to view the parent/child relationships between items in
the FeatureManager design tree. When you hover over
a feature with references in the FeatureManager design
tree, arrows display showing the relationships. If a
reference cannot be shown because a feature is not
expanded, the arrow points to the feature that contains
the reference and the actual reference appears in a text
box to the right of the arrow.

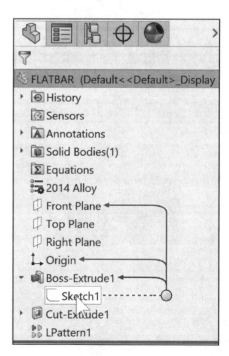

Use Dynamic reference visualization for a part, assembly and mates.

To display the Dynamic Reference Visualization, click **View ➢ User Interface ➢ Dynamic Reference Visualization (Parent/Child)** from the Main menu bar.

Mouse Movements

A mouse typically has two buttons: a primary button (usually the left button) and a secondary button (usually the right button). Most mice also include a scroll wheel between the buttons to help you scroll through documents and to Zoom in, Zoom out and rotate models in SOLIDWORKS. It is highly recommended that you use a mouse with at least a Primary, Scroll and Secondary button.

Single-click

To click an item, point to the item on the screen, and then press and release the primary button (usually the left button). Clicking is most often used to select (mark) an item or open a menu. This is sometimes called single-clicking or left-clicking.

Double-click

To double-click an item, point to the item on the screen, and then click twice quickly. If the two clicks are spaced too far apart, they might be interpreted as two individual clicks rather than as one double-click. Double-clicking is most often used to open items on your desktop. For example, you can start a program or open a folder by double-clicking its icon on the desktop.

Right-click

To right-click an item, point to the item on the screen, and then press and release the secondary button (usually the right button). Right-clicking an item usually displays a list of things you can do with the item. Right-click in the open Graphics window or on a command in SOLIDWORKS, and additional pop-up context is displayed.

Scroll wheel

Use the scroll wheel to zoom-in or to zoom-out of the Graphics window in SOLIDWORKS. To zoom-in, roll the wheel backward (toward you). To zoom-out, roll the wheel forward (away from you).

Summary

The SOLIDWORKS (UI) is designed to make maximum use of the Graphics window for your model. Displayed toolbars and commands are kept to a minimum.

The SOLIDWORKS User Interface and CommandManager consist of the following main options: Menu bar toolbar, Menu bar menu, Drop-down menus, Context toolbars, Consolidated fly-out menus, System feedback icons, Confirmation Corner and Heads-up View toolbar.

The Part CommandManager controls the display of tabs: *Features*, *Sketch*, *Evaluate*, *MBD Dimensions* and various *SOLIDWORKS Add-Ins*.

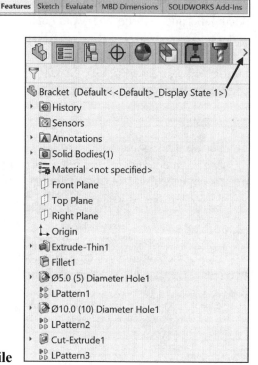

The FeatureManager consists of various tabs:

- *FeatureManager design tree* tab.

- *PropertyManager* tab.

- *ConfigurationManager* tab.

- *DimXpertManager* tab.

- *DisplayManager* tab.

- *CAM FeatureManager tree* tab.

- *CAM Operation tree* tab.

- *CAM Tools tree* tab.

Click the direction arrows to expand or collapse the FeatureManager design tree.

CommandManager, FeatureManager, and file folders will vary depending on system set-up and Add-ins.

You learned about creating a new SOLIDWORKS part and opening an existing SOLIDWORKS part along with using the Rollback bar to view the sketches and features.

If you modify a document property from an Overall drafting standard, a modify message is displayed as illustrated.

Overall drafting standard
ANSI-MODIFIED
Derived from: ANSI

Use the Search box, in the upper left corner of the Materials dialog box, to search through the entire materials library.

Templates are part, drawing and assembly documents which include user-defined parameters. Open a new part, drawing or assembly. Select a template for the new document.

In Chapter 2, obtain the working familiarity of the following SOLIDWORKS sketch and feature tools: *Line, Circle, Centerpoint Straight Slot, Smart Dimension, Extruded Boss/Base, Extruded Cut and Linear Pattern.*

Create three individual parts: AXLE, SHAFT-COLLAR and FLATBAR.

Create the assembly, LINKAGE using the three created parts and the downloaded sub-assembly - AirCylinder.

The book provides information on creating and storing special Part templates, Assembly templates and Drawing templates in the MY-TEMPLATES folder. The MY-TEMPLATES folder is added to the New SOLIDWORKS Document dialog box. Talk to your IT department *before you set* any new locations on a network system. The procedure in the book is designed for your personal computer.

If you do not create the MY-TEMPLATE tab or the special part, drawing, or assembly templates, use the standard SOLIDWORKS default template and apply all of the needed document properties and custom properties.

Notes:

Chapter 2
Parts and Assembly Creation

Below are the desired outcomes and usage competencies based on the completion of Chapter 2.

Desired Outcomes:	Usage Competencies:

• Create three parts:
 	
    ○ AXLE
 	
    ○ SHAFT-COLLAR
 	
    ○ FLATBAR
	• Create 2D sketch profiles on the correct Sketch plane.
	
	• Apply the following 3D features: Extruded Boss/Base, Extruded Cut and Linear Pattern.

• Create an assembly:
 	
    ○ LINKAGE assembly
 	
    ○ Apply Standard mate types
 	
    ○ Motion Study
	• Understand the Assembly toolbar.
	
	• Insert components into an assembly.
	
	• Apply the following Standard mates: Concentric, Coincident and Parallel.
	
	• Comprehend a Simple Motion Study.

Notes:

Chapter 2 - Parts and Assembly Creation

Chapter Objective

SOLIDWORKS is a design software application used to model and create 2D and 3D sketches, 3D parts, 3D assemblies and 2D drawings. The chapter objective is to obtain the working familiarity of the following SOLIDWORKS sketch and feature tools: Line, Circle, Centerpoint, Straight Slot, Smart Dimension, Extruded Boss/Base, Extruded Cut, and Linear Pattern.

Create three individual parts: AXLE, SHAFT-COLLAR, and FLATBAR.

Create the assembly, LINKAGE using the three created parts and the downloaded sub-assembly; AirCylinder.

On the completion of this chapter, you will be able to:

- Set overall drafting standard, units, precision, and dimensioning standards for a SOLIDWORKS Part and Assembly document.

- Create a 2D sketch.

- Understand and apply the following Sketch tools:

  o Circle, Smart dimension and Centerpoint Straight Slot.

- Identify the correct Sketch plane for the 2D sketch:

  o Front, Top, Right.

- Insert geometric relations and dimensions:

  o Equal, Vertical, Horizontal, Parallel, Perpendicular, Coincident and MidPoint.

- Modify sketch dimensions and geometric relations.

- Create a 3D model.

- Understand and apply the following SOLIDWORKS features:

  o Extruded Boss/Base, Extruded Cut and Linear Pattern.

- Download an assembly document (with references) in SOLIDWORKS.

- Create an assembly document.

- Insert components:

  o Download and insert components.

- Understand and apply various Assembly feature tools.

- Apply the following Standard mates: Coincident, Concentric and Parallel.

- Utilize Standard and the Quick mate procedure.

Chapter Overview

SOLIDWORKS is a 3D solid modeling CAD software package used to produce and model parts, assemblies, and drawings.

SOLIDWORKS provides design software to create 3D models and 2D drawings.

Create three parts in this chapter:

- AXLE

- SHAFT-COLLAR

- FLATBAR

Create the LINKAGE assembly. Download the AirCylinder assembly from the Pneumatic Components folder.

Combine the created parts with the AirCylinder sub-assembly to create the LINKAGE assembly.

Illustrations in the book display the default SOLIDWORKS user interface for 2023 SP0.

AXLE

SHAFT-COLLAR

FLATBAR

AirCylinder assembly

LINKAGE assembly

AXLE Part

The AXLE is a cylindrical rod. The AXLE supports the two FLATBAR parts.

The AXLE part is manufactured from 6061 alloy.

Tangent edges and origins are displayed for educational purposes in this book.

The AXLE rotates about its axis. The dimensions for the AXLE are determined from other components in the LINKAGE assembly.

AXLE

FLATBAR

Start a new SOLIDWORKS session.

Create the AXLE part.

Apply the feature to create the part. Features are the building blocks that add or remove material.

Utilize the Extruded Boss/Base tool from the Features toolbar to create a Boss-Extrude1 feature. The Extruded Boss/Base feature adds material. The Base feature (Boss-Extrude1) is the first feature of the part. The Base feature is the foundation of the part. Keep the Base feature <u>simple!</u>

AXLE

The Base feature geometry for the AXLE is a simple extrusion. How do you create a solid Extruded Boss/Base feature for the AXLE?

Origin

- Select the Front Plane as the Sketch plane.

- Sketch a circular 2D profile on the Front Plane, centered at the Origin as illustrated.

- Apply the Extruded Boss/Base Feature. Extend the profile perpendicular (⊥) to the Front Plane.

Utilize symmetry. Extrude the sketch with the Mid Plane End Condition in Direction 1. The Extruded Boss/Base feature is centered on both sides of the Front Plane.

Origin

Create a new part. Select File, New from the Main menu toolbar or click New ⬜ from the Main menu. There are two options for new documents: **Novice** and **Advanced**. Select the Advanced option. Select the default Part document.

> **Activity: Start a SOLIDWORKS Session. Create a New Part.**

Start a SOLIDWORKS 2023 session.

1) Click the **SOLIDWORKS 2023** application (or if available, **double-click** the SOLIDWORKS icon on the Desktop). View the Welcome dialog box. The Welcome dialog box provides a convenient means to open documents, view folders, access SOLIDWORKS resources, and stay updated on SOLIDWORKS news.

2) **Close** the SOLIDWORKS Welcome dialog box. The SOLIDWORKS program is displayed.

3) **View** your options.

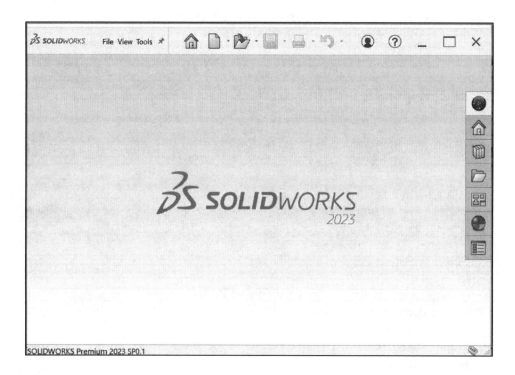

There are two default modes in the New SOLIDWORKS Document dialog box: Novice and Advanced. The *Novice* option is the default option with three templates. The *Advanced* mode contains access to additional templates and tabs that you create in system options. Use the *Advanced* mode in this book.

Create a new part.

4) Click **New** ⬜ from the Main menu. The New SOLIDWORKS Document dialog box is displayed.

Select Advanced Mode.

5) Click the **Advanced** button to display the New SOLIDWORKS Document dialog box in Advanced mode.

6) Click the **Templates** tab.

7) Click **Part**. Part is the default template from the New SOLIDWORKS Document dialog box.

8) Click **OK**.

The *Advanced* mode remains selected for all new documents in the current SOLIDWORKS session. When you exit SOLIDWORKS, the *Advanced* mode setting is saved.

The default SOLIDWORKS installation contains three tabs in the New SOLIDWORKS Document dialog box: *Templates, Tutorial and MBD*. The *Templates* tab corresponds to the default SOLIDWORKS templates. The *Tutorial* tab corresponds to the templates utilized in the SOLIDWORKS Tutorials. The *MBD* tab corresponds to the templates utilized in the SOLIDWORKS (Model Based Definition).

💡 The book is written using the ANSI (US) overall drafting standard and Third Angle Projection for all drawing documents.

Part1 is displayed in the FeatureManager and is the name of the document. Part1 is the default part window name.

The Part Origin ⤴ is displayed in blue in the center of the Graphics window. The Origin represents the intersection of the three default reference planes: *Front Plane*, *Top Plane* and *Right Plane*. The positive X-axis is horizontal and points to the right of the Origin in the Front view. The positive Y-axis is vertical and points upward in the Front view. The FeatureManager contains a list of features, reference geometry, and settings utilized in the part.

💡 Edit the document units directly from the Graphics window as illustrated.

💡 CommandManager, FeatureManager, Task Pane tabs and tree folders will vary depending on system setup and Add-ins.

Activity: Set Document Properties - AXLE Part

Set Document Properties. Set drafting standard, units, and precision.

9) Click **Options** from the Main menu. The System Options General dialog box is displayed.

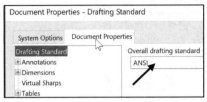

10) Click the **Document Properties** tab.

11) Select **ANSI** from the Overall drafting standard drop-down menu.

The Overall drafting standard determines the display of dimension text, arrows, symbols, and spacing.

Units are the measurement of physical quantities. Millimeter dimensioning and decimal inch dimensioning are the two most common unit types specified for engineering parts and drawings.

The primary units in this book are provided in IPS (inch, pound, second). The optional secondary units are provided in MMGS (millimeters, grams, second) and are indicated in brackets [ ].

Set document units. Set length and angle precision.

12) Click **Units**.

13) Click **IPS** (inch, pound, second) **[MMGS]** for Unit system.

14) Select **.123, [.12]** (three decimal places) for Length basic units.

15) Select **None** for Angle decimal places.

16) Click **OK** from the Document Properties - Units dialog box. The Part FeatureManager is displayed.

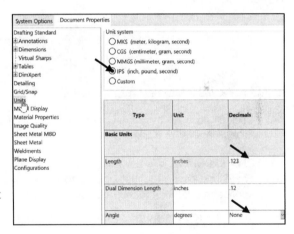

💡 The first sketch of a part is called the Base sketch. The Base sketch is the foundation for the 3D model. The book focuses on 2D sketches and 3D features.

💡 If you modify a document property from an Overall drafting standard, a modify message is displayed as illustrated.

Overall drafting standard

ANSI-MODIFIED

Derived from: ANSI

Activity: AXLE Part - Base Sketch - Extruded Base Feature

Create the Base sketch for the Extruded Base feature.

17) Right-click **Front Plane** from the FeatureManager. This is your Sketch plane. The Context toolbar is displayed.

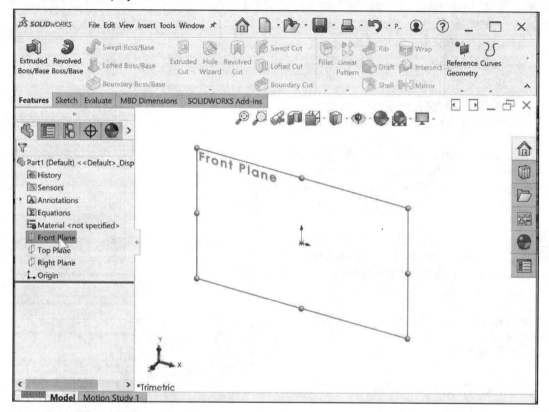

18) Click **Sketch** from the Context toolbar as illustrated. The Sketch toolbar is displayed. Front Plane is your Sketch plane.

19) Click the **Circle** tool from the Sketch toolbar. The Circle PropertyManager is displayed. The Circle-based tool uses a Consolidated Circle PropertyManager. The SOLIDWORKS application defaults to the last used tool type.

20) Move the **mouse pointer** into the Graphics window. The cursor displays the Circle icon.

21) Click the **Origin** of the circle. The cursor displays the Coincident icon.

22) Drag the **mouse pointer** to the right of the Origin to create the circle as illustrated. The center point of the circle is positioned at the Origin.

23) Click a **position** directly to the right to create the circle. The activated circle is displayed in blue.

Circle

R = 11.73

Sketches a circle. Click to place the center of the circle, then drag and release to define its radius.

Add a dimension. Use the Smart dimension tool.

24) Click **Smart Dimension** from the Sketch toolbar. The cursor displays the Smart Dimension icon.

25) Click the **circumference** of the circle.

26) Click a **position** diagonally above the circle in the Graphics window.

27) Enter **.188**in, [**4.78**] in the Modify dialog box. The Dimension Modify dialog box provides the ability to select a unit drop-down menu to modify units in a sketch or feature.

28) Click the **Green Check mark** in the Modify dialog box. The diameter of the circle is .188 inches.

The circular sketch is centered at the Origin. The dimension indicates the diameter of the circle. Sketch Relations are displayed in the model.

Press the f key to fit the part document to the Graphics window.

Add relations, then dimensions. This keeps the user from having too many unnecessary dimensions. This also helps to show the design intent of the model. Dimension what geometry you intend to modify or adjust.

The primary units in this book are provided in IPS (inch, pound, second). The optional secondary units are provided in MMGS (millimeters, grams, second) and are indicated in brackets [ ].

When you create a new part or assembly, the three default Planes (Front, Right and Top) are aligned with specific views.

Extrude the sketch to create the Base Feature.

29)   Click the **Features** tab from the CommandManager.

30)   Click the **Extruded Boss/Base** Features tool.
The Boss-Extrude PropertyManager is displayed.
Blind is the default End Condition in Direction 1.

31)   Select **Mid Plane** for End Condition in Direction 1.
The Mid Plane End Condition in the Direction 1 box
extrudes the sketch equally on both sides of the
Sketch plane. The depth defines the extrude
distance.

32)   Enter **1.375**in, **[34.93]** for Depth in Direction 1.
Accept the default conditions.

33)   Click **OK** ✔ from the Boss-Extrude
PropertyManager. Boss-Extrude1 is displayed in the
FeatureManager.

Fit the model to the Graphics window.
34)   Press the **f** key. Note the location of the Origin in the model.

Use Symmetry. When possible and if it makes sense, model objects symmetrically about
the origin.

The Boss-Extrude1 feature name is displayed in the FeatureManager. The FeatureManager lists the features, planes, and other geometry that constructs the part. Extrude features add material. Extrude features require the following: *Sketch Plane*, *Sketch* and *depth*.

Activity: AXLE Part - Save

Save the AXLE part. Enter part name. Enter part description. Save all parts to the SW-TUTORIAL 2023 folder.

35) Click **Save As** 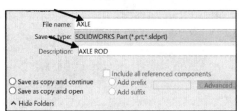 from the drop-down menu.

36) Double-click the **SW-TUTORIAL 2023** folder. In this book all models, assemblies and drawings are saved to the SW-TUTORIAL 2023 folder.

37) Enter **AXLE** for the File name.

38) Enter **AXLE ROD** for the Description.

39) Click **Save**. The AXLE FeatureManager is displayed.

Activity: AXLE Part - Edit Appearance

Modify the color of the AXLE part.

40) Right-click the **AXLE** icon at the top of the FeatureManager.

41) Click the **Appearances** drop-down arrow.

42) Click **AXLE** as illustrated. The Color PropertyManager is displayed. AXLE is displayed in the Selection box.

43) Select a **light blue** color from the Color box. View your options.

44) Click **OK** ✔ from the Color PropertyManager. View the AXLE in the Graphics window.

Use the Appearances PropertyManager to apply colors, material appearances, and transparency to parts and assembly components. For sketches or curves only, use the Sketch/Curve Color PropertyManager to apply colors.

The Advanced tab includes the Illumination and Surface Finish tabs, and additional options in the Color/Image and Mapping tabs. To display the simplified Color/Image or Mapping interfaces, click the Basic tab.

Download all needed model files (SW-TUTORIAL 2023 folder) from the SDC Publication website (www.SDCpublications.com/downloads/978-1-63057-556-4).

Sketching in SOLIDWORKS is the basis for creating features. Features are the basis for creating parts, which can be put together into assemblies.

The sketch status appears in the window status bar and in the FeatureManager. Colors indicate the state of individual sketch entities. Sketches are generally in one of the following states:

1.) *(+) Over defined.* The sketch is displayed in orange with a warning ⚠ icon.

2.) *(-) Under defined.* The sketch is displayed in blue.

3.) *No prefix.* The sketch is fully defined. This is the ideal sketch state. A fully defined sketch has complete information (manufacturing and inspection) and is displayed in black.

Activity: AXLE Part - View Modes

Orthographic projection is the process of projecting views onto Parallel planes with ⊥ projectors.

The default reference planes are the Front, Top and Right Planes.

The Isometric view displays the part in 3D with two equal projection angles.

💡 Click **View**, **Hide/Show**, **Origins** from the Main menu menu to display the Origin in the Graphics window.

Origin

Display the various view modes using the Heads-up View toolbar.

45) **Deactivate** the View Selector. Hide the in-context View Selector box.

Display a Front view.

46) Click **Front view** ⬜ from the Heads-up View toolbar.

Display a Top view.

47) Click **Top view** ⬜ from the Heads-up View toolbar.

Display a Right view.

48) Click **Right view** ⬜ from the Heads-up View toolbar.

Display an Isometric view.

49) Click **Isometric view** from the Heads-up View toolbar.

View modes manipulate the model in the Graphics window.

Zoom out/Zoom in.

50) Press the lower-case **z** key to zoom out. Note: you can also use the middle mouse button.

51) Press the upper-case **Z** key to zoom in. Note: you can also use the middle mouse button.

View the available view tools in the Graphics window.

52) **Right-click** in the Graphics window. View the available view tools.

Deselect the view menu.

53) Click **inside** the Graphics window.

Rotate the model.

54) Click the **middle mouse** button and move your mouse. The model rotates. The Rotate icon ⟳ is displayed.

View the various Display Styles.

55) Click **Isometric view** from the Heads-up View toolbar.

56) Click the **drop-down arrow** from the Display Styles box from the Heads-up Views toolbar as illustrated. SOLIDWORKS provides five key Display Styles:

- *Shaded* . Displays a shaded view of the model with no edges.

- *Shaded With Edges* . Displays a shaded view of the model, with edges.

- *Hidden Lines Removed* . Displays only those model edges that can be seen from the current view orientation.

- *Hidden Lines Visible* . Displays all edges of the model. Edges that are hidden from the current view are displayed in a different color or font.

- *Wireframe* . Displays all edges of the model.

Return to Shaded With Edges. Save the AXLE part.

57) Click the **Shaded With Edges** icon.

58) Click **Save** . The AXLE part is complete. Later, apply material to the part.

Review the AXLE Part.

The AXLE part utilized the Extruded Boss/Base feature. The Extruded Boss/Base feature adds material. The Extruded feature required a Sketch Plane, sketch and depth. The AXLE Sketch plane was the Front Plane. The 2D circle was sketched centered at the Origin. A dimension defined the overall size of the sketch based on the dimensions of mating parts in the LINKAGE assembly.

The default name of the Base feature is Boss-Extrude1. Boss-Extrude1 utilized the Mid Plane End Condition. The Boss-Extrude1 feature is symmetrical about the Front Plane.

The Edit Color option modified the part color. Select the Part icon in the FeatureManager to modify the color of the part. Color and a prefix define the sketch status. A blue sketch is under defined. A black sketch is fully defined. A red sketch is over defined.

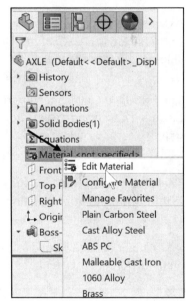

The default Reference planes are the Front, Top, and Right Planes. Utilize the Heads-up View toolbar to display the principle views of a part. The View Orientation and Display Style tools manipulate the model in the Graphics windows.

Later, add material to the AXLE part. To add material, right-click the **Material** folder in the FeatureManager. Click **Edit Material**. The Material dialog box is displayed. Select **material type**. Click **Apply**. Click **Close**. Return to the Graphics window.

When you create a new part or assembly, the three default Planes (Front, Right and Top) are aligned with specific views. The Plane you select for the Base sketch determines the orientation of the part.

SHAFT-COLLAR Part

The SHAFT-COLLAR part is fastened to the AXLE part.

The SHAFT-COLLAR part is manufactured from 6061 alloy.

Two SHAFT-COLLAR parts are used to position the two FLATBAR parts on the AXLE.

Create the SHAFT-COLLAR part.

Utilize the Extruded Boss/Base feature. The Extruded Boss/Base feature requires a 2D circular profile.

Utilize symmetry. Sketch a circle on the Front Plane centered at the Origin.

Extrude the sketch with the Mid Plane End Condition. The Extruded Boss/Base feature (Boss-Extrude1) is centered on both sides of the Front Plane.

SHAFT-COLLAR

The Extruded Cut feature removes material. Utilize an Extruded Cut feature to create a hole. The Extruded Cut feature requires a 2D circular profile. Sketch a circle on the front face centered at the Origin.

The Through All End Condition extends the Extruded Cut feature from the front face through all existing geometry.

The book is designed to expose the new SOLIDWORKS user to many different tools, techniques and procedures.

Activity: SHAFT-COLLAR Part - Extruded Boss/Base Feature

Create a new part.

59) Click **New** ⬜ from the Main menu. The New SOLIDWORKS Document dialog box is displayed.

60) Click the **Templates** tab. Part is the default template from the New SOLIDWORKS Document dialog box.

61) Double-click **Part**. The Part FeatureManager is displayed.

New SOLIDWORKS Document

Templates   Tutorial   MBD

Part
Assembly
Drawing

Save the part. Enter part name. Enter part description. Save in the SW-TUTORIAL 2023 folder.

62) Click **Save As** from the drop-down menu.

63) Double-click the **SW-TUTORIAL 2023** folder. In this book all models, assemblies and drawings are saved to the SW-TUTORIAL 2023 folder.

64) Enter **SHAFT-COLLAR** for File name in the SW-TUTORIAL 2023 folder.

65) Enter **SHAFT-COLLAR** for Description.

66) Click **Save**. The SHAFT-COLLAR FeatureManager is displayed.

Set Document Properties. Set drafting standard, units, and precision.

67) Click **Options** from the Main menu.

68) Click the **Document Properties** tab.

69) Select **ANSI** from the Overall drafting standard drop-down menu.

70) Click **Units**.

71) Click **IPS** (inch, pound, second), [**MMGS**] for Unit system.

72) Select **.123**, [**.12**] (three decimal places) for Length units Decimal places.

73) Select **None** for Angular units Decimal places.

74) Click **OK** from the Document Properties - Units dialog box.

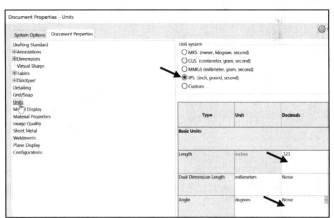

When you create a new part or assembly, the three default Planes (Front, Right and Top) are aligned with specific views. The Plane you select for the Base sketch determines the orientation of the part.

Insert a new sketch for the Extruded Base feature.

75) Right-click **Front Plane** from the FeatureManager. This is the Sketch plane. The Context toolbar is displayed.

76) Click **Sketch** from the Context toolbar as illustrated. The Sketch toolbar is displayed.

77) Click the **Circle** tool from the Sketch toolbar. The Circle PropertyManager is displayed.

78) Click the **Origin** . The cursor displays the Coincident to point feedback symbol.

79) Drag the **mouse pointer** to the right of the Origin as illustrated.

80) Click a **position** to create the circle.

Add a dimension.

81) Click **Smart Dimension** from the Sketch toolbar.

82) Click the **circumference** of the circle. The cursor displays the diameter feedback symbol.

83) Click a **position** diagonally above the circle in the Graphics window.

84) Enter **.4375**in, **[11.11]** in the Modify dialog box.

85) Click the **Green Check mark** in the Modify dialog box. The black sketch is fully defined.

Note: Three decimal places are displayed. The diameter value .4375 rounds to .438.

Extrude the sketch to create the Base feature.
86) Click the **Features** tab from the CommandManager.

87) Click the **Extruded Boss/Base** features tool. The Boss-Extrude PropertyManager is displayed. Select **Mid Plane** for End Condition in Direction 1.

88) Enter .**250**in, [**6.35**] for Depth. Accept the default conditions. View the location of the Origin.

89) Click **OK** ✔ from the Boss-Extrude PropertyManager.

Fit the model to the Graphics window. Display a trimetric view.
90) Press the **f** key.

91) Click **Trimetric** from the Heads-Up View toolbar.

Save the model.
92) Click **Save** .

Activity: SHAFT-COLLAR Part - Extruded Cut Feature

Insert a new sketch for the Extruded Cut feature.
93) Right-click the **front circular face** of the Boss-Extrude1 feature for the Sketch plane. The mouse pointer displays the face feedback icon.

94) Click **Sketch** from the Context toolbar as illustrated. The Sketch toolbar is displayed. This is your Sketch plane.

💡 View the mouse pointer feedback icon for the correct geometry: line, face, point, or vertex.

95) Click **Hidden Lines Removed** from the Heads-up View toolbar.

96) Click the **Circle** tool from the Sketch toolbar. The Circle PropertyManager is displayed.

97) Click the red **Origin**. The cursor displays the Coincident to point feedback symbol.

Hidden Lines Visible
Displays all edges of the model. Edges that are hidden from the current view are displayed in a different color or font.

98) Drag the **mouse pointer** to the right of the Origin.

99) Click a **position** to create the circle as illustrated.

Add a dimension.

100) Click the **Smart Dimension** Sketch tool.

101) Click the **circumference** of the circle.

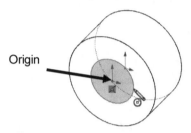

Origin

102) Click a **position** diagonally above the circle in the Graphics window.

103) Enter **.190**in, [**4.83**] in the Modify dialog box.

104) Click the **Green Check mark** in the Modify dialog box.

Insert an Extruded Cut feature.
105) Click the **Features** tab from the CommandManager.

106) Click **Extruded Cut** from the Features toolbar. The Cut-Extrude PropertyManager is displayed.

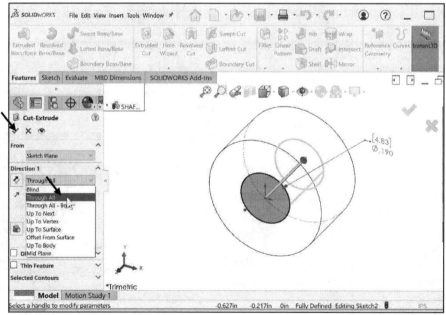

107) Select **Through All** for End Condition in Direction 1. The direction arrow points to the back. If needed, click the Reverse Direction button. Accept the default conditions.

108) Click **OK** ✔ from the Cut-Extrude PropertyManager. Cut-Extrude1 is displayed in the FeatureManager.

The Extruded Cut feature is named Cut-Extrude1. The Through All End Condition removes material from the Front Plane through the Boss-Extrude1 geometry.

 Model about the origin; this provides a point of reference.

Activity: SHAFT-COLLAR - Modify Dimensions and Edit Color

Modify the dimensions.

109) Click the **Z** key a few times to Zoom in.

110) Double-click the **outside cylindrical face** of the SHAFT-COLLAR. The Boss-Extrude1 dimensions are displayed. Sketch dimensions are displayed in black. The Extrude depth dimensions are displayed in blue.

111) Double-click the **.250**in, [**6.35**] depth dimension.

112) Enter **.500**in, [**12.70**].

Rebuild the model.

113) Click **Rebuild** 🌡 from the Main menu.

The Boss-Extrude1 feature and Cut-Extrude1 feature are modified.

Return to the original dimensions.

114) Click the **Undo** ↺ tool from the Main menu.

Display Shaded With Edges.

115) Click **Shaded With Edges** 🧊 from the Heads-up View toolbar. Modify the part color.

116) Right-click the **SHAFT-COLLAR Part** icon at the top of the FeatureManager.

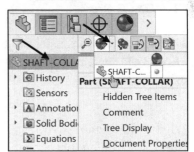

117) Click the **Appearances** drop-down arrow.

118) Click **SHAFT-COLLAR** as illustrated. The Color PropertyManager is displayed. SHAFT-COLLAR is displayed in the Selection box.

119) Click the **Basic** tab.

120) Select a **light green** color from the Color box.

121) Click **OK** ✔ from the Color PropertyManager.

122) **View** the SHAFT-COLLAR in the Graphics window.

Save the SHAFT-COLLAR part.

123) Click **Save** 💾. The SHAFT-COLLAR part is complete. All sketches should be fully defined. Later, apply material to the part.

      Review the SHAFT-COLLAR Part.

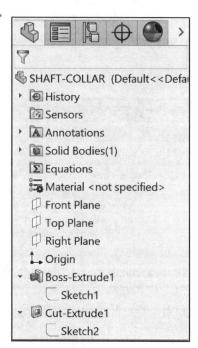

The SHAFT-COLLAR utilized an Extruded Boss/Base 🗐 feature and an Extruded Cut 🗐 feature. The Extruded Boss/Base feature adds material. An Extruded Boss/Base feature required a Sketch Plane, sketch and depth.

The Sketch plane was the Front Plane. The 2D circle was sketched centered at the Origin. A dimension fully defined the overall size of the sketch. The default name of the feature was Boss-Extrude1. Boss-Extrude1 utilized the Mid Plane End Condition. The Boss-Extrude1 feature was symmetric about the Front Plane.

The Extruded Cut 🗐 feature removed material. An Extruded Cut feature required a Sketch Plane, sketch and depth. The Sketch plane was the front face of the Boss-Extrude1 feature. A dimension fully defined the overall size of the sketch.

FLATBAR Part

The FLATBAR part fastens to the AXLE. The FLATBAR contains nine, ⌀.190in holes spaced 0.5in apart.

The FLATBAR part is manufactured from .090inch 2014 alloy. Create the FLATBAR part. Utilize the Centerpoint Straight Slot Sketch 🔲 tool with an Extruded Boss/Base 🗐 feature.

The Extruded feature requires a 2D profile sketched on the Front Plane.

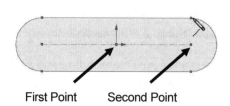

First Point   Second Point

The Centerpoint Straight Slot Sketch tool automatically applies design symmetry (Midpoint and Equal geometric relations). Create the 2D profile centered about the Origin. Relations control the size and position of entities with constraints.

Utilize an Extruded Cut feature to create the first hole. This is the seed feature for the Linear Pattern.

Utilize a Linear Pattern feature to create the remaining holes. A Linear Pattern creates an array of features in a specified direction.

Add relations, then dimensions. This keeps the user from having too many unnecessary dimensions. This also helps to show the design intent of the model.

Activity: FLATBAR Part - Extruded Base Feature

Create a new part.

124) Click **New** ⬜ from the Main menu. The New SOLIDWORKS Document dialog box is displayed.

125) Click the **Templates** tab.

126) Double-click **Part**. The Part FeatureManager is displayed.

Save the part. Enter part name. Enter part description. Save the part in the SW-TUTORIAL 2023 folder.

127) Click **Save As** 🖼 from the drop-down menu.

128) Enter **FLATBAR** for File name in the SW-TUTORIAL 2023 folder.

129) Enter **FLAT BAR 9-HOLES** for Description.

130) Click **Save**. The FLATBAR FeatureManager is displayed.

Set Document Properties. Set drafting standard, units, and precision.

131) Click **Options** ⚙ from the Main menu.

132) Click the **Document Properties** tab from the dialog box.

133) Select **ANSI** from the Overall drafting standard drop-down menu.

134) Click **Units**.

135) Click **IPS**, **[MMGS]** for Unit system.

136) Select **.123**, **[.12]** for Length units Decimal places.

137) Select **None** for Angular units Decimal places.

138) Click **OK** to set the document units.

Insert a new sketch for the Extruded Base feature.

139) Right-click **Front Plane** from the FeatureManager. This is the Sketch plane.

140) Click **Sketch** 🔲 from the Context toolbar as illustrated. The Sketch toolbar is displayed.

Utilize the Consolidated Slot Sketch toolbar. Apply the Centerpoint Straight Slot Sketch tool.

The Straight Slot Sketch tool provides the ability to sketch a straight slot from a center point. In this example, use the origin as your center point.

141) Click the **Centerpoint Straight Slot** 🔲 tool from the Sketch toolbar. The Slot PropertyManager is displayed.

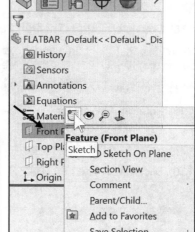

Create the Centerpoint Straight Slot with three points.

142) Click the **Origin**. This is your first point.

143) Click a **point** directly to the right of the Origin. This is your second point.

144) Click a **point** directly above the second point. This is your third point. The Centerpoint Straight Slot is displayed.

145) Click **OK** ✔ from the Slot PropertyManager.

First Point   Second Point

View Sketch relations.

146) Click **View**, **Hide/Show**, **Sketch Relations** from the Main menu. View the sketch relations in the Graphics window.

Deactivate Sketch relations.

147) Click **View**, **Hide/Show**, un-check **Sketch Relations** from the Main menu. The Straight Slot Sketch tool provides a midpoint relation with the Origin and Equal relations between the other sketch entities.

Add dimensions.

148) Click the **Smart Dimension** ✏ tool from the Sketch toolbar.

149) Click the **horizontal** centerline.

150) Click a **position** above the top horizontal line in the Graphics window.

151) Enter **4.000**in, [**101.6**] in the Modify dialog box.

152) Click the **Green Check mark** ✔ in the Modify dialog box.

153) Click the **right arc** of the FLATBAR.

154) Click a **position** diagonally to the right in the Graphics window.

155) Enter **.250**in, [**6.35**] in the Modify dialog box.

156) Click the **Green Check mark** ✔ in the Modify dialog box. The black sketch is fully defined.

💡 Model about the Origin; this provides a point of reference for your dimensions to fully define the sketch.

[101.60]
4.000

[6.35]
R.250

Extrude the sketch to create the Base (Boss-Extrude1) feature.

157) Click **Extruded Boss/Base** from the Features toolbar. The Boss-Extrude
PropertyManager is displayed.

158) Enter .**090**in, [**2.29**] for Depth. Accept the default conditions.

159) Click **OK** ✔ from the Boss-Extrude PropertyManager. Boss-Extrude1 is displayed in the
FeatureManager.

Fit the model to the Graphics window.
160) Press the **f** key.

Save the FLATBAR part.

161) Click **Save** 💾.

Activity: FLATBAR Part - Extruded Cut Feature

Insert a new sketch for the Extruded Cut Feature.
162) Right-click the **front face** of the Boss-Extrude1
feature in the Graphics window. This is the Sketch
plane. Boss-Extrude1 is highlighted in the
FeatureManager.

163) Click **Sketch** ✏️ from the Context toolbar as
illustrated. The Sketch toolbar is displayed.

Display a Front view - Hidden Lines Removed.
164) Click **Front view** 🔲 from the Heads-up View
toolbar.

165) Click **Hidden Lines Removed** 🔲 from the
Heads-up View toolbar.

The process of placing the mouse pointer over
an existing arc to locate its center point is called
"wake up."

Wake up the Center point.

166) Click the **Circle** ⊙ Sketch tool from the Sketch toolbar. The Circle PropertyManager is displayed.

167) Place the **mouse pointer** on the left arc. Do not click. The center point of the slot arc is displayed.

168) Click the **center point** of the arc. Click a **position** to the right of the center point to create the circle as illustrated.

Add a dimension.

169) Click the **Smart Dimension** ✏ Sketch tool.

170) Click the **circumference** of the circle.

171) Click a **position** diagonally above and to the left of the circle in the Graphics window.

172) Enter **.190**in, [**4.83**] in the Modify box.

173) Click the **Green Check mark** ✔ in the Modify dialog box.

Display an Isometric view - Shaded With Edges.

174) Click **Isometric view** 🔷 from the Heads-up View toolbar.

175) Click **Shaded With Edges** 🔷 from the Heads-up View toolbar.

Insert an Extruded Cut feature.

176) Click the **Features** tab from the CommandManager.

177) Click **Extruded Cut** 📷 from the Features toolbar. The Cut- Extrude PropertyManager is displayed.

178) Select **Through All** for End Condition in Direction 1. The direction arrow points to the back. Accept the default conditions.

179) Click **OK** ✔ from the Cut-Extrude PropertyManager. The Cut-Extrude1 feature is displayed in the FeatureManager.

Save the FLATBAR part.

180) Click **Save** 💾.

💡 Think design intent. When do you use various End Conditions? What are you trying to do with the design? How does the component fit into an Assembly?

The blue Cut-Extrude1 icon indicates that the feature is selected.

Select features by clicking their icons in the FeatureManager or by selecting their geometry in the Graphics window.

🔆 When you create a new part or assembly, the three default Planes (Front, Right and Top) are aligned with specific views. The Plane you select for the Base sketch determines the orientation of the part.

Activity: FLATBAR Part - Linear Pattern Feature

Create a Linear Pattern feature.

181) Click the **Linear Pattern** tool from the Features toolbar. The Linear Pattern PropertyManager is displayed. Cut-Extrude1 is displayed in the Features to Pattern box. Note: If Cut-Extrude1 is not displayed, click inside the Features to Pattern box. Click Cut-Extrude1 from the fly-out FeatureManager.

182) Click **inside** the Pattern Direction box.

183) Click the **top edge** of the Boss-Extrude1 feature for Direction1 in the Graphics window. Edge<1> is displayed in the Pattern Direction box.

184) Enter **0.5**in, [**12.70**] for Spacing.

185) Enter **9** for Number of Instances. Instances are the number of occurrences of a feature.

186) The Direction arrow should point to the right. Click the **Reverse Direction** button if required.

187) Check **Geometry pattern** from the Options box.

188) Click **OK** from the Linear Pattern PropertyManager. The LPattern1 feature is displayed in the FeatureManager.

Save the FLATBAR part.

189) Click **Save** 🖫. The FLATBAR part is complete. Later, apply material to the part.

Close all documents.
190) Click **Windows, Close All** from the Main menu.

 Review the FLATBAR Part.

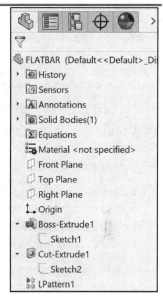

The FLATBAR part utilized an Extruded Boss/Base 🗍
feature as the first feature. The Sketch plane was the Front
Plane. The 2D sketch utilized the Straight Slot Sketch tool to
create the slot profile.

You added linear and radial dimensions to define your sketch.
You applied the Extruded Boss/Base feature with a Blind End
Condition in Direction 1. Boss-Extrude1 was created.

You created a circle sketch for the Extruded Cut feature on
the front face of Boss-Extrude1. The front face was your
Sketch plane for the Extruded Cut feature. The Extruded

Cut 🗍 feature removed material to create the hole. The
Extruded Cut feature default name was Cut-Extrude1. The Through All End Condition
option in Direction 1 created the Cut-Extrude1 feature. The Cut-Extrude1 feature is the
seed feature for the Linear Pattern of holes.

The Linear Pattern ⬚ feature created an array of 9 holes, equally spaced along the
length of the FLATBAR part. Enter material later in an exercise.

Later, add material to the FLATBAR part. To add material, right-click the **Material**
folder in the FeatureManager. Click **Edit Material**. The Material dialog box is displayed.
Select **material type**. Click **Apply**. Click **Close**. Return to the Graphics window.

LINKAGE Assembly

An assembly is a document that contains two or more parts. An assembly inserted into
another assembly is called a sub-assembly. A part or sub-assembly inserted into an
assembly is called a component. The LINKAGE assembly consists of the following
components: AXLE, SHAFT-COLLAR, FLATBAR, and AirCylinder sub-assembly.

Establishing the correct component relationship in an assembly requires forethought on
component interaction. Mates are geometric relationships that align and fit components in
an assembly. Mates remove degrees of freedom from a component.

Mate Types

Mates provide the ability to create geometric relationships between assembly components. Mates define the allowable directions of rotational or linear motion of the components in the assembly. Move a component within its degrees of freedom in the Graphics window to view the behavior of an assembly.

Standard Mates:

The Standard mates tab is the default tab. The mate tab provides the ability to insert a *Standard*, *Advanced*, *or Mechanical* mate or to address *Analysis*. Standard mate types:

- **Coincident**: Locate the selected faces, edges, points or planes so they use the same infinite line. A Coincident mate positions two vertices for contact.

- **Parallel**: Locate the selected items to lie in the same direction and to remain a constant distance apart. A parallel mate permits only translational motion of a single part with respect to another. No rotation is allowed.

- **Perpendicular**: Locate the selected items at a 90 degree angle to each other. The perpendicular mate allows both translational and rotational motion of one part with respect to another.

- **Tangent**: Locates the selected items in a tangent mate. At least one selected item must be either a conical, cylindrical, or spherical face.

- **Concentric**: Locates the selected items so they can share the same center point. You can prevent the rotation of components that are mated with concentric mates by selecting the Lock Rotation option. Locked concentric mates are indicated by an icon in the FeatureManager design tree.

- **Lock**: Maintains the position and orientation between two components.

- **Distance**: Locates the selected items with a specified distance between them. Use the drop-down arrow box or enter the distance value directly.

- **Angle**: Locates the selected items at the specified angle to each other. Use the drop-down arrow box or enter the angle value directly.

- **Mate Alignment** options:

    - **Aligned**: Positions the components so that the normal vectors from the selected faces point in the same direction.

- **Anti-Aligned**: Positions the components so that the normal vectors from the selected faces point in opposite directions.

- **Mate Options**: The following options are available: **Add to new folder**, **Show popup dialog**, **Show preview**, **Use for positioning only**, **Make first selection transparent**.

Advanced Mates:

Advanced mate types:

- **Profile Center**: Mate to center automatically center-aligns common component types such as rectangular and circular profiles to each other.

- **Symmetric**: Force two similar entities to be symmetric about a planar face or plane.

- **Width**: Center a tab within the width of a groove. The Width mate is used to replace the Symmetric mate where components have tolerance and a gap rather than a tight fit.

- **Path**: Enable any point on a component to be set to follow a defined path.

- **Linear/Linear Coupler**: Establish a relationship between the translation of one component and the translation of another component.

- **Limit**: Allow components to move within a range of values for distance and angle. Select the angle and distance from the provided boxes. Specify a starting distance or angle as well as a maximum and minimum value.

- **Distance**: Locate the selected items with a specified distance between them. Use the drop-down arrow box or enter the distance value directly (Limit mate).

- **Angle**: Locate the selected items at the specified angle to each other. Use the drop-down arrow box or enter the angle value directly.

- **Mate Alignment** options:

  - **Aligned**: Positions the components so that the normal vectors from the selected faces point in the same direction.

  - **Anti-Aligned**: Positions the components so that the normal vectors from the selected faces point in opposite directions.

Mechanical Mates:

Mechanical mate types:

- **Cam**: Forces a plane, cylinder, or point to be tangent or coincident to a series of tangent extruded faces.

- **Slot**: Mate bolts to straight or arced slots and you can mate slots to slots. Select an axis, cylindrical face, or a slot to create slot mates.

- **Hinge**: Limit the movement between two components to one rotational degree of freedom. It has the same effect as adding a Concentric mate plus a Coincident mate. You can also limit the angular movement between the two components.

- **Gear**: Force two components to rotate relative to one another around selected axes. The Gear mate provides the ability to establish gear type relations between components without making the components physically mesh. Align the components before adding the Mechanical gear mate.

- **Rack and Pinion**: Linear translation of one component (the rack) causes circular rotation in another component (the pinion), and vice versa. You can mate any two components to have this type of movement relative to each other. The components do not need to have gear teeth.

- **Screw**: Constrain two components to be concentric, and adds a pitch relationship between the rotation of one component and the translation of the other. Translation of one component along the axis causes rotation of the other component according to the pitch relationship. Likewise, rotation of one component causes translation of the other component. Align the components before adding the Mechanical screw mate.

- **Universal Joint**: Permit the transfer of rotations from one rigid body to another. This mate is particularly useful to transfer rotational motion around corners, or to transfer rotational motion between two connected shafts that are permitted to bend at the connection (drive shaft on an automobile).

Example: Utilize a Concentric mate between the AXLE cylindrical face and the FLATBAR Extruded Cut feature (hole). Utilize a Coincident mate between the SHAFT-COLLAR back face and the FLATBAR front flat face. The LINKAGE assembly requires the AirCylinder assembly. The AirCylinder assembly is located in the Pneumatic Components folder.

Activity: AirCylinder Assembly - Open and Save As option

Download all needed folders and files to your local hard drive.

191) Download the SW-TUTORIAL 2023 folder. Work from a local hard drive.

Create a new assembly.

192) Click **New** ⬜ from the Main menu. The New SOLIDWORKS Document dialog box is displayed.

193) Click the **Templates** tab.

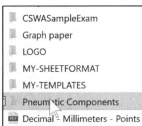

194) Double-click **Assembly** from the New SOLIDWORKS Document dialog box. The Begin Assembly PropertyManager is displayed. Open models are displayed in the Open documents box. If there are no Open SOLIDWORKS models, the Window Open dialog is also displayed.

195) Browse to the **SW-TUTORIAL 2023\Pneumatic Components** folder.

196) Double-click the **AirCylinder** assembly.

197) Click **OK** ✓ from the Begin Assembly PropertyManager. The AirCylinder assembly is displayed in the Graphics window. The (f) symbol is placed in front of the AirCylinder name in the FeatureManager. It is fixed to the origin.

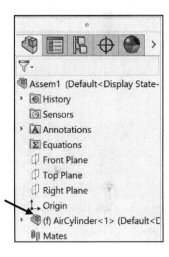

Mate to the first component added to the assembly. If you mate to the first component or base component of the assembly and decide to change its orientation later, all the components will move with it.

🔅 Determine the static and dynamic behavior of mates in each sub-assembly before creating the top-level assembly.

198) If required, click **Yes** to Rebuild.

199) Click **Save As** 🖫.

200) Select the **SW-TUTORIAL 2023** for Save in folder.

201) Enter **LINKAGE** for file name.

202) Click the **Include all referenced components** box.

203) Click the **Advanced** button. An assembly and drawing document has part\component references. You must save the assembly document and all of the reference components to the same folder. Note your options.

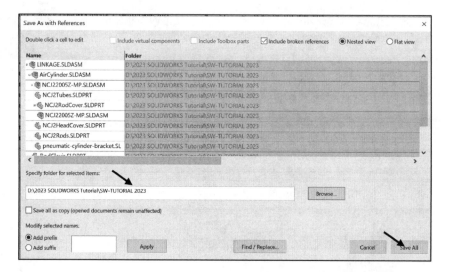

204) Click the **Browse** button from the Save As with References folder.

205) Select the **SW-TUTORIAL 2023** folder.

206) Click **Save All**. The LINKAGE assembly FeatureManager is displayed.

207) **View** that all components and the assembly are in the folder.

You can also use the Pack and Go option to save an assembly or drawing document with references, toolbox components, studies, etc.

Display the RodClevis component in the FeatureManager.
208) **Expand** the AirCylinder assembly in the FeatureManager.

209) Click **RodClevis<1>** from the FeatureManager. The RodClevis is highlighted in the Graphics window.

If required hide the Origins.
210) Click **View**, **Hide/Show**, un-check **Origins** from the Main menu.

The AirCylinder is the first component in the LINKAGE assembly and is fixed (f) to the LINKAGE assembly Origin.

Display an Isometric view.
211) Click **Isometric view** .

Insert the AXLE part.
212) Click the **Assembly** tab in the CommandManager.

213) Click the **Insert Components** Assembly tool. The Insert Component PropertyManager is displayed. Note: Open SOLIDWORKS models are displayed in the Open documents box. If there are no Open models, the Window Open dialog is also displayed.

214) Browse to the **SW-TUTORIAL 2023** folder. Note if AXLE is active, double-click AXLE from the Open documents box.

215) Double-click **AXLE** from the SW-TUTORIAL 2023 folder.

216) Click a **position** to the front of the AirCylinder assembly as illustrated.

Cut-Extrude5 of RodClevis<1>

Enlarge the view.
217) **Zoom in** on the RodClevis and the AXLE.

Insert a Concentric mate. Use the Quick mate procedure.
218) Click the **inside front hole face** of the RodClevis. The cursor displays the face feedback symbol.

219) Hold the **Ctrl** key down.

220) Click the **long cylindrical face** of the AXLE. The cursor displays the face feedback symbol.

221) Release the **Ctrl** key. The Mate pop-up menu is displayed.

Concentric

Lock rotation

222) Click **Concentric** from the Mate pop-up menu. The AXLE is positioned concentric to the RodClevis hole. Note your options (Lock rotation, Flip Mate Alignment). Do not select at this time.

Move the AXLE.

223) Click and drag the **AXLE** left to right. The AXLE translates in and out of the RodClevis holes.

Position the mouse pointer in the middle of the face to select the entire face. Do not position the mouse pointer near the edge of the face. If the wrong face or edge is selected, click inside the Graphics window to deselect the entity.

Display a Top view.

224) Click **Top view** .

Insert a Coincident mate. Expand the LINKAGE assembly and components in the FeatureManager.

225) **Expand** the LINKAGE assembly FeatureManager.

226) **Expand** the AirCylinder assembly.

227) Click the **Front Plane** of the AirCylinder assembly.

228) **Expand** the AXLE part from the FeatureManager.

229) Hold the **Ctrl** key down.

230) Click the **Front Plane** of the AXLE part.

231) Release the **Ctrl** key. The Mate pop-up menu is displayed.

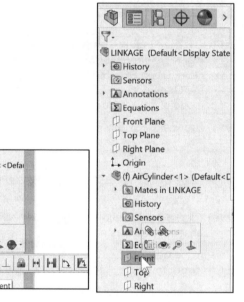

232) Click **Coincident mate** from the pop-up menu. The AirCylinder Front Plane and the AXLE Front Plane are Coincident. The AXLE is centered in the RodClevis.

If you delete a mate and then recreate it, the mate numbers will be in a different order.

To fix the first component to the Origin in an assembly, click OK from the Begin Assembly PropertyManager or click the Origin in the Graphics window.

Display an Isometric view.

233) Click **Isometric view** .

Display the Mates in the folder.
234) **Expand** the Mates folder in the FeatureManager. View the created mates.

Activity: LINKAGE Assembly - Insert FLATBAR Part

Insert the FLATBAR part.

235) Click the **Insert Components** Assembly tool. The Insert Component PropertyManager is displayed. Note: Open SOLIDWORKS models are displayed in the Open documents box. If there are no Open models, the Window Open dialog is also displayed.

236) Double-click **FLATBAR** from the SW-TUTORIAL 2023 folder.

Place the component in the assembly.
237) Click a **position** in the Graphics window as illustrated.

Enlarge the view.
238) **Zoom in** on the AXLE and the left side of the FLATBAR to enlarge the view.

Insert a Concentric mate.
239) Click the inside **left front hole face** of the FLATBAR.

240) Hold the **Ctrl** key down.

241) Click the **long cylindrical face** of the AXLE.

242) Release the **Ctrl** key. The Mate pop-up menu is displayed.

243) Click **Concentric** from the Mate pop-up menu.

Fit the model to the Graphics window.
244) Press the **f** key.

Move the FLATBAR.
245) Click and drag the **FLATBAR**. The FLATBAR translates and rotates along the AXLE.

Insert a Coincident mate.
246) Click the **front face** of the FLATBAR.

247) **Rotate** the model to view the back face of the RodClevis.

248) Hold the **Ctrl** key down.

249) Click the **back face** of the RodClevis as illustrated.

250) Release the **Ctrl** key. The Mate pop-up menu is displayed.

251) Click **Coincident** from the Mate pop-up menu.

Display an Isometric view.
252) Click **Isometric view** .

Insert the second FLATBAR component.

253) Click the **Insert Components** Assembly tool.

254) Browse to the **SW-TUTORIAL 2023** folder.

255) Double-click **FLATBAR**.

256) Click a **position** to the front of the AirCylinder in the Graphics window as illustrated.

Enlarge the view.
257) **Zoom in** on the second FLATBAR and the AXLE.

Insert a Concentric mate.
258) Click the **left front inside hole face** of the second FLATBAR.

259) Hold the **Ctrl** key down.

260) Click the **long cylindrical face** of the AXLE.

261) Release the **Ctrl** key. The Mate pop-up menu is displayed.

262) Click **Concentric** from the Mate pop-up menu.

263) Click and drag the **second FLATBAR** to the front.

Fit the model to the Graphics window.
264) Press the **f** key.

Insert a Coincident mate.

265) **Rotate** the model to view the back face of the second FLATBAR.

266) Click the **back face** of the second FLATBAR.

267) **Rotate** the model to view the front face of the RodClevis.

268) Hold the **Ctrl** key down.

269) Click the **front face** of the RodClevis.

270) Release the **Ctrl** key. The Mate pop-up menu is displayed.

271) Click **Coincident** from the Mate pop-up menu.

Insert a Parallel mate.

272) Press the **Shift-z** keys to Zoom in on the model.

273) Click the **top narrow face** of the first FLATBAR.

274) Hold the **Ctrl** key down.

275) Click the **top narrow face** of the second FLATBAR.

276) Release the **Ctrl** key.

277) Click **Parallel** ⟍ from the Mate pop-up menu.

Display an Isometric view.

278) Click **Isometric view** .

Move the two FLATBAR parts.

279) Click and drag the **second FLATBAR**. Both FLATBAR parts move together.

View the Mates folder.

280) **Expand** the Mates folder from the FeatureManager. View the created mates.

Determine the static and dynamic behavior of mates in each sub-assembly before creating the top-level assembly.

▸ 🔩 (f) AirCylinder<1> (Default<Default_Display State-1>)
▸ 🔩 (-) AXLE<1> (Default<<Default>_Display State 1>)
▸ 🔩 (-) FLATBAR<1> (Default<<Default>_Display State 1>)
▸ 🔩 (-) FLATBAR<2> (Default<<Default>_Display State 1>)
▾ 🔩 Mates
   ◎ Concentric1 (AirCylinder<1>,AXLE<1>)
   ⟋ Coincident1 (AirCylinder<1>,AXLE<1>)
   ◎ Concentric2 (AXLE<1>,FLATBAR<1>)
   ⟋ Coincident2 (AirCylinder<1>,FLATBAR<1>)
   ◎ Concentric3 (AXLE<1>,FLATBAR<2>)
   ⟋ Coincident3 (AirCylinder<1>,FLATBAR<2>)
   ⟍ Parallel1 (FLATBAR<1>,FLATBAR<2>)

Activity: LINKAGE Assembly - Insert SHAFT-COLLAR Part

Insert the first SHAFT-COLLAR.

281) Click the **Insert Components** Assembly tool.

282) Double-click **SHAFT-COLLAR** from the SW-TUTORIAL 2023 folder.

283) Click a **position** to the back of the AXLE as illustrated.

Enlarge the view.

284) Click the **Zoom to Area** tool.

285) Zoom-in on the SHAFT-COLLAR and the AXLE component.

Deactivate the tool.

286) Click the **Zoom to Area** tool.

Insert a Concentric mate.

287) Click the **inside hole face** of the SHAFT-COLLAR.

288) Hold the **Ctrl** key down.

289) Click the **long cylindrical face** of the AXLE.

290) Release the **Ctrl** key. The Mate pop-up menu is displayed.

291) Click **Concentric** from the Mate pop-up menu.

Insert a Coincident mate.

292) Press the **Shift-z** keys to Zoom in on the model.

293) Click the **front face** of the SHAFT-COLLAR as illustrated.

294) Rotate the model to view the back face of the first FLATBAR.

295) Hold the **Ctrl** key down.

296) Click the **back face** of the first FLATBAR.

297) Release the **Ctrl** key. The Mate pop-up menu is displayed.

298) Click **Coincident** from the Mate pop-up menu.

Display an Isometric view.

299) Click **Isometric view** .

Insert the second SHAFT-COLLAR.

300) Click the **Insert Components** Assembly tool.

301) Double-click the **SHAFT-COLLAR** part from the SW-TUTORIAL 2023 folder.

302) Click a **position** near the AXLE as illustrated.

Enlarge the view.

303) Click the **Zoom to Area** tool.

304) **Zoom-in** on the second SHAFT-COLLAR and the AXLE to enlarge the view.

305) Click the **Zoom to Area** tool to deactivate the tool.

Insert a Concentric mate.
306) Click the **inside hole face** of the second SHAFT-COLLAR.

307) Hold the **Ctrl** key down.

308) Click the **long cylindrical face** of the AXLE.

309) Release the **Ctrl** key. The Mate pop-up menu is displayed.

310) Click **Concentric** from the Mate pop-up menu.

Insert a Coincident mate.
311) Click the **back face** of the second SHAFT-COLLAR.

312) **Rotate** to view the front face of the second FLATBAR.

313) Hold the **Ctrl** key down.

314) Click the **front face** of the second FLATBAR.

315) Release the **Ctrl** key. The Mate pop-up menu is displayed.

316) Click **Coincident** from the Mate pop-up menu.

View the created Mates.

317) Expand the Mates folder. View the created mates.

Display an Isometric view.

318) Click **Isometric view** .

Fit the model to the Graphics window.

319) Press the **f** key.

Save the LINKAGE assembly.

320) Click **Save** . The LINKAGE assembly is complete.

 Use the Pack and Go option to save an assembly or drawing with references. The Pack and Go tool saves either to a folder or creates a zip file to e-mail.

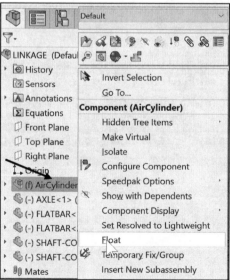

Review the LINKAGE Assembly.

An assembly is a document that contains two or more parts. A part or sub-assembly inserted into an assembly is called a component. You created the LINKAGE assembly. The AirCylinder sub-assembly was the first component inserted into the LINKAGE assembly. The AirCylinder assembly was obtained from the Pneumatic Components folder.

The AirCylinder assembly was fixed to the Origin. The Concentric and Coincident mates added Geometric relationships between the inserted components in the LINKAGE assembly.

To remove the fixed state, Right-click a component name in the FeatureManager. Click Float. The component is free to move.

To fix the first component to the Origin in an assembly, click OK from the Begin Assembly PropertyManager or click the Origin in the Graphics window.

The AXLE part was the second component inserted into the LINKAGE assembly. The AXLE required a Concentric mate between the two cylindrical faces and a Coincident mate between the two Front Planes.

The FLATBAR part was the third component inserted into the LINKAGE assembly. The FLATBAR required a Concentric mate between the two cylindrical faces and a Coincident mate between the two flat faces.

A second FLATBAR was inserted into the LINKAGE assembly. A Parallel mate was added between the two FLATBARs.

Two SHAFT-COLLAR parts were inserted into the LINKAGE assembly. Each SHAFT-COLLAR required a Concentric mate between the two cylindrical faces and a Coincident mate between the two flat faces.

Motion Study - Basic Motion Tool

Motion Studies are graphical simulations of motion for assembly models. You can incorporate visual properties such as lighting and camera perspective into a motion study. Motion studies do not change an assembly model or its properties. They simulate and animate the motion you prescribe for your model. Use SOLIDWORKS mates to restrict the motion of components in an assembly when you model motion.

Create a Motion Study. Select the Basic Motion option from the MotionManager. The Basic Motion option provides the ability to approximate the effects of motors, springs, collisions and gravity on your assembly. Basic Motion takes mass into account in calculating motion. Note: The Animation option does not.

Activity: LINKAGE Assembly - Basic Motion

Insert a Rotary Motor using the Motion Study tab.

321) Click the **Motion Study 1** tab located in the bottom left corner of the Graphics window. The MotionManager is displayed.

322) Select **Basic Motion** for Type of study from the MotionManager drop-down menu as illustrated.

323) Click **Motor** from the
MotionManager. The Motor
PropertyManager is
displayed.

324) Click the **Rotary Motor** box.

325) Click the **FLATBAR front
face** as illustrated. A red
Rotary Motor icon is
displayed. The red direction
arrow points
counterclockwise.

326) Enter **150 RPM** for speed in
the Motion box.

327) Click **OK** ✔ from the Motor PropertyManager.

Record the Simulation.

328) Click **Calculate** ⚙. The FLATBAR rotates in a
counterclockwise direction for a set period of time.

329) Click **Play** ▶ . View the simulation.

Save the Animation file to the SW-TUTORIAL 2023 folder.
330) Click **Save Animation**.

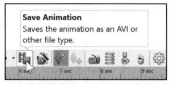

331) Select **Save as type**.

332) Select **Save in folder**.

333) Click **Save** from the Save Animation to File dialog box.
View your options.

334) Click **OK** from the Video Compression box.

Close the Motion Study and return to SOLIDWORKS.
335) Click the **Model** tab location in the bottom left corner of the Graphics window.

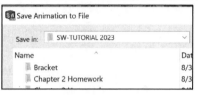

Fit the assembly to the Graphics window.
336) Press the **f** key.

Save the LINKAGE assembly.

337) Click **Save** 💾.

Exit SOLIDWORKS.
338) Click **Windows**, **Close All** from the Main menu.

The LINKAGE assembly chapter is complete.

Rename a feature or sketch for clarity. Slowly click the feature or sketch name twice and enter the new name when the old one is highlighted.

💡 Use the Pack and Go option to save an assembly or drawing with references. The Pack and Go tool saves either to a folder or creates a zip file to e-mail.

 Review the Motion Study.

The Rotary Motor Basic Motion tool combined Mates and Physical Dynamics to rotate the FLATBAR components in the LINKAGE assembly. The Rotary Motor was applied to the front face of the FLATBAR. You utilized the Calculate option to play the simulation. You saved the simulation in a file.

Chapter Summary

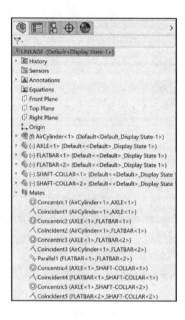

In this chapter, you created three parts (AXLE, SHAFT-COLLAR and FLATBAR), downloaded the AirCylinder assembly from the Pneumatic Components folder and created the LINKAGE assembly using the Standard mates.

You developed an understanding of the SOLIDWORKS User Interface: Menus, Toolbars, Task Pane, CommandManager, FeatureManager, System feedback icons, Document Properties, Parts, and Assemblies.

You created 2D sketches and addressed the three key states of a sketch: *Fully Defined*, *Over Defined* and *Under Defined*. Note: Always review your FeatureManager for the proper sketch state.

You obtained the knowledge of the following SOLIDWORKS features: Extruded Boss/Base, Extruded Cut, and Linear Pattern. Features are the building blocks of parts. The Extruded Boss/Base feature required a Sketch plane, sketch, and depth.

The Extruded Boss/Base feature added material to a part. The Boss-Extruded1 feature was utilized in the AXLE, SHAFT-COLLAR and FLATBAR parts.

The Extruded Cut feature removed material from the part. The Extruded Cut feature was utilized to create a hole in the SHAFT-COLLAR and FLATBAR parts. Note: Both were Through All holes. We will address the Hole Wizard later in the book. The Linear Pattern feature was utilized to create an array of holes in the FLATBAR part.

When parts are inserted into an assembly, they are called components. You created the LINKAGE assembly by inserting the AirCylinder assembly, AXLE, SHAFT-COLLAR and FLATBAR parts.

Mates are geometric relationships that align and fit components in an assembly. Concentric, Coincident and Parallel mates were utilized to assemble the components.

You created a Motion Study. The Rotary Motor Basic Motion tool combined Mates and Physical Dynamics to rotate the FLATBAR components in the LINKAGE assembly.

Note: Use the Select Other tool (See SOLIDWORKS Help) to select faces and edges that are hidden in an assembly.

Questions

1. Explain the steps in starting a SOLIDWORKS session.

2. Describe the procedure to begin a new 2D sketch.

3. Explain the steps required to modify units in a part document from inches to millimeters.

4. Describe the procedure to create a simple 3D cylinder on the Front Plane with an Extruded Boss/Base (Boss-Extrude1) feature.

5. Identify the three default Reference planes in SOLIDWORKS.

6. Describe a Base feature. Provide an example from this chapter.

7. Describe the differences between an Extruded Boss/Base feature and an Extruded Cut feature.

8. The sketch color black indicates a sketch is _____ defined.

9. The sketch color blue indicates a sketch is _____ defined.

10. The sketch color red indicates a sketch is _____ defined.

11. Describe the procedure to "wake up" a center point in a sketch.

12. Define a Geometric relation. Provide three examples.

13. Describe the procedure to create a Linear Pattern feature.

14. Describe an assembly or sub-assembly.

15. What are mates and why are they important in assembling components?

16. In an assembly, each component has_____# degrees of freedom. Name them.

17. True or False. A fixed component cannot move in an assembly.

18. Review the Design Intent section in the book. Identify how you incorporated design intent into a part or assembly document.

Exercises

Exercise 2.1: Identify the Sketch plane for the Boss-Extrude1 (Base) feature as illustrated. Simplify the number of features.

A: Top Plane

B: Front Plane

C: Right Plane

D: Left Plane

Origin

Correct answer _____.

Create the part. Dimensions are arbitrary.

Exercise 2.2: Identify the Sketch plane for the Boss-Extrude1 (Base) feature as illustrated. Simplify the number of features.

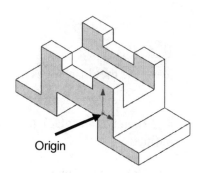

A: Top Plane

B: Front Plane

C: Right Plane

D: Left Plane

Origin

Correct answer _____.

Create the part. Dimensions are arbitrary.

Exercise 2.3: Identify the Sketch plane for the Boss-Extrude1 (Base) feature as illustrated. Simplify the number of features.

A: Top Plane

B: Front Plane

C: Right Plane

D: Left Plane

Origin

Correct answer _____.

Create the part. Dimensions are arbitrary.

Exercise 2.4: FLATBAR - 3HOLE Part

Create an ANSI, IPS FLATBAR - 3HOLE part document.

- Utilize the Front Plane for the Sketch plane. Insert an Extruded Base (Boss-Extrude1) feature. Do not display Tangent Edges.

- Create an Extruded Cut feature. This is your seed feature. Apply the Linear Pattern feature. The FLATBAR - 3HOLE part is manufactured from 0.06in., [1.5mm] 6061 Alloy.

Exercise 2.5: FLATBAR - 5HOLE Part

Create an ANSI, IPS, FLATBAR - 5HOLE part as illustrated.

- Utilize the Front Plane for the Sketch plane. Insert an Extruded Base (Boss-Extrude1) feature.

- Create an Extruded Cut feature. This is your seed feature. Apply the Linear Pattern feature. The FLATBAR - 5HOLE part is manufactured from 0.06in, [1.5mm] 6061 Alloy.

- Calculate the required dimensions for the FLATBAR - 5HOLE part. Use the following information: Holes are .500in. on center, Radius is .250in., and hole diameter is .190in.

- Do not display Tangent edges in the final model.

Think design intent. When do you use the various End Conditions and Geometric sketch relations? What are you trying to do with the design? How does the component fit into an assembly?

Exercise 2.6: Simple Block Part

Create the illustrated ANSI part. Note the location of the Origin in the illustration.

- Calculate the overall mass of the illustrated model.

- Apply the Mass Properties tool.

- Think about the steps that you would use to build the model.

- Review the provided information carefully.

- Units are represented in the IPS (inch, pound, second) system.

- A = 3.50in, B = .70in

Given:
A = 3.50
B = .70
Material: 1060 Alloy
Density = 0.0975 lb/in^3
Units: IPS
Decimal places: 2

Origin

Exercise 2.7: Simple Block Part

Create the illustrated ANSI part. Note the location of the Origin in the illustration.

Create the sketch symmetric about the Front Plane. The Front Plane in this problem is **not** your Sketch Plane. Utilize the Blind End Condition in Direction 1.

Given:
A = 3.00
B = .75
Material: Copper
Density = 0.321 lb/in^3
Units: IPS
Decimal places: 2

- Calculate the overall mass of the illustrated model.

- Apply the Mass Properties tool.

- Think about the steps that you would use to build the model.

- Review the provided information carefully. Units are represented in the IPS (inch, pound, second) system.

- A = 3.00in, B = .75in

Note: Sketch1 is symmetrical.

Origin

Exercise 2.8: Simple Block Part

Create an ANSI part from the illustrated model. Note the location of the Origin in the illustration.

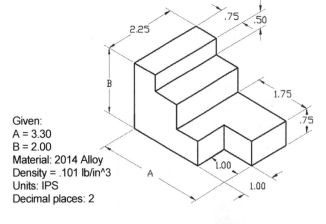

- Calculate the volume of the part and locate the Center of mass with the provided information.

Given:
A = 3.30
B = 2.00
Material: 2014 Alloy
Density = .101 lb/in^3
Units: IPS
Decimal places: 2

- Apply the Mass Properties tool.

- Think about the steps that you would use to build the model.

- Review the provided information carefully.

Origin

Exercise 2.9: Simple Block Part

Create an ANSI, MMGS part from the illustrated drawing: Front, Top, Right and Isometric views.

Note the location of the Origin in the illustration. The drawing views are displayed in Third Angle Projection.

- Apply 1060 Alloy for material.

- Calculate the Volume of the part.

- Locate the Center of mass.

Think about the steps that you would use to build the model. The part is symmetric about the Front Plane.

Origin

Exercise 2.10: Simple Block Part

Create the ANSI, MMGS part from the illustrated drawing: Front, Top, Right and Isometric views.

- Apply 1060 Alloy for material.

- The part is symmetric about the Front Plane.

- Calculate the Volume of the part and locate the Center of mass.

Think about the steps that you would use to build the model.

The drawing views are displayed in Third Angle Projection.

Origin

Exercise 2.11: LINKAGE-2 Assembly

Create the LINKAGE-2 assembly.

- Copy the LINKAGE assembly from the Chapter 2 Homework folder to your local hard drive.

- Open the assembly.

- Select Save As from the drop-down Main menu.

- Check the Save as copy and open check box.

- Enter LINKAGE-2 for file name. LINKAGE-2 ASSEMBLY for description.

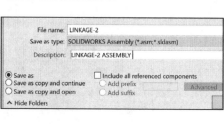

The FLATBAR-3HOLE part was created in
Exercise 2.4. Utilize two AXLE parts, four
SHAFT COLLAR parts, and two FLATBAR-
3HOLE parts to create the LINKAGE-2
assembly as illustrated.

- Insert the first AXLE part.

- Insert a Concentric mate.

- Insert a Coincident mate.

- Insert the first FLATBAR-3HOLE part.

- Insert a Concentric mate.

- Insert a Coincident mate.

- Perform the same procedure for the second
  FLATBAR-3HOLE part.

- Insert a Parallel mate between the 2
  FLATBAR-3HOLE parts. Note: The 2
  FLATBAR-3HOLE parts move together.

🔆 When a component is in the Lightweight

🧊 state, only a subset of its model data is
loaded into memory. The remaining model data
is loaded on an as-needed basis.

🔆 When a component is fully resolved, all its
model data is loaded into memory.

- Insert the second AXLE part.

- Insert a Concentric mate.

- Insert a Coincident mate.

- Insert the first SHAFT-COLLAR part.

- Insert a Concentric mate.

- Insert a Coincident mate.

- Perform the same tasks to insert
  the other three required SHAFT-
  COLLAR parts as illustrated.

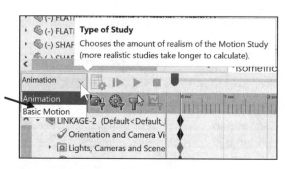

Exercise 2.12: LINKAGE-2 Assembly Motion Study

Create a Motion Study using the
LINKAGE-2 Assembly that was created
in the previous exercise.

- Create a Basic Motion Study.

- Apply a Rotary Motor to the front
  FLATBAR-3HOLE as illustrated.

- Play and Save the Simulation.

Exercise 2.13: **ROCKER Assembly**

Create a ROCKER assembly. The
ROCKER assembly consists of two AXLE
parts, two FLATBAR-5HOLE parts, and
two FLATBAR-3HOLE parts.

The FLATBAR-3HOLE parts are linked
together with the FLATBAR-5HOLE.

The three parts rotate clockwise and
counterclockwise, above the Top Plane.
Create the ROCKER assembly.

- Insert the first FLATBAR-5HOLE part. The
  FLATBAR-5HOLE is fixed to the Origin of the
  ROCKER assembly.

- Insert the first AXLE part.

- Insert a Concentric mate.

- Insert a Coincident mate.

- Insert the second AXLE part.

- Insert a Concentric mate.

- Insert a Coincident mate.

- Insert the first FLATBAR-3HOLE part.

- Insert a Concentric mate.

- Insert a Coincident mate.

- Insert the second FLATBAR-3HOLE part.

- Insert a Concentric mate.

- Insert a Coincident mate.

- Insert the second FLATBAR-5HOLE part.

- Insert the required mates.

Note: The end holes of the second FLATBAR-5HOLE are concentric with the end holes of the FLATBAR-3HOLE parts.

Note: In mechanical design, the ROCKER assembly is classified as a mechanism. A Four-Bar Linkage is a common mechanism comprised of four links.

Link1 is called the Frame.

The AXLE part is Link1.

Link2 and Link4 are called the Cranks.

The FLATBAR-3HOLE parts are Link2 and Link4. Link3 is called the Coupler. The FLATBAR-5HOLE part is Link3.

💡 If an assembly or component is loaded in a Lightweight state, right-click the assembly name or component name from the FeatureManager. Click Set Lightweight to Resolved.

💡 Determine the static and dynamic behavior of mates in each sub-assembly before creating the top-level assembly.

Exercise 2.14: 4 Bar linkage

Create the 4-bar linkage assembly as illustrated.
Create the five components. Assume
dimensions.

In an assembly, fix (f) the first component to
the origin or fully define it to the three default
reference planes.

Insert all needed mates to simulate the
movement of a 4-bar linkage assembly.

Read the section on Coincident, Concentric and
Distance mates in SOLIDWORKS help and in
the SOLIDWORKS 2023 Tutorial book.

Create a base with text for extra credit.

Below are a few sample models
from my Freshman Engineering
class.

Note the different designs to
maintain the proper movement
of a 4-bar linkage using a base.

Exercise 2.15: Fill Pattern 2-1

Create a Polygon Layout Fill Pattern feature. Apply the seed cut option.

1. Open **Fill Pattern 2-1** from the Chapter 2 Homework folder.

2. Click the **Fill Pattern** Features tool. The Fill Pattern PropertyManager is displayed.

3. Click the **Front face** of Boss-Extrude1. Face<1> is displayed in the Fill Boundary box. The direction arrow points to the right.

4. Click **Polygon** for Pattern Layout.

5. Click **Target spacing**.

6. Enter **15**mm for Loop Spacing.

7. Enter **6** for Polygon sides.

8. Enter **10**mm for Instances Spacing.

9. Enter **10**mm for Margins. View the direction arrow.

10. Click **Create seed cut**.

11. Click **Circle** for Features to Pattern.

12. Enter **4**mm for Diameter.

13. Click **OK** from the Fill Pattern PropertyManager. Fill Pattern1 is created and is displayed in the FeatureManager.

14. **View** the results.

Direction 1

Exercise 2.16: Fill Pattern 2-2

Create a Circular Layout Fill Pattern feature.

1. **Copy** the Chapter 2 Homework folder to your hard drive

2. Open **Fill Pattern 2-2** from the Chapter 2 Homework folder.

3. Click the **Fill Pattern** 🖼 Features tool. The Fill Pattern PropertyManager is displayed.

4. Click the **top face** of Boss-Extrude1. Face<1> is displayed in the Fill Boundary box. The direction arrow points to the back.

5. Click **Circular** for Pattern Layout.

6. Enter **.10**in for Loop Spacing.

7. Click **Target spacing**.

8. Enter .**10**in Instance Spacing.

9. Enter .**05**in for Margins. Edge<1> is selected for Pattern Direction.

10. Click **inside** the Features to Pattern box.

11. Click **Boss-Extrude2** from the fly-out FeatureManager. Boss-Extrude2 is displayed in the Features to Pattern box.

12. Click **OK** ✔ from the Fill Pattern PropertyManager. Fill Pattern1 is created and is displayed in the FeatureManager.

13. **View** the results.

Exercise 2.17: Tangent Mate

Create a Tangent mate between the roller-wheel and the cam on both sides to simulate movement.

1. **Copy** the Chapter 2 Homework folder to your hard drive.

2. Open **Tangent** from the Chapter 2 Homework\Tangent folder. The assembly is displayed.

Create a Tangent mate between the roller-wheel and the cam on both sides. The cam was created with imported geometry (curve).

💡 A Tangent mate places the selected items tangent to each other. At least one selected item must be either a conical, cylindrical, spherical face. An Angle mate places the selected items at the specified angle to each other.

3. Click the **Mate** 🖇 tool from the Assembly tab.

4. Click the **Standard** tab.

5. **Pin** the Mate PropertyManager.

Create the first Tangent mate.

6. Click the **contact face** of the left cam as illustrated.

7. Click the **contact face** of the left roller-wheel as illustrated.

8. Click **Tangent** ○ from the Mate Pop-up menu.

9. Click **Add/Finish** mate ✔ from the Pop-up menu.

You can set an option in the Mate PropertyManager so that the first component you select from becomes transparent. Then selecting from the second component is easier, especially if the second component is behind the first. The option is supported for all mate types except those that might have more than one selection from the first component (width, symmetry, linear coupler, cam, and hinge).

Create the second Tangent mate.

10. Click the **contact face** of the right cam.

11. Click the **contact face** of the right roller-wheel.

12. Click **Tangent** ○ from the Mate Pop-up menu.

13. Click **Add/Finish mate** ✔ from the Pop-up menu.

14. **Un-Pin** the Mate PropertyManager.

15. Click **OK** ✔ from the Mate PropertyManager.

16. Display an **Isometric** view.

17. **Move** the cam. View the results.

Notes:

Chapter 3

FRONT-SUPPORT Assembly

Below are the desired outcomes and usage competencies based on the completion of Chapter 3.

Desired Outcomes:	**Usage Competencies**:

• Create four parts:
 	
   o HEX-STANDOFF
 	
   o ANGLE-13HOLE
 	
   o TRIANGLE
 	
   o SCREW
	• Apply the following model features: Extruded Boss/Base, Extruded Thin, Extruded Cut, Revolved Boss/Base, Hole Wizard, Linear Pattern, Circular Pattern, Mirror, Fillet and Chamfer.
	
	• Apply sketch techniques with various sketch tools and Construction geometry.

• Create an assembly: o FRONT-SUPPORT assembly	• Comprehend the assembly process and insert the following mate types: Concentric, Coincident, Parallel, Distance, Gear, Screw, Hinge, and Cam.

Notes:

Chapter 3 - FRONT-SUPPORT Assembly

Chapter Objective

Create four new parts utilizing the Top, Front and Right Planes. Determine the Sketch plane for each feature. Obtain the knowledge of the following SOLIDWORKS features: Extruded Boss/Base, Extruded Thin, Extruded Cut, Revolved Boss/Base, Hole Wizard, Linear Pattern, Circular Pattern, Fillet, and Chamfer.

Apply sketch techniques with various Sketch tools: Line, Circle, Corner Rectangle, Centerline, Dynamic Mirror, Straight Slot, Trim Entities, Polygon, Tangent Arc, Sketch Fillet, Offset Entities, and Convert Entities.

Create four new parts:

1. HEX-STANDOFF
2. ANGLE-13HOLE
3. TRIANGLE
4. SCREW

Create the FRONT-SUPPORT assembly.

On the completion of this chapter, you will be able to:

- Select the correct Sketch plane.

- Generate a 2D sketch.

- Insert the required dimensions and Geometric relations.

- Apply the following SOLIDWORKS features:
  - Extruded Boss/Base
  - Extruded Cut
  - Extruded Thin
  - Revolved Base
  - Linear and Circular Pattern
  - Mirror
  - Fillet
  - Hole Wizard
  - Chamfer
- Apply the following mate types: Concentric, Coincident, Parallel, Distance, Gear, and Cam.

Chapter Overview

The FRONT-SUPPORT assembly supports various pneumatic components and is incorporated into the PNEUMATIC-TEST-MODULE.

Create four new parts in this chapter:

1. HEX-STANDOFF

2. ANGLE-13HOLE

3. TRIANGLE

4. SCREW

ANGLE-13HOLE

HEX-STANDOFF

Create the FRONT-SUPPORT assembly using the four new created parts.

The FRONT-SUPPORT assembly is used in the exercises at the end of this chapter and in later chapters of the book.

TRIANGLE

Simplified SCREW

Think design intent. When do you use various End Conditions and Geometric sketch relations? What are you trying to do with the design? How does the component fit into an assembly?

Download all needed model files (SW-TUTORIAL 2023 folder) from the SDC Publication website (www.SDCpublications.com/downloads/978-1-63057-556-4).

FRONT-SUPPORT

Reference Planes and Orthographic Projection

The three default ⊥ Reference planes represent infinite 2D planes in 3D space:

- Front
- Top
- Right

Planes have no thickness or mass.

Orthographic projection is the process of projecting views onto parallel planes with ⊥ projectors.

The default ⊥ datum planes are:

- Primary
- Secondary
- Tertiary

These are the planes used in manufacturing:

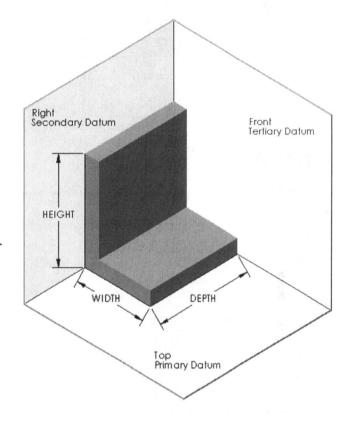

- Primary datum plane contacts the part at a minimum of three points.
- Secondary datum plane contacts the part at a minimum of two points.
- Tertiary datum plane contacts the part at a minimum of one point.

The part view orientation depends on the Base feature Sketch plane.

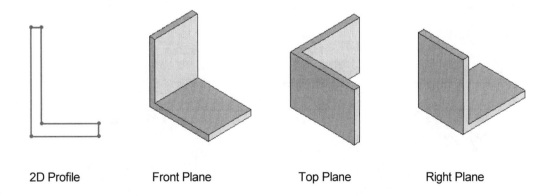

2D Profile   Front Plane   Top Plane   Right Plane

The part view orientation is dependent on the Base feature Sketch plane. Compare the available default Sketch planes in the FeatureManager: Front, Top and Right Plane.

Each Boss-Extrude1 feature above was created with an L-shaped 2D Sketch profile. The six principle views of Orthographic projection listed in the ASME Y14.3standard are:

- Top

- Front

- Right side

- Bottom

- Rear (Back)

- Left side

SOLIDWORKS Standard view names correspond to these Orthographic projection view names.

ASME Y14.3M Principle View Name:	SOLIDWORKS Standard View:
Front	Front
Top	Top
Right side	Right
Bottom	Bottom
Rear	Back
Left side	Left

The standard drawing views in Third Angle Orthographic projection are:

- Front

- Top

- Right

- Isometric

There are two Orthographic projection drawing systems. The first Orthographic projection system is called the Third Angle projection. The second Orthographic projection system is called the First Angle projection. The systems are derived from positioning a 3D object in the third or first quadrant.

Third Angle Projection

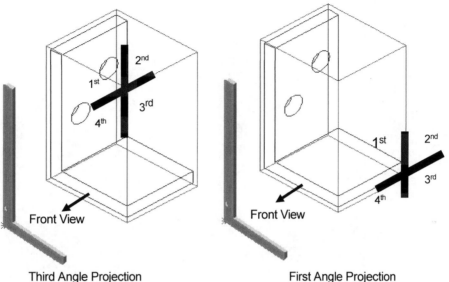

Third Angle Projection   First Angle Projection

The part is positioned in the third quadrant in third angle projection. The 2D projection planes are located between the viewer and the part. The projected views are placed on a drawing.

First Angle Projection

The part is positioned in the first quadrant in First Angle projection. Views are projected onto the planes located behind the part. The projected views are placed on a drawing. First Angle projection is primarily used in Europe and Asia.

Third Angle projection is primarily used in the U.S. & Canada and is based on the ASME Y14.3M Multi and Sectional View Drawings standard. Designers should have knowledge and understanding of both systems.

There are numerous multi-national companies.
Example: A part is designed in the U.S., manufactured in Japan and destined for a European market.

Third Angle projection is used in this text. A truncated cone symbol appears on the drawing to indicate the Projection System:

Third Angle Projection Symbol

First Angle Projection Symbol

Select the Sketch plane based on symmetry and orientation of the part in the FRONT-SUPPORT assembly. Utilize the standard views: *Front*, *Back*, *Right*, *Left*, *Top*, *Bottom* and *Isometric* to orient the part. Create the 2D drawings for the parts in Chapter 4.

HEX-STANDOFF Part

The HEX-STANDOFF part is a hexagonal shaped part utilized to elevate components in the FRONT-SUPPORT assembly. Machine screws are utilized to fasten components to the HEX-STANDOFF.

Create the HEX-STANDOFF part with the Extruded Boss/Base feature. The Sketch plane for the HEX-STANDOFF Boss-Extrude1 feature is the Top Plane.

Create the HEX-STANDOFF in the orientation utilized by the FRONT-SUPPORT assembly.

HEX-STANDOFF

Origin   Origin   Origin

Note the location of the origin in the model. When you create a new part, the three default Planes (Front, Right and Top) are aligned with specific views. The Plane you select for the Base sketch determines the orientation of the part.

All sketches should be fully defined in the FeatureManager.

Insert Geometric relations first then dimensions in a sketch to maintain design intent. Display geometric relations in the sketch.

The Boss-Extrude1 feature sketch consists of two profiles. The first sketch is a circle centered at the Origin on the Top Plane.

The second sketch is a polygon with 6 sides centered at the Origin. The polygon utilizes an inscribed circle to construct the geometry.

Geometric relations are constraints that control the size and position of the sketch entities. Apply a Horizontal relation in the polygon sketch.

Extrude the sketch perpendicular to the Top Plane. Utilize the Edit Sketch tool to modify the sketch.

The Hole Wizard feature creates complex and simple Hole features. Utilize the Hole Wizard feature to create a Tapped Hole.

The Tapped Hole depth and diameter are based on drill size and screw type parameters. Apply a Coincident relation to position the Tapped Hole aligned with the Origin.

Activity: HEX-STANDOFF Part - Extruded Boss/Base Feature

Create a new part.

1) Click **New** from the Main menu.

2) Click the **Templates** tab.

3) Double-click **Part**. The Part FeatureManager is displayed.

Set Document Properties. Set drafting standard.

4) Click **Options** from the Main menu.

5) Click the **Document Properties** tab from the dialog box. Select **ANSI** from the Overall drafting standard box.

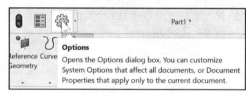

Set document units. Set Length and Angle precision.

6) Click **Units**.

7) Select **IPS**, [**MMGS**] for Unit system.

8) Select **.123**, [**.12**] for Length units Decimal places.

9) Select **None** for Angular units Decimal places.

10) Click **OK**. The Part FeatureManager is displayed.

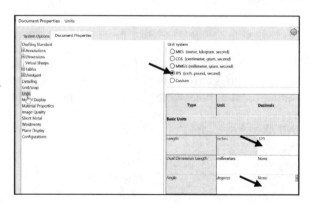

The primary units are provided in IPS (inch, pound, seconds). The optional secondary units are provided in MMGS (millimeter, gram, second) and are indicated in brackets [ ]. Illustrations are provided in inches and millimeters.

Save the part. Enter part name. Enter part description.

11) Click **Save As** 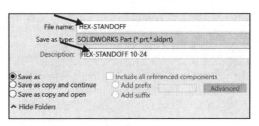.

12) Select the **SW-TUTORIAL 2023** folder.

13) Enter **HEX-STANDOFF** for File name.

14) Enter **HEX-STANDOFF 10-24** for Description. Click **Save**. The HEX-STANDOFF FeatureManager is displayed.

Select the Sketch plane.
15) Right-click **Top Plane** from the FeatureManager. This is the Sketch plane.

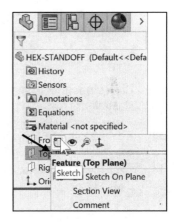

Insert a Boss-Extrude1 feature (Base feature) sketched on the Top Plane. Note: A plane is an infinite 2D area. The blue boundary is for visual reference.

Insert a new sketch.
16) Click **Sketch** from the Context toolbar. The Sketch toolbar is displayed.

17) Click the **Circle** ⊙ Sketch tool. The Circle PropertyManager is displayed.

18) Drag the **mouse pointer** into the Graphics window. The cursor displays the Circle icon.

19) Click the **Origin** ↳.

20) Drag the **mouse pointer** to the right of the Origin.

21) Click a **position** to create the circle as illustrated.

Insert a Polygon.
22) Click the **Polygon** ⊙ Sketch tool. The Polygon PropertyManager is displayed. The cursor displays the Polygon icon.

23) Click the **Origin** ↳ as illustrated.

24) Drag the **mouse pointer** horizontally to the right.

25) Click a **position** to the right of the circle to create the hexagon as illustrated.

Add a dimension.

26) Click the **Smart Dimension** Sketch tool.

27) Click the **circumference** of the first circle.

28) Click a **position** diagonally above the hexagon to locate the dimension.

29) Enter **.150**in, [**3.81**] in the Modify box.

30) Click the **Green Check mark** in the Modify dialog box.

31) Click the **circumference** of the inscribed circle.

32) Click a **position** to locate the dimension as illustrated.

33) Enter **.313**in, [**7.95**] in the Modify box.

34) Click the **Green Check mark** in the Modify dialog box.

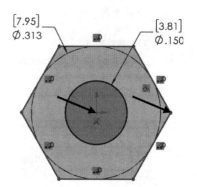

Exit Smart Dimension.

35) Right-click **Select**.

Add a Horizontal relation to fully defined the sketch.

36) Click the **Origin** .

37) Hold the **Ctrl** key down.

38) Click the right most **point** of the hexagon. The Properties PropertyManager is displayed.

39) Release the **Ctrl** key.

40) Click **Horizontal** from the Add Relations box.

41) Click **OK** from the Properties PropertyManager. The black sketch is fully defined.

Extrude the sketch.

42)   Click the **Features** tab in the CommandManager.

43)   Click **Extruded Boss/Base** from the Features toolbar. The Boss-Extrude PropertyManager is displayed. Blind is the default End Condition in Direction1. The direction arrow points upward.

44)   Enter **.735**in, [**18.67**] for Depth.

45)   Click **OK** from the Boss-Extrude PropertyManager. Boss-Extrude1 is displayed in the FeatureManager.

Fit the model to the Graphics window.

46)   Press the **f** key.

The Boss-Extrude1 feature (Base feature) was sketched on the Top Plane. Changes occur in the design process. Edit the sketch of Boss-Extrude1. Delete the circle and close the sketch. Apply the Hole Wizard feature to create a Tapped Hole.

Edit Sketch1.

47)   **Expand** Boss-Extrude1 in the FeatureManager.

48)   Right-click **Sketch1** in the FeatureManager.

49)   Click **Edit Sketch** from the Context toolbar.

Delete the inside circle.

50)   Click the **circumference** of the inside circle as illustrated. The Circle PropertyManager is displayed.

51)   Press the **Delete** key.

52)   Click **Yes** to the Sketcher Confirm Delete message. Both the circle geometry and its dimension are deleted.

Save and close the sketch.

53)   Click **Save**.

54)   Click **Rebuild and Save**. The Boss-Extrude1 feature is updated. Note: Sketch1 is fully defined.

Fit the model to the Graphics window.

55)   Press the **f** key.

Activity: HEX-STANDOFF Part - Hole Wizard Feature

Insert a Tapped Hole with the Hole Wizard feature tool. Create a 2D Sketch.

56) Click **Hidden Lines Visible** ⊞ from the Heads-up View toolbar.

57) Click **Hole Wizard** 🕳 from the Features toolbar. The Hole Specification PropertyManager is displayed.

Note: For metric, utilize ANSI Metric and M5x0.8 for size.

58) Click **Straight Tap** for Hole Specification.

59) Select **ANSI Inch**, [**ANSI Metric**] for Standard.

60) Select **Bottoming Tapped Hole** for Type.

61) Select **#10-24**, [**Ø5**] for Size.

62) Select **Through All** for End Condition.

63) Click the **Cosmetic thread** box. Accept the default conditions.

64) Click the **Positions** tab.

65) Click the **top face** of Boss-Extrude1 to the right of the origin. Do not select the center of the part.

66) Click the **center of the part** to locate the center point of the hole as illustrated. The Straight Tapped hole is displayed in yellow. Yellow is a preview color. The Point Sketch tool is automatically selected. No other holes are required. You created a Coincident relation to the center point of the Tapped Hole in the Top view.

67) Right-click **Select** in the Graphics window to deselect the Point Sketch tool. **Do not** right-click on the model.

68) Click **OK** ✅ from the Hole Position PropertyManager.

The #10-24 Tapped Hole1 feature is displayed in the FeatureManager. Sketch2 determines the center point location of the Tapped Hole. Sketch3 is the profile of the Tapped Hole.

Save the HEX-STANDOFF part.

69) Click **Shaded With Edges** 🟦 from the Heads-up View toolbar.

70) Click **inside** the Graphics window.

71) Click **Save** 💾. The HEX-STANDOFF is complete. Later, add material. View the threads.

💡 To view the thread, right-click the Annotations folder, and click Details. Check the Cosmetic thread box and the Shaded cosmetic threads box. Click OK.

 Review the HEX-STANDOFF Part.

The HEX-STANDOFF part utilized the Extruded Boss/Base feature. The Boss-Extrude1 feature required a sketch on the Top Plane. The first profile was a circle centered at the Origin on the Top Plane. The second profile used the Polygon Sketch tool. You utilized the Edit Sketch tool to modify the Sketch profile and to delete the circle.

The Hole Wizard feature created a Tapped Hole. The Hole Wizard feature required the Boss-Extrude1 top face as the Sketch plane.

A Coincident relation located the center point of the Tapped Hole aligned with respect to the Origin.

CommandManager and FeatureManager tabs and file folders will vary depending on system setup and Add-ins.

ANGLE-13HOLE Part

The ANGLE-13HOLE part is an L-shaped support bracket. The ANGLE-13HOLE part is manufactured from 0.090in, [2.3] aluminum.

The ANGLE-13HOLE part contains fillets, holes, and slot cuts.

ANGLE-13HOLE

Simplify the overall design into seven features. Utilize symmetry and Linear Patterns.

The open L-Shaped profile is sketched on the Right Plane.

Utilize an Extruded Thin feature with the Mid Plane option to locate the part symmetrical to the Right Plane.

Insert the first Extruded Cut feature for the first hole. This is the seed feature for the Linear Pattern. The hole sketch is located on the top face of the Extruded Thin feature.

Insert a Linear Pattern feature to create an array of 13 holes along the bottom horizontal edge.

Insert a Fillet feature to round the four corners.

Model about the origin; this provides a point of reference.

Insert the second Extruded Cut feature on the front face of the Extruded Thin feature. This is the seed feature for the second Linear Pattern.

Insert a Linear Pattern feature to create an array of 3 holes along the top horizontal edge.

Utilize the Sketch Mirror tool to create the slot profile. Use the Slot Sketch tool.

Insert the third Extruded Cut feature to create the two slots.

Select the Sketch plane for the Base feature that corresponds to the parts orientation in the assembly.

Activity: ANGLE-13HOLE Part - Documents Properties

Create a new part.

72) Click **New** from the Main menu.

73) Double-click **Part**. The Part FeatureManager is displayed.

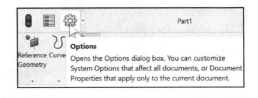

Set Document Properties. Set drafting standard, units, and precision.

74) Click **Options** from the Main menu.

75) Click the **Document Properties** tab from the dialog box.

76) Select **ANSI** from the Overall drafting standard drop-down menu.

77) Click **Units**.

78) Select **IPS**, [**MMGS**] for Unit system.

79) Select **.123**, **[.12]** for Length units Decimal places. Select **None** for Angular units Decimal places. Click **OK**.

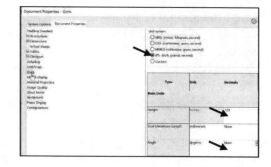

Save the part. Enter part name. Enter part description. Save the part in the SW-TUTORIAL 2023 folder.

80) Click **Save As** .

81) Select the **SW-TUTORIAL 2023** folder. Enter **ANGLE-13 HOLE** for File name.

82) Enter **ANGLE BRACKET-13 HOLE** for Description. Click **Save**. The ANGLE-13 Hole FeatureManager is displayed.

Activity: ANGLE-13HOLE Part - Extruded Thin Feature

Insert an Extruded Thin feature sketched on the Right Plane. Select the Sketch plane.

83) Right-click **Right Plane** from the FeatureManager.

Sketch a horizontal line.

84) Click **Sketch** from the Context toolbar. The Sketch toolbar is displayed.

85) Click the **Line** Sketch tool from the Sketch toolbar.

86) Click the **Origin** as illustrated. Click a **position** to the right of the Origin.

Sketch a vertical line.

87) Click a **position** directly above the right end point.

De-select the Line Sketch tool.

88) Right-click **Select** in the Graphics window.

Add an Equal relation.

89) Click the **vertical** line. Hold the **Ctrl** key down.

90) Click the **horizontal line**. Release the **Ctrl** key.

91) Right-click **Make Equal** ═ from the Context toolbar.

92) Click **OK** from the Properties PropertyManager.

Add a dimension.

93)   Click the **Smart Dimension** ✎ Sketch tool.

94)   Click the **horizontal** line.

95)   Click a **position** below the profile.

96)   Enter .**700**in, [**17.78**] in the Modify box.

97)   Click the **Green Check mark** ✔. The black sketch is fully defined.

💡 Save rebuild time. Add relations and dimensions to fully define a sketch. Fully defined sketches are displayed in black.

Extrude the sketch.

98)   Click **Extruded Boss/Base** 🗔 from the Features toolbar. The Boss-Extrude PropertyManager is displayed.

99)   Select **Mid Plane** for End Condition in Direction 1.

100) Enter **7.000**in, [**177.8**] for Depth. Note: Thin Feature is checked.

101) If needed, click the **Reverse Direction Arrow** button for One-Direction. Material thickness is created **above** the Origin.

102) Enter .**090**in, [**2.3**] for Thickness.

103) Check the **Auto-fillet corners** box.

104) Enter .**090**in, [**2.3**] for Fillet Radius.

105) Click **OK** ✔ from the Boss-Extrude PropertyManager. Extrude-Thin1 is displayed in the FeatureManager. Sketch1 is fully defined.

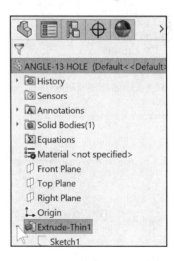

Fit the model to the Graphics window.
106) Press the **f** key.

💡 Think design intent. When do you use the various End Conditions and Geometric sketch relations? What are you trying to do with the design? How does the component fit into an assembly?

Clarify the Extrude-Thin1 feature direction and thickness options. Utilize multiple view orientations and Zoom In before selecting OK ✔ from the Boss-Extrude PropertyManager.

Modify feature dimensions.

107) Click **Extrude-Thin1** in the FeatureManager. **Drag** the dimensions off the model.

108) Click the **7.000**in, **[177.80]** dimension in the Graphics window.

109) Enter **6.500**in, **[165.10]**.

110) Click **inside** the Graphics window.

Save the model.

111) Click **Save** 💾.

Activity: ANGLE-13HOLE Part - Extruded Cut Feature

Insert a new sketch for the Extruded Cut feature.

112) Right-click the **top face** of Extrude-Thin1 as illustrated. This is the Sketch plane.

113) Click **Sketch** ▦ from the Context toolbar. The Sketch toolbar is displayed.

114) Click **Top view** ▱ from the Heads-up View toolbar.

115) Click the **Circle** ⊙ Sketch tool.

116) Sketch a **circle** on the left side of the Origin as illustrated.

Add dimensions.

117) Click the **Smart Dimension** ⟋ Sketch tool.

118) Click the **Origin** ⌖.

119) Click the **center point** of the circle.

120) Click a **position** below the horizontal profile line.

121) Enter **3.000**in, [**76.2**].

122) Click the **Green Check mark** ✔.

123) Click the **bottom horizontal** line.

124) Click the **center point** of the circle.

125) Click a **position** to the left of the profile.

126) Enter **.250**in, [**6.35**]. Click the **Green Check mark** ✔.

127) Click the **circumference** of the circle.

128) Click a **position** diagonally above the profile. Enter **.190**in, [**4.83**].

129) Click the **Green Check mark** ✔.

Insert an Extruded Cut Feature.

130) Click **Extruded Cut** 🔲 from the Features toolbar. The Cut-Extrude PropertyManager is displayed.

131) Select **Through All** for End Condition in Direction 1. Accept the default conditions.

132) Click **OK** ✔ from the Cut-Extrude PropertyManager. Cut-Extrude1 is displayed in the FeatureManager. Cut-Extrude1 is the seed feature for the Linear Pattern feature of holes.

Display an Isometric view.

133) Click **Isometric view** 🔾 from the Heads-up View toolbar.

Save the model.

134) Click **Save** 💾.

Activity: ANGLE-13HOLE Part - Linear Pattern Feature

Insert a Linear Pattern feature.

135) Click **Top View** from the Heads-up View toolbar.

136) Click **Linear Pattern** from the Features toolbar. The Linear Pattern PropertyManager is displayed. Cut-Extrude1 is displayed in the Features to Pattern box.

137) Click the **bottom horizontal edge** of the Extrude-Thin1 feature for Direction1. Edge<1> is displayed in the Pattern Direction box. The direction arrow points to the right. If required, click the Reverse Direction button.

138) Enter **0.5**in, [**12.70**] for Spacing.

139) Enter **13** for Number of Instances. If needed, click inside the **Features and Faces** box.

140) Click **Cut-Extrude1** from the fly-out FeatureManager.

141) Check the **Geometry pattern** box under Options.

142) Click **OK** from the Linear Pattern PropertyManager. LPattern1 is displayed in the FeatureManager.

Display an Isometric view.

143) Click **Isometric view** from the Heads-up View toolbar.

Save the ANGLE-13HOLE part.

144) Click **Save**. Note: All sketches should be fully defined in the FeatureManager.

When possible and if it makes sense, model objects symmetrically about the origin. Even if the part is not symmetrical, the way it attaches or is manufactured could have symmetry.

Activity: ANGLE-13HOLE Part - Fillet Feature

Insert a Constant Radius Fillet Feature. Use the Pop-up Content menu.
145) Zoom in on the right top edge as illustrated.

146) Click the **right top edge** of the
Extrude-Thin1 feature.

147) Click **Fillet** from the Features
toolbar. The Fillet PropertyManager
is displayed.

148) Click the **Manual** tab.

149) Click **Constant Radius Fillet** Type.

150) Enter **.250 [6.35]** for Radius.

151) Click the **right bottom edge**. The
Fillet Pop-up Content menu is
displayed.

152) Click **Connected to start virtual
loop, 2 Edges**. Edge<1> thru
Edge<4> are displayed in the Items
To Fillet box.

153) Click **OK** from the Fillet
PropertyManager. Fillet1 is displayed in the
FeatureManager.

Fit the model to the Graphics window.
154) Press the **f** key.

Display an Isometric view.

155) Click **Isometric view** 🔲 from the Heads-up View toolbar.

Save the ANGLE-13HOLE part.

156) Click **Save** 💾.

The book is designed to expose the new SOLIDWORKS user to many different tools, techniques and procedures. It may not always use the most direct tool or process.

Activity: ANGLE-13HOLE Part - Second Extruded Cut/Linear Pattern

Insert a new sketch for the second Extruded Cut feature.

157) Right-click the **front face** of the Extrude-Thin1 feature in the Graphics window. The front face is the Sketch plane.

158) Click **Sketch** 📝 from the Context toolbar. The Sketch toolbar is displayed.

159) Click **Front view** 🔲 from the Heads-up View toolbar.

160) Click **Wireframe** 🔲 from the Heads-up View toolbar to display LPattern1.

Note: **Do not** align the center point of the circle with the center point of the LPattern1 feature. **Do not** align the center point of the circle with the center point of the Fillet radius. Control the center point position with dimensions.

161) Click the **Circle** ⊙ Sketch tool. The cursor displays the Circle icon.

162) Sketch a **circle** on the left side of the Origin between the two LPattern1 holes as illustrated.

Wireframe
Displays all edges of the model.

Add dimensions.

163) Click the **Smart Dimension** Sketch tool.

164) Click the **Origin** .

165) Click the **center point** of the circle.

166) Click a **position** below the horizontal profile line.

167) Enter **3.000**in, [**76.20**].

168) Click the **Green Check mark** .

169) Click the **top horizontal line**.

170) Click the **center point** of the circle.

171) Click a **position** to the left of the profile.

172) Enter **.250**in, [**6.35**].

173) Click the **Green Check mark** .

174) Click the **circumference** of the circle.

175) Click a **position** above the profile.

176) Enter **.190**in, [**4.83**]. Click the **Green Check mark** .

Insert an Extruded Cut Feature.

177) Click **Extruded Cut** from the Features toolbar. The Cut-Extrude PropertyManager is displayed.

178) Select **Through All** for End Condition in Direction1. Accept the default conditions.

179) Click **OK** from the Cut-Extrude PropertyManager. Cut-Extrude2 is displayed in the FeatureManager.

Create the second Linear Pattern Feature.

180) Click **Linear Pattern** from the Features toolbar. The Linear Pattern FeatureManager is displayed. Cut-Extrude2 is displayed in the Features to Pattern box.

181) Click inside the **Pattern Direction** box.

182) Click the **top horizontal edge** of the Extrude-Thin1 feature for Direction1. Edge<1> is displayed in the Pattern Direction box. The Direction arrow points to the right. If required, click the **Reverse Direction** button.

183) Enter **3.000**in, [**76.20**] for Spacing.

184) Enter **3** for Number of Instances. Note: Cut-Extrude2 (Seed feature) is displayed in the Features to Pattern box. Check the **Geometry pattern** box under Options.

185) Click **OK** ✔ from the Linear Pattern PropertyManager. LPattern2 is displayed in the FeatureManager.

Display an Isometric view - Shaded With Edges.

186) Click **Isometric view** 🔲 from the Heads-up View toolbar.

187) Click **Shaded With Edges** 🔲 from the Heads-up View toolbar.

Save the ANGLE-13HOLE part.

188) Click **Save** 💾.

Activity: ANGLE-13HOLE Part - Third Extruded Cut

Insert a new sketch for the third Extruded Cut Feature.

189) Select the Sketch plane. Right-click the **front face** of Extrude-Thin1.

190) Click **Sketch** 🔲 from the Context toolbar. The Sketch toolbar is displayed.

191) Click **Front view** 🔲 from the Heads-up View toolbar.

192) Click **Hidden Lines Removed** 🔲 from the Heads-up View toolbar.

Sketch a vertical centerline.

193) Click the **Centerline** ✏ Sketch tool. The Insert Line PropertyManager is displayed.

194) Click the **Origin** ⌙.

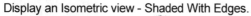

195) Click a **vertical position** above the top horizontal line.

196) Click **Tools**, **Sketch Tools**, **Dynamic Mirror** from the Main menu.

197) Click the **centerline** in the Graphics window.

Sketch a rectangle.

198) Click **Wireframe** ⬠ from the Heads-up View toolbar.

Apply the Straight Slot Sketch tool. Sketch a straight slot using two end points and a point for height.

199) Click the **Straight Slot** ⬭ Sketch tool from the Consolidated Slot toolbar. Do not align the rectangle first point and second point to the center points of Lpattern1.

200) Click the **first point** of the rectangle to the left of the Origin as illustrated.

201) Click the **second point** directly to the right.

202) Click the **third point** to create the slot as illustrated. View the two slots on the model.

203) Click **OK** ✅ from the Slot PropertyManager.

Deactivate the Dynamic Mirror tool.

204) Click **Tools**, **Sketch Tools**, **Dynamic Mirror** from the Main menu.

Add a Concentric relation.
205) **Zoom in** on the centerline.

206) Click the **top end point** of the centerline.

207) Hold the **Ctrl** key down.

208) Click the **circumference** of the circle.

209) Release the **Ctrl** key.

210) Click **Concentric** from the Add Relations box.

The endpoint of the centerline is positioned in the center of the circle. Note: Right-click Clear Selections to remove selected entities from the Add Relations box.

Add an Equal relation.
211) Click the **circumference** of the circle.

212) Hold the **Ctrl** key down.

213) Click the **left arc** of the first rectangle.

214) Release the **Ctrl** key.

215) Click **Equal** ⁼ from the Add Relations box. The arc radius is equal to the circle radius.

Add a Horizontal relation.
216) Click the **top end point** of the centerline as illustrated.

217) Hold the **Ctrl** key down.

218) Click the **center point** of the left arc of the first rectangle.

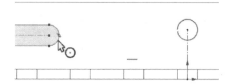

219) Release the **Ctrl** key.

220) Click **Horizontal** ⎯ from the Add Relations box.

221) Click **OK** ✔ from the Properties PropertyManager.

The right arc is horizontally aligned to the left arc due to symmetry from the Sketch Mirror tool.

Add dimensions. Dimension the distance between the two slots.

222) Click the **Smart Dimension** ✎ Sketch tool.

223) Click the **right arc center point** of the left slot.

224) Click the **left arc center point** of the right slot.

225) Click a **position** above the top horizontal line.

226) Enter **1.000**in, **[25.40]** in the Modify dialog box.

227) Click the **Green Check mark** ✔.

228) Click the **left center point** of the left arc.

229) Click the **right center point** of the left arc.

230) Click a **position** above the top horizontal line.

231) Enter **2.000**in, **[50.80]** in the Modify dialog box.

232) Click the **Green Check mark** ✔. The black sketch is fully defined.

Display an Isometric view.
233) Click **Isometric view** 🧊.

💡 Origin and Tangent edges are displayed for educational purposes.

Insert an Extruded Cut Feature.

234) Click **Extruded Cut** ⬚ from the Features toolbar. The Cut-Extrude PropertyManager is displayed.

235) Select **Through All** for End Condition in Direction1. The direction arrow points to the back.

236) Click **OK** ✅ from the Extrude PropertyManager. Cut-Extrude3 is displayed in the FeatureManager.

Display Shaded With Edges.

237) Click **Shaded With Edges** ⬚ from the Heads-up View toolbar.

Save the ANGLE-13HOLE part.

238) Click **Save** 💾. The ANGLE-13HOLE is complete. All sketches in the FeatureManager should be fully defined. Add material later.

💡 Model about the origin; it provides a point of reference.

The dimension between the two slots is over-defined if the arc center points are aligned to the center points of the LPattern1 feature. An over-defined sketch is displayed in red, in the FeatureManager.

The mouse pointer displays a blue dashed line when horizontal and vertical sketch references are inferred.

💡 Add relations, then dimensions. This will keep the user from having too many unnecessary dimensions. This helps to show the design intent of the model.

Review the ANGLE-13HOLE Part.

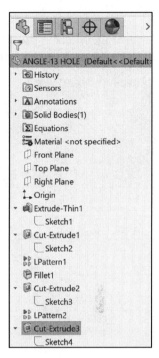

The ANGLE-13HOLE part utilized an open L-Shaped profile sketched on the Right Plane. The Extruded Thin feature with the Mid Plane option located the part symmetrical to the Right Plane. The first Extruded Cut feature created the first hole sketched on the top face of the Extruded Thin feature.

The first Linear Pattern feature created an array of 13 holes along the bottom horizontal edge. The Fillet feature rounded the four corners. The second Extruded Cut feature created a hole on the Front face. The second Linear Pattern feature created an array of 3 holes along the top horizontal edge. The third Extruded Cut feature created two slot cuts using the Straight Slot Sketch tool.

CommandManager, FeatureManager, Task Pane tabs and file folders will vary depending on system setup and Add-ins.

TRIANGLE Part

The TRIANGLE part is a multipurpose supporting plane.

The TRIANGLE is manufactured from .090in, [2.3] aluminum. The TRIANGLE contains numerous features.

Utilize symmetry and Sketch tools to simplify the geometry creation.

The center points of the slots and holes locate key geometry for the TRIANGLE.

Utilize sketched construction geometry to locate the center points.

Construction geometry is not calculated in the extruded profile.

Utilize the Sketch Offset tool and Sketch Fillet to create the sketch profile for the Extruded Boss/Base 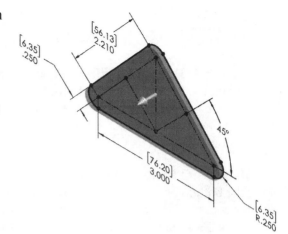 (Boss-Extrude1) feature.

Utilize the Dynamic Mirror Sketch tool and the Circle Sketch tool to create the first Extruded Cut feature.

Utilize the Corner Rectangle, Trim, and the Tangent Arc Sketch tools to create the second Extruded Cut feature left bottom slot.

Note: You could use the Slot sketch tool and eliminate the need for the Trim and Tangent Arc sketch tools. A goal of this book is to expose the new user to different tools and methods.

Utilize the Mirror feature to create the right bottom slot.

Utilize the Straight Slot Sketch tool to create the third Extruded Cut feature.

Utilize the Circular Pattern feature to create the three radial slot cuts.

Activity: TRIANGLE Part - Mirror, Offset and Fillet Sketch Tools
---

Create a new part.

239) Click **New** ⬜ from the Main menu.

240) Click the **Templates** tab.

241) Double-click **Part**. The Part FeatureManager is displayed.

Save the part. Enter part name. Enter part description. Save the part in the SW-TUTORIAL 2023 folder.

242) Click **Save As** 📋 .

243) Select **SW-TUTORIAL 2023** for the Save in file folder.

244) Enter **TRIANGLE** for File name. Enter **TRIANGLE** for Description. Click **Save**. The TRIANGLE FeatureManager is displayed.

Set the Document Properties. Set drafting standard, units, and precision.

245) Click **Options** ⚙ ˅ from the Main menu.

246) Click the **Document Properties** tab from the dialog box. Select **ANSI** from the Overall drafting standard box.

247) Click **Units**. Click **IPS**, [**MMGS**] for Unit system.

248) Select **.123**, [**.12**] for Length units Decimal places.

249) Select **None** for Angular units Decimal places.

250) Click **OK** to set the document units.

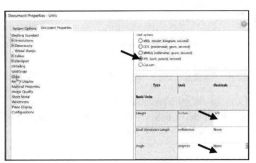

Insert a new sketch for the Extruded Base feature.
251) Right-click **Front Plane** from the FeatureManager.

252) Click **Sketch** ⬚ from the Context toolbar. The Sketch toolbar is displayed.

Sketch a vertical centerline.
253) Click the **Centerline** ⟋ Sketch tool.

254) Click the **Origin** ↳.

255) Click a vertical **position** above the Origin as illustrated.

Deselect the Centerline Sketch tool.
256) Right-click **Select**.

Sketch a Mirrored profile.
257) Click **Tools**, **Sketch Tools**, **Dynamic Mirror** from the Main menu.

258) Click the **centerline** in the Graphics window.

259) Click the **Centerline** ⟋ Sketch tool.

260) Click the **Origin** ↳.

261) Click a **position** to the left of the Origin to create a horizontal line.

262) Click the **top end point** of the vertical centerline to complete the triangle as illustrated.

263) Right-click **End Chain** to end the line segment. The Centerline tool is still active.

264) Click the **Origin** ↳.

265) Click a **position** coincident with the right-angled centerline. *Do not select the Midpoint.*

266) Right-click **End Chain** to end the line segment.

Deactivate the Dynamic Mirror Sketch tool.
267) Click **Tools**, **Sketch Tools**, **Dynamic Mirror** from the Main menu.

Add a dimension.

268) Click **Smart Dimension** from the Sketch toolbar.

269) Click the **right horizontal** centerline.

270) Click the **inside right** centerline.

271) Click a **position** between the two lines.

272) Enter **45**deg in the Modify dialog box for the angular dimension.

273) Click the **Green Check mark** in the Modify dialog box.

274) Click the **horizontal centerline**.

275) Click a position **below** the centerline.

276) Enter **3.000**in, [**76.20**].

277) Click the **Green Check mark** in the Modify dialog box.

278) Click the **left angled** centerline.

279) Click **position** aligned to the left angled centerline.

280) Enter **2.120**in, [**53.85**].

281) Click the **Green Check mark** in the Modify dialog box.

Offset the sketch.

282) Right-click **Select** in the Graphics window.

283) Hold the **Ctrl** key down.

284) Click the **three outside centerlines**.

285) Release the **Ctrl** key.

286) Click the **Offset Entities** Sketch tool.

287) Enter **.250**in, [**6.35**] for Offset Distance. The yellow Offset direction is outward.

288) Click **OK** ✔ from the Offset Entities PropertyManager.

289) Pull the **dimensions** off the model.

Three profile lines are displayed. The centerlines are on the inside.

Insert the Sketch Fillet.

290) Click **Sketch Fillet** ⌐| from the Sketch toolbar. The Sketch Fillet PropertyManager is displayed.

291) Enter **.250**in, [**6.35**] for Radius.

292) Click the **three outside corner** points.

293) Click **OK** ✔ from the Sketch Fillet PropertyManager.

Activity: TRIANGLE Part - Extruded Boss/Base Feature

Extrude the sketch. Create the Boss-Extrude1 feature.

294) Click **Extruded Boss/Base** 🗔 from the Features toolbar. Blind is the default End Condition in Direction1.

295) Enter **.090**in, [**2.3**] for Depth in Direction 1. The direction arrow points to the front.

296) Click **OK** ✔ from the Boss-Extrude PropertyManager. Boss-Extrude1 is displayed in the FeatureManager.

Display an Isometric view. Save the TRIANGLE part.

297) Click **Isometric view** 🧊 from the Heads-up View toolbar.

298) Click **Save** 💾.

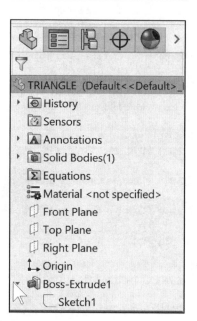

🔅 Insert centerlines and relations to build sketches that are referenced by multiple features.

Display Sketch1.

299) Expand Boss-Extrude1 in the FeatureManager.

300) Right-click **Sketch1**.

301) Click **Show**.

Activity: TRIANGLE Part - First Extruded Cut Feature

Insert a new sketch for the first Extruded Cut.

302) Right-click the **front face** of Boss-Extrude1. This is your Sketch plane.

303) Click **Sketch** from the Context toolbar. The Sketch toolbar is displayed.

304) Click **Front view** from the Heads-up View toolbar.

305) Display **Hidden Lines Removed**.

306) Click the **Circle** Sketch tool. The Circle PropertyManager is displayed.

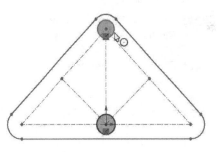

307) Sketch a **circle** centered at the Origin.

308) Sketch a **circle** centered at the endpoint of the vertical centerline as illustrated.

Sketch a vertical centerline.

309) Click the **Centerline** Sketch tool.

310) Click the **Origin**.

311) Click the **center point** of the top circle.

Deselect the Centerline Sketch tool.

312) Right-click **Select**.

Sketch a Mirrored profile.

313) Click **Tools**, **Sketch Tools**, **Dynamic Mirror** from the Main menu.

314) Click the **centerline** in the Graphics window.

315) Click the **Circle** Sketch tool. The Circle PropertyManager is displayed.

316) Sketch a **circle** on the left side of the centerline, coincident with the left centerline, in the lower half of the triangle.

317) Sketch a **circle** on the left side of the centerline, coincident with the left centerline, in the upper half of the triangle. Right-click **Select.**

318) Deactivate the Dynamic Mirror tool. Click **Tools, Sketch Tools, Dynamic Mirror** from the Main menu.

Add an Equal relation.
319) Click a **circle**. Hold the **Ctrl** key down.

320) Click the five other **circles**. Release the **Ctrl** key.

321) Click **Equal** ＝ . Click **OK** ✔ from the PropertyManager.

Add a dimension.

322) Click the **Smart Dimension** ✎ Sketch tool.

323) Click the circumference of the **top circle**.

324) Click a **position** off the TRIANGLE.

325) Enter **.190**in, [**4.83**].

326) Click the **Green Check mark** ✔ in the Modify dialog box.

Create the aligned dimensions.
327) Click the bottom **left point**. Click the **center point** of the bottom left circle.

328) Click a **position** aligned to the angled centerline.

329) Enter **.710**in, [**18.03**]. Click the **Green Check mark** ✔ in the Modify dialog box.

330) Click the **bottom left point**. Click the **center point** of the top left circle.

331) Click a **position** aligned to the angled centerline. Enter **1.410**in, [**35.81**].

332) Click the **Green Check mark** ✔ in the Modify dialog box. The sketch is fully defined.

Insert an Extruded Cut Feature.

333) Click **Extruded Cut** from the Features toolbar. The Cut-Extrude PropertyManager is displayed.

334) Select **Through All** for the End Condition in Direction1.

335) Click **OK** ✔ from the Cut-Extrude PropertyManager. Cut-Extrude1 is displayed in the FeatureManager.

Display an Isometric view. Save the model.

336) Click **Isometric view** from the Heads-up View toolbar.

337) Click **Save** .

Activity: TRIANGLE Part - Second Extruded Cut Feature

Insert a new slot sketch for the second Extruded Cut.

338) Right-click the **front face** of Boss-Extrude1 for the Sketch plane. Click **Sketch** from the Context toolbar. The Sketch toolbar is displayed.

339) Click **Front view** from the Heads-up View toolbar.

340) Click the **Corner Rectangle** tool from the Consolidated Sketch toolbar. Note: The purpose of this book is to teach you different tools and methods. You can also apply the Straight Slot Sketch tool and eliminate many of the next steps.

341) Sketch a **rectangle** to the left of the Origin as illustrated.

Trim the vertical lines.

342) Click the **Trim Entities** Sketch tool. The Trim PropertyManager is displayed. Click **Trim to closest** in the Options box.

The Trim to closest icon is displayed.

343) Click the **left vertical line** of the rectangle.

344) Click the **right vertical line** of the rectangle.

345) Click **OK** ✔ from the Trim PropertyManager.

First Point

Sketch the Tangent Arcs.

346) Click the **Tangent Arc** Sketch tool from the Consolidated Sketch toolbar. The Arc PropertyManager is displayed.

347) Sketch a **180º arc** on the left side.

348) Sketch a **180º arc** on the right side.

349) Click **OK** from the Arc PropertyManager.

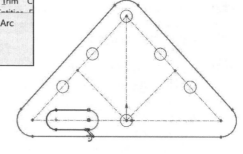

Add an Equal relation.
350) Click the **right arc**. Hold the **Ctrl** key down.

351) Click the **left arc**. Click the **bottom center circle**. Release the **Ctrl** key.

352) Click **Equal** = . Click **OK** from the Properties PropertyManager.

Add a Coincident relation.
353) Press the **f** key to fit the model to the Graphics window. Click the **center point** of the left arc.

354) Hold the **Ctrl** key down.

355) Click the **left lower point**.

356) Release the **Ctrl** key.

357) Click **Coincident** from the Add Relations box.

Add a Tangent relation.
358) Click the **bottom horizontal line**.

359) Hold the **Ctrl** key down.

360) Click the **first arc tangent**.

361) Release the **Ctrl** key.

362) Click **Tangent** from the Add Relations box.

363) Click **OK** from the Properties PropertyManager.

Add a dimension.

364) Click the **Smart Dimension** Sketch tool.

365) Click the **left center point** of the left arc.

366) Click the **right center point** of the right arc.

367) Click a **position** below the horizontal line.

368) Enter **1.000**in, [**25.40**] in the Modify dialog box.

369) Click the **Green Check mark** .

Insert an Extruded Cut feature.

370) Click **Isometric view** .

371) Click **Extruded Cut** from the Features toolbar. The Cut-Extrude PropertyManager is displayed.

372) Select **Through All** for End Condition in Direction 1.

373) Click **OK** from the Cut-Extrude PropertyManager.

Save the model.

374) Click **Save** . Cut-Extrude2 is highlighted in the FeatureManager.

Activity: TRIANGLE Part - Mirror Feature

Mirror the Cut-Extrude2 feature. Note: The new feature to have a Secondary Mirror Face/Plane.

375) Click **Mirror** from the Features toolbar. The Mirror PropertyManager is displayed. Cut-Extrude2 is displayed in the Feature to Mirror box.

376) Click **Right Plane** from the fly-out TRIANGLE FeatureManager. Right Plane is displayed in the Mirror Face/Plane box.

377) Check the **Geometry Pattern** box.

378) Click **OK** from the Mirror PropertyManager. Mirror1 is displayed in the FeatureManager.

Activity: TRIANGLE Part - Third Extruded Cut Feature

Insert a new sketch for the third Extruded Cut feature.

379) Right-click the **front face** of Boss-Extrude1 for the Sketch plane.

380) Click **Sketch** from the Context toolbar. The Sketch toolbar is displayed.

381) Click **Front view** .

Sketch a Straight Slot.

382) Click the **Straight Slot** Sketch tool from the Consolidated Rectangle toolbar. The Straight Slot icon is displayed.

383) Click a **position** coincident with the left angled centerline as illustrated.

Sketch the second point.

384) Click a **position** aligned to the centerline. A dashed line is displayed.

Sketch the third point.

385) Click a **position** above the inside left centerline. The Straight Slot sketch is displayed.

Click the Question mark ⑦ in the PropertyManager to obtain additional information on the tool.

386) Click **OK** from the Slot PropertyManager.

Add an Equal relation.

387) Click the **left arc**.

388) Hold the **Ctrl** key down.

389) Click the **bottom circle**.

390) Release the **Ctrl** key.

391) Click **Make Equal** from the Pop-up menu.

392) Click **OK** from the PropertyManager.

Add a dimension.

393) Click the **Smart Dimension** Sketch tool.

394) Click the **left center point** of the left arc.

395) Click the **right center point** of the right arc.

396) Click a **position** below the horizontal line.

397) Enter **.560**in, [**14.22**] in the Modify dialog box.

398) Click the **Green Check mark** . The sketch is fully defined.

Display an Isometric view. Insert an Extruded Cut Feature.

399) Click **Isometric view** .

400) Click **Extruded Cut** from the Features toolbar. The Cut-Extrude PropertyManager is displayed.

401) Select **Through All** for the End Condition in Direction 1.

402) Click **OK** from the Cut-Extrude PropertyManager. Cut-Extrude3 is displayed in the FeatureManager.

Save the model.

403) Click **Save** .

Display the Temporary Axis.

404) Click **View**, **Hide/Show**, check **Temporary Axes** from the Main menu.

The book is designed to expose the new SOLIDWORKS user to many different tools, techniques and procedures. It may not always use the most direct tool or process.

Activity: TRIANGLE Part - Circular Pattern Feature

Insert a Circular Pattern feature.

405) Click **Circular Pattern** from the Features Consolidated toolbar. The Circular Pattern PropertyManager is displayed. Cut-Extrude3 is displayed in the Features to Pattern box.

406) Click the **Temporary Axis** displayed through the center hole located at the Origin. The Temporary Axis is displayed as Axis <1> in the Pattern Axis box.

407) Enter **90**deg for Angle.

408) Enter **3** for Number of Instances.

409) Check the **Equal spacing** box. If required, click **Reverse Direction**.

410) Click the **Geometry pattern** box.

411) Click **OK** from the PropertyManager.

Hide Sketch1.

412) **Expand** Boss-Extrude1 from the FeatureManager.

413) Right-click **Sketch1**.

414) Click **Hide**.

Display an Isometric view. Deactivate the Temporary Axes. Save the TRIANGLE part.

415) Click **Isometric view** .

416) Click **View**, **Hide/Show**, un-check **Temporary Axes** from the Main menu.

417) Display Shaded with Edges.

418) Click **Save** . The TRIANGLE part is complete. Add material later.

Review the TRIANGLE Part.

The TRIANGLE part utilized a Boss-Extrude1 feature. A triangular shape profile was sketched on the Front Plane. Symmetry and construction geometry sketch tools located center points for slots and holes.

The Sketch Fillet tool created rounded corners for the profile. The Sketch Mirror and Circle Sketch tools were utilized to create the Extruded Cut features.

The Corner Rectangle, Sketch Trim, and Tangent Arc tools were utilized to create the second Extruded Cut feature, left bottom slot. The Mirror feature was utilized to create the right bottom slot. **Note:** The next feature to have a Secondary Mirror Face/Plane.

The Parallelogram and Tangent Arc Sketch tools were utilized to create the third Extruded Cut feature. The Circular Pattern feature created the three radial slot cuts. The following Geometric relations were utilized: *Equal*, *Parallel*, *Coincident* and *Tangent*.

Note: Use the Slot sketch tool and eliminate the need for the Trim and Tangent Arc sketch tools. A goal of this book is to expose the new user to different tools and methods.

SCREW Part

The SCREW part is a simplified model of a 10-24 x 3/8 Machine screw. Screws, nuts and washers are classified as fasteners. An assembly contains hundreds of fasteners. Utilize simplified versions to conserve model and rebuild time.

Simplified version

Machine screws are described in terms of the following:

- Nominal diameter - Size 10.

- Threads per inch - 24.

- Length - 3/8.

Screw diameter, less than ¼ inch, is represented by a size number. Size 10 refers to a diameter of .190 inch. Utilize the SCREW part to fasten components in the FRONT-SUPPORT assembly.

The SCREW part utilizes a Revolved Base feature
to add material. The Revolved Boss/Base feature
requires a centerline and sketch on a Sketch plane. A
Revolved feature requires an angle of revolution. The
sketch is revolved around the centerline.

Sketch a centerline on the Front Sketch plane.

Sketch a closed profile.

Revolve the sketch 360 degrees.

Utilize the Edit Sketch tool to modify the
sketch. Utilize the Sketch Trim and Tangent
Arc tool to create a new profile.

Utilize an Extruded Cut feature sketched
on the Front Plane. This is the seed feature for
the Circular Pattern.

Utilize the Circular Pattern feature to create four
instances.

Apply the Fillet feature to round edges and faces.
Utilize the Fillet feature to round the top edge.

Apply the Chamfer feature to bevel
edges and faces. Utilize a Chamfer feature
to bevel the bottom face.

Note: Utilize an M5 Machine screw for
metric units.

Activity: SCREW Part-Documents Properties

Create a new part.

419) Click **New** ⬜ from the Main menu.

420) Click the **Templates** tab. Double-click **Part**.

Save the part. Enter name. Enter description.

421) Click **Save As** .

422) Select **SW-TUTORIAL 2023** for the Save in folder.

423) Enter **SCREW** for File name.

424) Enter **MACHINE SCREW 10-24x3/8** for Description.

425) Click **Save**. The SCREW FeatureManager is displayed.

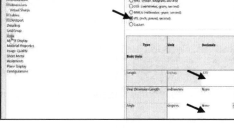

Set Document Properties. Set drafting standard, units, and precision.

426) Click **Options** ⚙ ˙ from the Main menu.

427) Click the **Documents Properties** tab from the dialog box. Select **ANSI** from the Overall drafting standard box.

428) Click **Units**. Select **IPS**, [**MMGS**] for Unit system. Select **.123**, [**.12**] for Length units Decimal places. Select **None** for Angular units Decimal places. Click **OK**.

Activity: SCREW Part - Revolved Feature

Insert a Revolved feature sketched on the Front Plane. The Front Plane is the default Sketch plane.

Insert a new sketch.

429) Right-click **Front Plane** from the FeatureManager.

430) Click **Sketch** ⌐ from the Context toolbar. The Sketch toolbar is displayed.

431) Click the **Centerline** ⌿ Sketch tool. The Insert Line PropertyManager is displayed.

Origin

Sketch a vertical centerline.

432) Click the **Origin** ⤬. Click a **position** directly above the Origin as illustrated.

433) Right-click **End Chain** to end the centerline.

Add a dimension.

434) Click the **Smart Dimension** Sketch tool.

435) Click the **centerline**. Click a **position** to the left.

436) Enter .**500**in, [**12.70**]. Click the **Green Check mark** .

Fit the sketch to the Graphics Window.
437) Press the **f** key.

Sketch the profile.
438) Click the **Line** Sketch tool.

Sketch the first horizontal line.

439) Click the **Origin** .

440) Click a **position** to the right of the Origin.

Sketch the first vertical line.
441) Click a position **above** the horizontal line endpoint.

Sketch the second horizontal line.

442) Sketch the **second vertical line**. The top point of the vertical line is collinear with the top point of the centerline.

443) Sketch the **third horizontal line**. The left endpoint of the horizontal line is coincident with the top point of the centerline.

444) Right-click **Select** to deselect the Line Sketch tool.

Add a Horizontal relation.
445) Click the **top** most right point. Hold the **Ctrl** key down.

446) Click the **top** most left point. Release the **Ctrl** key. Click **Horizontal** from the Add Relations box.

447) Click **OK** from the Properties PropertyManager.

Add a dimension.
448) Click the **Smart Dimension** Sketch tool. Create the first diameter dimension.

449) Click the **centerline** in the Graphics window.

450) Click the **first vertical line**.

451) Click a **position** to the left of the Origin to create a diameter dimension.

452) Enter **.190**in, **[4.83]**.

453) Click the **Green Check mark** .

A diameter dimension for a Revolved sketch requires a centerline, profile line, and a dimension position to the left of the centerline.

A dimension position directly below the bottom horizontal line creates a radial dimension.

Create the second diameter dimension.
454) Click the **centerline** in the Graphics window.

455) Click the **second vertical line**.

456) Click a **position** to the bottom left of the Origin to create a diameter dimension.

457) Enter **.373**in, **[9.47]**.

Create a vertical dimension.
458) Click the **first vertical line**.

459) Click a **position** to the right of the line.

460) Enter **.375**in, **[9.53]**.

Center the dimension text.
461) Click the **.190**in, **[4.83]** dimension. Drag the **text** between the two extension lines.

462) Click the **.373**in, **[9.47]** dimension.

463) Drag the **text** between the two extension lines. If required, click the **blue arrow dots** to flip the arrows inside the extension lines.

464) Right-click **Select** to deselect the Smart Dimension Sketch tool.

Select the Centerline for axis of revolution.
465) Click the **vertical centerline** as illustrated.

Revolve the sketch.
466) Click **Revolved Boss/Base** from the Features toolbar.

467) Click **Yes**. The Revolve PropertyManager is displayed.

The "Yes" button causes a vertical line to be automatically sketched from the top left point to the Origin. The Graphics window displays the Isometric view and a preview of the Revolved Base feature.

The Revolve PropertyManager displays 360 degrees for the Angle of Revolution.

468) Click **OK** ✔ from the Revolve PropertyManager.

The FeatureManager displays the Revolve1 name for the first feature. The Revolved Boss/Base feature requires a centerline, sketch, and an angle of revolution. A solid Revolved Boss/Base feature requires a closed sketch. Draw the sketch on one side of the centerline.

The SCREW requires a rounded profile. Edit the Revolved Base sketch. Insert a Tangent Arc.

Edit the Revolved Base sketch.
469) Right-click **Revolve1** in the FeatureManager.

470) Click **Edit Sketch** 🖉 from the Context toolbar.

471) Click **Front view** 🔲.

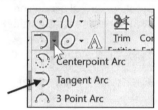

472) Click the **Tangent Arc** ⌒ Sketch tool.

473) Click the **top centerline** point as illustrated.

474) Drag the **mouse pointer** to the right and downward.

475) Click a **position** collinear with the right vertical line, below the midpoint. The arc is displayed tangent to the top horizontal line.

Deselect the Tangent Arc Sketch tool.
476) Right-click **Select**.

Delete unwanted geometry.
477) Click the **Trim Entities** ✂ Sketch tool. The Trim PropertyManager is displayed.

478) Click **Trim to closest** from the Options box. The Trim to closest icon is displayed.

479) Click the **right top vertical line** as illustrated.

480) Click the **top horizontal line** as illustrated. The two lines are removed.

481) Click **OK** ✔ from the Trim PropertyManager. Note: You may still view lines until you exit the sketch.

Add a dimension.
482) Click the **Smart Dimension** ✎ Sketch tool.

483) Click the **arc**.

484) Click a **position** above the profile.

485) Enter **.304**in, **[7.72]**.

486) Click the **Green Check mark** ✔ . The sketch should be fully defined; if not, add a .05 inch dimension as illustrated.

Exit the Sketch. Save the model.
487) Click **Exit Sketch** from the Sketch toolbar.

488) Click **Save** 🖫.

The SCREW requires an Extruded Cut feature on the Front Plane. Utilize the Convert Entities Sketch tool to extract the Revolved Base top arc edge for the profile of the Extruded Cut.

Activity: SCREW Part - Extruded Cut Feature

Insert a new sketch for the Extruded Cut feature.
489) Right-click **Front Plane** from the FeatureManager.

490) Click **Sketch** ⊏ from the Context toolbar. The Sketch toolbar is displayed.

491) Click the **top arc** as illustrated. The mouse pointer displays the silhouette edge icon for feedback.

492) Click the **Convert Entities** ⬦ Sketch tool.

493) Click the **Line** ✎ Sketch tool.

494) Sketch a **vertical line**. The top endpoint of the line is coincident with the arc, vertically aligned to the Origin.

495) Sketch a **horizontal line**. The right end point of the line is coincident with the arc. Do not select the arc midpoint. Right-click **Select**.

496) Click **Isometric view** .

497) Click the **Trim Entities** Sketch tool. The Trim PropertyManager is displayed.

498) Click **Power trim** from the Options box.

499) Click and drag the mouse pointer to **intersect** the **right** arc line as illustrated. If needed, click and drag the mouse pointer to **intersect** the **left** arc line.

500) Click **OK** from the Trim PropertyManager.

Add a dimension.
501) Click the **Smart Dimension** Sketch tool.

502) Click the **vertical line**.

503) Click a **position** to the right of the profile.

504) Enter .030in, [0.76].

505) Click the **Green Check mark** .

Insert an Extruded Cut Feature.
506) Click **Isometric view** .

507) Click **Extruded Cut** from the Features toolbar. The Cut-Extrude PropertyManager is displayed.

508) Select **Mid Plane** for the End Condition in Direction 1.

509) Enter .050in, [1.27] for Depth.

510) Click **OK** from the Cut-Extrude PropertyManager. Cut-Extrude1 is displayed in the FeatureManager.

Activity: SCREW Part - Circular Pattern Feature

Insert the Circular Pattern feature. Show Temporary Axes.

511) Click **View, Hide/Show,** check **Temporary Axes** from the Main menu. The Temporary Axis is required for the Circular Pattern feature.

512) Click **Circular Pattern** from the Features Consolidated toolbar. Cut-Extrude1 is displayed in the Features to Pattern box.

513) Click the **Temporary Axis** in the Graphics window. Axis<1> is displayed in the Pattern Axis box.

514) Check the **Equal spacing** box.

515) Enter **360**deg for Angle.

516) Enter **4** for Number of Instances.

517) Check the **Geometry pattern** box.

518) Click **OK** ✔ from the Circular Pattern PropertyManager. CirPattern1 is displayed in the FeatureManager.

Save the model.

519) Click **Save** 🖫.

Activity: SCREW Part - Fillet Feature

Insert a Constant Radius Fillet Type feature.

520) Click the **top circular edge** as illustrated.

521) Click **Fillet** 🗍 from the Features toolbar. The Fillet PropertyManager is displayed.

522) Click the **Manual** tab. Edge<1> is displayed in the Items to Fillet box.

523) Click the **Constant Radius Fillet** icon for Fillet Type.

524) Enter **.010**in, [**.25**] for Radius.

525) Click **OK** ✔ from the Fillet PropertyManager. Fillet1 is displayed in the FeatureManager.

Activity: SCREW Part - Chamfer Feature

Insert a 45 deg Angle Distance Chamfer feature.
526) Click the **bottom circular edge**.

527) Click **Chamfer** from the
Features Consolidated toolbar.
Edge<1> is displayed in the Items
to Chamfer box.

528) Click **Angle distance** for type.

529) Enter **.050**in, **[1.27]** for Distance.
View the direction of the arrow.

530) Click **OK** ✔ from the Chamfer PropertyManager. Chamfer1 is
displayed in the FeatureManager.

🔆 Simplify the part. Save rebuild time. Suppress features that
are not required in the assembly.

A suppressed feature is not displayed in the Graphics window. A
suppressed feature is removed from any rebuild calculations.

Suppress the Fillet and Chamfer feature.
531) Hold the **Ctrl** key down.

532) Click **Fillet1** and **Chamfer1** from the FeatureManager.

533) Release the **Ctrl** key.

534) Right-click **Suppress** ↓ from the Context toolbar. Note:
Suppressed features are displayed in light gray in the
FeatureManager.

Deactivate the Temporary Axes in the Graphics window.
535) Click **View**, **Hide/Show**, un-check **Temporary Axes** from the
Main menu.

Display an Isometric view. Save the SCREW part.
536) Click **Isometric view** from the Heads-up View toolbar.

537) Click **Save** .

Close all open documents.
538) Click **Window**, **Close All** from the Main menu.

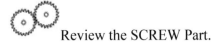 Review the SCREW Part.

The Revolved Boss/Base feature was utilized to create the simplified SCREW part. The Revolved Boss/Base feature required a centerline sketched on the Front Sketch plane and a closed profile. The sketch was revolved 360 degrees to create the Base feature for the SCREW part.

Edit Sketch was utilized to modify the sketch. The Sketch Trim and Tangent Arc tools created a new profile.

The Extruded Cut feature was sketched on the Front Plane. The Circular Pattern feature created four instances. The Fillet feature rounded the top edges.

The Chamfer feature beveled the bottom edge. The Fillet and Chamfer are suppressed to save rebuild time in the assembly. Note: suppressed features are displayed in light gray in the FeatureManager.

The Origin and Tangent edges are displayed for educational purposes.

Think design intent. When do you use the various End Conditions and Geometric sketch relations? What are you trying to do with the design? How does the component fit into an assembly?

Mate to the first part added to the assembly. If you mate to the first part or base part of the assembly and decide to change its orientation later, all the parts will move with it.

FRONT-SUPPORT Assembly

The FRONT-SUPPORT assembly consists
of the following parts:

- ANGLE-13HOLE

- TRIANGLE

- HEX-STANDOFF

- SCREW

Create the FRONT-SUPPORT assembly.
Insert the ANGLE-13HOLE part. The
ANGLE-13HOLE part is fixed to the
FRONT-SUPPORT Origin. Insert the first HEX-STANDOFF part.

Utilize Concentric and Coincident mates to assemble the HEX-STANDOFF to the left
hole of the ANGLE-13HOLE part. Insert the second HEX-STANDOFF part.

Utilize Concentric and Coincident mates to assemble the HEX-STANDOFF to the third
hole from the right side. Insert the TRIANGLE part. Utilize Concentric, Distance, and
Parallel mates to assemble the TRIANGLE. Utilize Concentric/Coincident SmartMates to
assemble the four SCREWS. Note: Other mate types can be used.

Determine the static and dynamic behavior of mates in each sub-assembly before
creating the top-level assembly.

Activity: FRONT-SUPPORT Assembly-Insert ANGLE-13HOLE

Create a new assembly.

539) Click **New** ▯ from the Main menu.

540) Double-click **Assembly** from the Templates tab. The
Begin Assembly PropertyManager is displayed. Open
models are displayed in the Open documents box. If
there are no Open SOLIDWORKS models, the
Window Open dialog is also displayed.

541) Double-click the **ANGLE-13HOLE** part from the
SW-TUTORIAL 2023 folder or from your stored file
folder.

Fix the first component to the Origin.

542) Click **OK** ✔ from the Begin Assembly PropertyManager. The first component is fixed to the Origin (f).

Save the assembly. Enter part name. Enter part description.

543) Click **Save As** 🖫.

544) Select **SW-TUTORIAL 2023** for the Save in file folder.

545) Enter **FRONT-SUPPORT** for File name. Enter **FRONT SUPPORT ASSEMBLY** for Description.

546) Click **Save**. The FRONT-SUPPORT assembly FeatureManager is displayed.

Set Document Properties. Set drafting standard.

547) Click **Options** ⚙ ˅ from the Main menu.

548) Click the **Document Properties** tab from the dialog box.

549) Select **ANSI** from the Overall drafting standard box.

Set document units. Set length and angle precision.
550) Click **Units**.

551) Select **IPS**, [**MMGS**] for Unit system.

552) Select **.123**, [**.12**] for Length units Decimal places. Select **None** for Angular units Decimal places.

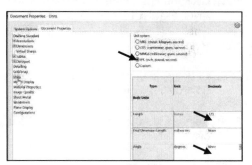

553) Click **OK**.

The ANGLE-13HOLE name in the FeatureManager displays an (f) symbol. The (f) symbol indicates that the ANGLE-13HOLE component is fixed to the FRONT-SUPPORT assembly Origin. The component cannot move or rotate.

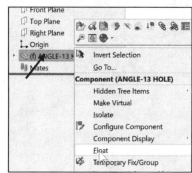

🔅 To remove the fixed state, Right-click a component name in the FeatureManager. Click **Float**. The component is free to move.

Display an Isometric view.

554) Click **Isometric view** from the Heads-up View toolbar.

Save the assembly.

555) Click **Save** .

Activity: FRONT-SUPPORT Assembly - Insert HEX-STANDOFF

Insert the HEX-STANDOFF part.

556) Click the **Insert Components** Assembly tool.

557) Double-click **HEX-STANDOFF** from the SW-TUTORIAL 2023 folder or your stored folder file.

558) Click a **position** near the left top front hole as illustrated.

Enlarge the view.
559) **Zoom in** on the front left side of the assembly.

Move the component.
560) Click and drag the **HEX-STANDOFF** component below the ANGLE-13HOLE left hole.

The HEX-STANDOFF name in the FeatureManager displays a (-) minus sign. The minus sign indicates that the HEX-STANDOFF part is free to move.

Insert a Concentric mate.
561) Click the **left inside cylindrical hole face** of the ANGLE-13HOLE component.

562) Hold the **Ctrl** key down.

563) Click inside the **cylindrical hole face** of the HEX-STANDOFF component.

564) Release the **Ctrl** key. The Mate pop-up menu is displayed.

565) Click **Concentric** from the Mate pop-up menu.

Cut-Extrude1 of ANGLE-13 HOLE<1>

Concentric

566) Click and drag the **HEX-STANDOFF** component below the ANGLE-13HOLE component until the top face is displayed.

Insert a Coincident mate.
567) Click the **HEX-STANDOFF top** face.

568) **Rotate** the assembly to view the bottom face of the ANGLE-13HOLE component.

569) Hold the **Ctrl** key down.

570) Click the **ANGLE-13HOLE bottom face**.

571) Release the **Ctrl** key. The Mate pop-up menu is displayed.

572) Click **Coincident** from the Mate pop-up menu.

Display an Isometric view.
573) Click **Isometric view** .

574) Click and drag the **HEX-STANDOFF** component. The HEX-STANDOFF rotates about its axis.

Insert a Parallel mate.
575) Click **Front view** .

576) Click the **HEX-STANDOFF front face**.

577) Hold the **Ctrl** key down.

578) Click the **ANGLE-13HOLE front face**.

579) Release the **Ctrl** key. The Mate pop-up menu is displayed.

580) Click **Parallel** from the Mate pop-up menu.

Display the created mates.
581) **Expand** the Mates folder in the FRONT-SUPPORT FeatureManager. Three mates are displayed between the ANGLE-13HOLE component and the HEX-STANDOFF component. The HEX-STANDOFF component is fully defined.

Save the FRONT-SUPPORT assembly.
582) Click **Save** 💾.

Insert the second HEX-STANDOFF part.
583) Hold the **Ctrl** key down.

584) Click and drag the **HEX-STANDOFF<1>** 🔲 HEX-STANDOFF<1> name from the FeatureManager into the FRONT-SUPPORT assembly Graphics window.

585) Release the **mouse pointer** below the far right hole of the ANGLE-13HOLE component.

586) Release the **Ctrl** key. HEX-STANDOFF<2> is displayed in the Graphics window and listed in the FeatureManager.

💡 The number <2> indicates the second instance or copy of the same component. The instance number increments every time you insert the same component. If you delete a component and then reinsert the component in the same SOLIDWORKS session, the instance number increments by one.

Enlarge the view.
587) **Zoom in** on the right side of the assembly.

Insert a Concentric mate.
588) Click inside the **cylindrical hole face** of the second HEX-STANDOFF component.

589) Hold the **Ctrl** key down.

590) Click the **third hole cylindrical face** from the right ANGLE-13HOLE component.

591) Release the **Ctrl** key. The Mate pop-up menu is displayed.

592) Click **Concentric** from the Mate pop-up menu.

Move the second HEX-STANDOFF part.
593) Click and drag the **HEX-STANDOFF** component below the ANGLE-13HOLE component until its top face is displayed.

Insert a Coincident mate.
594) Click the **second HEX-STANDOFF** top face.

595) **Rotate** the assembly to view the bottom face of the ANGLE-13HOLE component.

596) Hold the **Ctrl** key down.

597) Click the **ANGLE-13HOLE bottom face**.

598) Release the **Ctrl** key.

599) Click **Coincident** from the Mate pop-up menu.

Insert a Parallel mate.
600) Click **Front view** .

601) Click the **front face** of the second HEX-STANDOFF.

602) Hold the **Ctrl** key down.

603) Click the **front face** of the ANGLE-13HOLE.

604) Release the **Ctrl** key. The Mate pop-up menu is displayed.

605) Click **Parallel** from the Mate pop-up menu.

Display the created mates.
606) **Expand** the Mates folder in the FRONT-SUPPORT FeatureManager. Three mates are displayed between the ANGLE-13HOLE component and the second HEX-STANDOFF component. The second HEX-STANDOFF is fully defined.

Activity: FRONT-SUPPORT Assembly - Insert the TRIANGLE

Insert the TRIANGLE part.
607) Click **Isometric view** .

608) Click the **Insert Components** Assembly tool.

609) Double-click **TRIANGLE** from the SW-TUTORIAL 2023 folder.

610) Click a **position** in back of the ANGLE-13HOLE component as illustrated.

Enlarge the view.
611) Zoom in on the right side of the TRIANGLE and the ANGLE-13HOLE.

Insert a Concentric mate.
612) Click the **inside right arc face** of the TRIANGLE.

613) Hold the **Ctrl** key down.

614) Click the **inside right arc face** of the ANGLE-13HOLE slot.

615) Release the **Ctrl** key. The Mate pop-up menu is displayed.

616) Click **Concentric** from the Mate pop-up menu.

Fit the model to the Graphics window.
617) Press the **f** key.

Insert a Distance mate.
618) Click the **front face** of the TRIANGLE.

619) Rotate the model to view the back face of the ANGLE-13HOLE component.

620) Hold the **Ctrl** key down.

621) Click the **back face** of the ANGLE-13HOLE component.

622) Release the **Ctrl** key. The Mate pop-up menu is displayed.

623) Click **Distance** ⊢⊣ from the Mate pop-up menu.

624) Enter **0**.

625) Click the **Green Check mark** ✔.

 A Distance Mate of 0 provides additional flexibility compared to a Coincident mate. A Distance mate value can be modified.

Insert a Parallel mate.

626) Click **Bottom view** .

627) Click the **narrow bottom face** of the TRIANGLE.

628) Hold the **Ctrl** key down.

629) Click the **bottom face** of the ANGLE-13HOLE.

630) Release the **Ctrl** key.

631) Click **Parallel** ⟍ from the Mate pop-up menu.

Display the Isometric view.

632) Click **Isometric view** ◈.

View the created mates.

633) Expand the Mates folder. View the created mates.

Save the FRONT-SUPPORT assembly.

634) Click **Save** 🖫.

Assemble the four SCREW parts with SmartMates.

A SmartMate is a mate that automatically occurs when a component is placed into an assembly.

The mouse pointer displays a SmartMate feedback symbol when common geometry and relationships exist between the component and the assembly.

SmartMates are Concentric, Coincident, or Concentric and Coincident.

A Concentric SmartMate assumes that the geometry on the component has the same center as the geometry on an assembled reference.

Mating entities	Type of mate	Pointer
2 linear edges	Coincident	
2 planar faces	Coincident	
2 vertices	Coincident	
2 conical faces, or 2 temporary axes , or 1 conical face and 1 temporary axis	Concentric	
2 circular edges (peg-in-hole SmartMates). The edges do not have to be complete circles.	Concentric (conical faces) - and - Coincident (adjacent planar faces)	
2 circular patterns on flanges (flange SmartMates).	Concentric and coincident	
Origins and coordinate systems	Coincident	

As the component is dragged into place, the mouse pointer provides various feedback icons. The SCREW utilizes a Concentric and Coincident SmartMate. Assemble the first SCREW. The circular edge of the SCREW mates Concentric and Coincident with the circular edge of the right slot of the TRIANGLE.

Activity: FRONT-SUPPORT Assembly - Insert the SCREW

Insert the SCREW part.

635) Click **Open** from the Main menu.

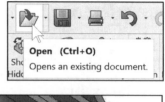

636) Double-click **SCREW** from the SW-TUTORIAL 2023 folder. The SCREW PropertyManager is displayed.

Display the SCREW part and the FRONT-SUPPORT assembly.
637) Click **Window**, **Tile Horizontally** from the Main menu.

638) **Zoom in** on the right side of the FRONT-SUPPORT assembly. Note: Work between the two tile windows.

Insert the first SCREW.
639) Click and drag the **circular edge** of the SCREW part into the FRONT-SUPPORT assembly Graphic window.

640) Release the mouse pointer on the **top 3rd circular hole edge** of the ANGLE-13HOLE. The mouse pointer displays the Coincident/Concentric circular edges icon.

Use the Pack and Go option to save an assembly or drawing with references. The Pack and Go tool saves either to a folder or creates a zip file to email. View SOLIDWORKS help for additional information.

Insert the second SCREW.

641) **Zoom in** on the right side of the FRONT-SUPPORT assembly. **Rotate** the model to view the right arc edge of the ANGLE-13HOLE.

642) Click and drag the **circular edge** of the SCREW part into the FRONT-SUPPORT assembly Graphic window.

643) Release the mouse pointer on the **right arc edge** of the

ANGLE-13HOLE. The mouse pointer displays the Coincident/Concentric circular edges icon.

Insert the third SCREW part.

644) **Zoom in** on the left side of the FRONT-SUPPORT assembly. **Rotate** the model if needed.

645) Click and drag the **circular edge** of the SCREW part into the FRONT-SUPPORT Assembly Graphic window.

646) Release the mouse pointer on the **left arc edge** of the

ANGLE-13HOLE. The mouse pointer displays the Coincident/Concentric circular edges icon.

Insert the fourth SCREW part.

647) **Zoom in** on the bottom circular edge of the ANGLE-13HOLE. **Rotate** the model if needed.

648) Click and drag the **circular edge** of the SCREW part into the FRONT-SUPPORT Assembly Graphic window.

649) Release the mouse pointer on the **bottom circular edge** of

the ANGLE-13HOLE. The mouse pointer displays the Coincident/Concentric circular edges icon.

650) **Close** the SCREW part window.

651) **Maximize** the FRONT-SUPPORT assembly window.

Display an Isometric view.

652) Click **Isometric view** . View the results in the Graphics window.

Save the FRONT-SUPPORT assembly.

653) Click **Save** .

Select the Ctrl-Tab keys to quickly alternate between open SOLIDWORKS documents.

Close all open parts and assemblies.
654) Click **Windows**, **Close All** from the Main menu.

The FRONT-SUPPORT assembly is complete.

Display the Mates in the FeatureManager to check that the components and the mate types correspond to the design intent.

When a component is in the Lightweight state, only a subset of its model data is loaded in memory. The remaining model data is loaded on an as-needed basis.

Review the FRONT-SUPPORT Assembly.

The ANGLE-13HOLE part was the first part inserted into the FRONT-SUPPORT assembly. The ANGLE-13HOLE part was fixed to the FRONT-SUPPORT Origin.

Concentric, Coincident, and Parallel mates were utilized to assemble the HEX-STANDOFF to the ANGLE-13HOLE. *Concentric*, *Distance*, and *Parallel* mates were utilized to assemble the TRIANGLE to the ANGLE-13HOLE. The *Concentric/Coincident* SmartMate was utilized to mate the four SCREW parts to the FRONT-SUPPORT assembly.

Chapter Summary

In this chapter you created four parts, HEX-STANDOFF, ANGLE-13HOLE, TRIANGLE and SCREW, utilizing the Top, Front, Right Planes and the FRONT-SUPPORT assembly.

You obtained the knowledge of the following SOLIDWORKS features: Extruded Boss/Base, Extruded Thin, Extruded Cut, Revolved Boss/Base, Hole Wizard, Linear Pattern, Circular Pattern, Fillet, and Chamfer. You also applied sketch techniques with various Sketch tools: Line, Circle, Corner Rectangle, Centerline, Dynamic Mirror, Straight Slot, Trim Entities, Polygon, Tangent Arc, Sketch Fillet, Offset Entities, and Convert Entities.

You utilized centerlines as construction geometry to reference dimensions and relationships. You incorporated the four new parts to create the FRONT-SUPPORT assembly. Concentric, Distance, and Parallel mates were utilized to assemble the TRIANGLE to the ANGLE-13HOLE.

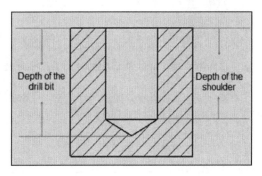

You can define the end condition of a hole to the depth of the drill tip or to the depth of the shoulder. The options are available for all Hole Wizard features (including the Hole Wizard Assembly feature) and Advanced Hole types with the following end conditions: Blind, Up to Vertex, Up to Surface, Offset to Surface. Previously, the end condition of a hole was calculated only up to the full diameter of the shoulder.

You applied the Quick mate procedure in the FRONT-SUPPORT assembly. Quick Mate is a procedure to mate components together. No command (click Mate from the Assembly CommandManager) is required. Hold the Ctrl key down, and make your selections. Release the Ctrl key, a Quick Mate pop-up is displayed below the context toolbar. Select your mate and you are finished.

The Concentric/Coincident SmartMate was utilized to mate the four SCREW parts to the FRONT-SUPPORT assembly.

Use Selection filters to select difficult individual features such as *faces*, *edges* and *points* when applying mates in an assembly. Right-click in the Assembly Graphics window, select Selection Filters. View your options.

Questions

1. Identify the three default Reference planes used in SOLIDWORKS.

2. True or False. Sketches are created only on the Front Plane.

3. Identify the sketch tool required to create a hexagon.

4. Describe the profile required for an Extruded Thin feature.

5. Mirror Entities, Offset Entities, Sketch Fillet, and Trim Entities are located in the _____ toolbar.

6. List the six standard principle orthographic views in a drawing: _____, _____, _____, _____, _____, _____,

7. Identify four types of Geometric relations that can be added to a sketch.

8. Describe the difference between a Circular Pattern feature and a Linear Pattern feature.

9. Describe the difference between a Fillet feature and a Chamfer feature.

10. Identify the function of the Hole Wizard feature.

11. Four 10-24x3/8 Machine Screws are required for an assembly. The diameter is _____. The threads per inch is _____. The length is _____.

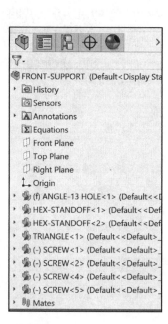

12. Describe the difference between a Distance mate and a Coincident mate.

13. True or False. A fixed component cannot move in an assembly.

14. Describe the procedure to remove the fix state (f) of a component in an assembly.

15. Determine the procedure to rotate a component in an assembly.

16. Describe the procedure to resolve a Lightweight component in an assembly.

Exercises

Exercise 3.1: HEX-NUT Part

Create an ANSI, IPS HEX-NUT part. Apply 6061 Alloy material. Apply the following dimensions:

- Depth: .125 in, [3.18].

- Inside hole diameter: .190in, [4.83].

- Outside diameter: .375in, [9.53].

Use the Top Plane as the Sketch plane.

Exercise 3.2: FRONT-SUPPORT-2 Assembly

Create an ANSI, IPS FRONT-SUPPORT-2 assembly.

- Name the new assembly FRONT-SUPPORT-2.

- Insert the FRONT-SUPPORT assembly. The FRONT-SUPPORT assembly was created in this Chapter. Note: The FRONT-SUPPORT assembly is provided in the Chapter 3 Homework folder.

- Fix the FRONT-SUPPORT assembly to the Origin.

- Insert the first HEX-NUT (Exercise 3.1) into the FRONT-SUPPORT-2 assembly.

- Insert a Concentric mate and Coincident mate.

- Insert the second HEX-NUT part.

- Insert a Concentric mate and Coincident mate.

You can also insert a Parallel mate between the HEX-NUT parts and the FRONT-SUPPORT assembly.

Exercise 3.3: Weight-Hook Assembly

Create an ANSI, IPS Weight-Hook assembly. The Weight-Hook assembly has two components: WEIGHT and HOOK.

- Create a new assembly document.

- Copy and insert the WEIGHT part from the Chapter 3 Homework folder.

- Fix the WEIGHT to the Origin in the Assem1 FeatureManager.

- Insert the HOOK part from the Chapter 3 Homework folder into the assembly.

- Insert a Concentric mate between the inside top cylindrical face of the WEIGHT and the cylindrical face of the thread. Concentric is the default mate.

- Insert the first Coincident mate between the top edge of the circular hole of the WEIGHT and the top circular edge of Sweep1, above the thread.

- Coincident is the default mate. The HOOK can rotate in the WEIGHT.

- Fix the position of the HOOK. Insert the second Coincident mate between the Right Plane of the WEIGHT and the Right Plane of the HOOK. Coincident is the default mate.

- Expand the Mates folder and view the created mates.

- Calculate the Mass and Volume of the assembly.

- Identify the Center of Mass for the assembly.

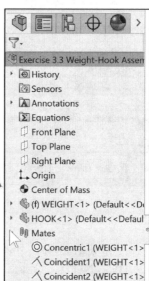

Exercise 3.4: Weight-Link Assembly

Create an ANSI, IPS Weight-Link assembly. The Weight-Link assembly has two components and a sub-assembly: Axle component, FLATBAR component, and the Weight-Hook sub-assembly that you created in a previous exercise.

- Create a new assembly document. Copy and insert the Axle part from the Chapter 3 Homework folder.

- Fix the Axle component to the Origin.

- Copy and insert the FLATBAR part from the Chapter 3 Homework folder. Rotate as needed.

- Insert a Concentric mate between the Axle cylindrical face and the FLATBAR inside face of the top circle.

- Insert a Coincident mate between the Front Plane of the Axle and the Front Plane of the FLATBAR.

- Insert a Coincident mate between the Right Plane of the Axle and the Top Plane of the FLATBAR. Position the FLATBAR as illustrated.

- Insert the Weight-Hook sub-assembly that you created in Exercise 3.3.

- Insert a Tangent mate between the inside bottom cylindrical face of the FLATBAR and the top circular face of the HOOK, in the Weight-Hook assembly. Tangent mate is selected by default. Click Flip Mate Alignment if needed.

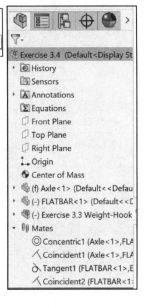

- Insert a Coincident mate between the Front Plane of the FLATBAR and the Front Plane of the Weight-Hook sub-assembly. Coincident mate is selected by default. The Weight-Hook sub-assembly is free to move in the bottom circular hole of the FLATBAR.

- Calculate the Mass and Volume and the Center of Mass for the assembly.

Exercise 3.5: Binder Clip

Create a simple Gem® binder clip.

Create an ANSI - IPS assembly. Create two components - BASE and HANDLE.

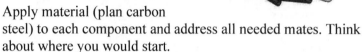

Apply material (plan carbon steel) to each component and address all needed mates. Think about where you would start.

Note: The cuts in the main body are made before folding the metal and strengthening the material during manufacturing. You do not need to model this exactly. The key to this exercise is to apply Advanced distance mates (Limit mates) to limit the movement of the handle.

What is the Base Sketch for each component?

What are the dimensions? Measure all dimensions (approximately) from a small or large Gem binder clip.

View SOLIDWORKS Help or Chapter 5 on the Swept Base feature to create the Binder Clip handle.

💡 Determine the static and dynamic behavior of mates in each sub-assembly before creating the top-level assembly.

Exercise 3.6: Limit Mate (Advanced Distance Mate)

- Copy all files from the Chapter 3 Homework\Limit Mate folder to your hard drive.

- Create the final assembly with all needed mates for proper movement.

- Insert a Distance (Advanced Limit Mate) to restrict the movement of the Slide Component - lower and upper movement.

- Use the Measure tool to obtain maximum and minimum distances.

- Use SOLIDWORKS Help for additional information.

A Distance (Limit Mate) is an Advanced Mate type. Limit mates allow components to move within a range of values for distance and angle. You specify a starting distance or angle as well as a maximum and minimum value.

- Save the model and move the slide to view the results in the Graphics window.

- Think about how you would use this mate type in other assemblies.

Exercise 3.7: Screw Mate (Mechanical Mate)

- Copy the Chapter 3 Homework\Screw Mate folder to your hard drive.

- Open the Screw Mate assembly.

- Insert a Screw mate between the inside Face of the Base and the Face of the vice and any other mates that are required. A Screw is a Mechanical Mate type.

- Calculate the total mass and volume of the assembly.

A Screw mate constrains two components to be concentric and also adds a pitch relationship between the rotation of one component and the translation of the other. Translation of one component along the axis causes rotation of the other component according to the pitch relationship. Likewise, rotation of one component causes translation of the other component. Use SOLIDWORKS Help if needed.

- Rotate the handle and view the results. Think about how you would use this mate type in other assemblies?

Use the Select Other tool (See SOLIDWORKS Help) to select faces and edges that are hidden in an assembly.

Exercise 3.8: Angle Mate

- Copy the Chapter 3 Homework\Angle Mate folder to your hard drive.

- Open the Angle Mate assembly.

- Move the Handle in the assembly. The Handle is free to rotate.

- Set the angle of the Handle.

- Insert an Angle mate (165 degree) between the Handle and the Side of the valve using Planes. An Angle mate places the selected items at the specified angle to each other. The Handle has a 165 degree Angle mate to restrict flow through the valve.

- Think about how you would use this mate type in other assemblies?

Exercise 3.8A: Angle Mate (Cont:)

Create two end caps (lids) for the ball valve using the Top-down Assembly method. Note: The Reference - In-Content symbols in the FeatureManager.

- Modify the Appearance of the body to observe the change, and enhance visualization.

- Apply the Select-other tool to obtain access to hidden faces and edges.

Exercise 3.9: Symmetric Mate (Advanced Mate)

- Copy the Chapter 3 Homework\Symmetric Mate folder to your hard drive.

- Open the Symmetric Mate assembly.

- Insert a Symmetric Mate between the two Guide Rollers.

- Apply LimitMates (Advanced Distance mate) to limit movement of the two Guide Rollers.

- Think about how you would use this mate type in other assemblies?

Exercise 3.10: Gear Mate (Mechanical mate)

View the ppt presentation and motion file on Gears located in the Chapter 3 Homework\Gears folder.

Create the Gear Mate assembly per the ppt presentation.
Create the **Base Plate** and **two Shafts** per the ppt presentation.

Use the SOLIDWORKS toolbox to insert the needed gear components. You create the **Base Plate**, and **Shafts**.

Insert all needed mates for proper movement.

Gear mate: Forces two components to rotate relative to one another around selected axes. The Gear mate provides the ability to establish gear type relations between components without making the components physically mesh.

SOLIDWORKS provides the ability to modify the gear ratio without changing the size of the gears. *Align the components* before adding the Mechanical gear mate.

Exercise 3.10A: Slot Mate (Center in Slot option)

Copy the Chapter 3 Homework\Slot Mate folder to your hard drive.

Open the GUIDE-ROD assembly. Create two Slot mates for the two flange bolts. You can mate bolts to straight or arced slots and you can mate slots to slots. Use the Quick Mate procedure.

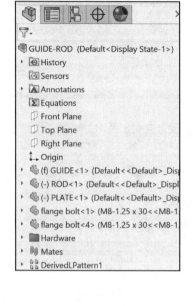

Insert a Slot mate (Center in Slot option) between the first flange bolt and the inside face of the right slot.

Insert the second slot mate (Center in Slot option) between the second flange bolt and the inside face of the left slot.

Insert a Coincident mate between the top face of the GUIDE and the bottom face of the second flange bolt cap.

Insert a Coincident mate between the top face of the GUIDE and the bottom face of the first flange bolt cap.

Insert a parallel mate between the front face of the first hex head flange bolt and the front of the GUIDE.

Inset a parallel mate between the front face of the second hex head flange bolt and the front face of the GUIDE.

Expand the Mates folder. View the created mates.

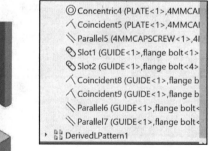

Exercise 3.11: Slider Part

Create the part from the illustrated A-ANSI - MMGS Third Angle Projection drawing below: Front, Top, Right and Isometric view.

Note: The location of the Origin (Shown in an Isometric view).

- Apply Cast Alloy steel for material.

- The part is symmetric about the Front Plane.

- Apply Mid Plane for End Condition in Boss-Extrude1.

- Apply Through All for End Condition in Cut-Extrude1.

- Apply Through All for End Condition in Cut-Extrude2.

- Apply Up to Surface for End Condition in Boss Extrude2

- Calculate the Volume of the part and locate the Center of mass.

Think about the steps that you would take to build the model. Do you need the Right view for manufacturing? Does it add any important information?

Exercise 3.12: Cosmetic Thread Part

Apply a Cosmetic thread: 1/4-20-1 UNC 2A. A cosmetic thread represents the inner diameter of a thread on a boss or the outer diameter of a thread.

Copy the Cosmetic thread part from the Chapter 3 Homework folder to your hard drive.

Open the part. Create a Cosmetic thread. Produce the geometry of the thread. Click the bottom edge of the part as illustrated.

Click Insert, Annotations, Cosmetic Thread from the Main menu. View the Cosmetic Thread PropertyManager. Edge<1> is displayed.

Select ANSI Inch for Standard.

Select Machine Threads for Type.

Select ¼-20 for Size.

End-Condition - Blind.

Enter 1.00 for depth.

Select 2A for Thread class. Click OK ✔ from the Cosmetic Thread FeatureManager.

Expand the FeatureManager. View the Cosmetic Thread feature. If needed, right-click the Annotations folder, click Details.

Check the Cosmetic threads and Shaded cosmetic threads box.

Click OK. View the cosmetic thread on the model.

☀ ¼-20-1 UNC 2A: ¼ inch major diameter - 20 threads / inch, 1 inch long, Unified National Coarse thread series, Class 2 (General Thread), A - External threads.

Exercise 3.12A: Stud Wizard

Create a stud on a Cylindrical Body. Create an ANSI Inch, Machine thread, ¼-20-2 UNC 2A: ¼ inch major diameter - 20 threads / inch, 2 inch long, Unified National Coarse thread series, Class 2 (General Thread), A -External threads.

Note: Class fit measures the looseness or tightness of external mating threads. You can retrieve and use this information for hole callouts in drawings.

Copy and open the Exercise 3.12A part from the Chapter 3 Homework folder.

Use the default setting for UnderCut. The UnderCut option creates an undercut at the end of the threaded portion of the stud to provide clearance. The default values are based on the selections under Standard, but you can override and customize these values.

Click the Creates Stud on a Cylindrical Body icon.

Click the top edge of the part as illustrated. Edge<1> is displayed.

Enter ANSI Inch for Standard.

Enter Machine Threads for Type.

Enter ¼-20 for Size.

Select End Condition: Blind.

Enter 2.00in for Depth.

Check the Thread class box.

Select 2A for Thread Class.

Click OK ✔ from the FeatureManager.

View the threads. Right-click the Annotations folder, and click Details.

Check the Cosmetic threads and Shaded cosmetic threads box.

Click OK from the Annotation Properties dialog box.

Exercise 3.12B: Stud Wizard

Create a stud on a Surface. Create an ANSI Inch, Machine thread, ¼-20-2 UNC 2A: ¼ inch major diameter - 20 threads / inch, 2 inch long, Unified National Coarse thread series, Class 2 (General Thread), A -External threads.

Copy and open Exercise 3.12B from the Chapter 3 Homework folder.

Use the default setting for UnderCut. The UnderCut option creates an undercut at the end of the threaded portion of the stud to provide clearance. The default values are based on the selections under Standard, but you can override and customize these values.

Click the Creates Stud on a Surface icon.

Enter ANSI Inch for Standard. Enter Machine Threads for Type.

Enter ¼-20 for Size. Select End Condition: Blind.

Enter 2.00in for Depth. Check the Thread class box.

Select 2A for Thread Class.

Click the Position tab.

Click Top view.

Click a position to the right on the top surface. Click the origin.

Click OK ✔ from the Stud Position PropertyManager. Display an Isometric view.

To view the threads, right-click the Annotations folder, and click Details.

Check the Cosmetic threads and Shaded cosmetic threads box.

Click OK from the Annotation Properties dialog box.

Exercise 3.13: Hole - Block

Create the Hole-Block ANSI - IPS part using the Hole Wizard feature as illustrated. Create the Hole-Block part on the Front Plane.

The Hole Wizard feature creates either a 2D or 3D sketch for the placement of the hole in the FeatureManager.

You can consecutively place multiple holes of the same type. The Hole Wizard creates 2D sketches for holes unless you select a non-planar face or click the 3D Sketch button in the Hole Position PropertyManager.

Hole Wizard creates two sketches. A sketch of the revolved cut profile of the selected hole type and the other sketch of the center placement for the profile. Both sketches should be fully defined.

Create a rectangular prism 2 inches wide by 5 inches long by 2 inches high. On the top surface of the prism, place four holes, 1 inch apart.

- Hole #1: Simple Hole Type: Fractional Drill Size, 7/16 diameter, End Condition: Blind, 0.75 inch deep.
- Hole #2: Counterbore hole Type: for 3/8 inch diameter Hex bolt, End Condition: Through All.
- Hole #3: Countersink hole Type: for 3/8 inch diameter Flat head screw, 1.5 inch deep.
- Hole #4: Tapped hole Type, Size ¼-20, End Condition: Blind -1.0 inch deep.

Exercise 3.14: Hole Wizard Part

Apply the 3D sketch placement method as illustrated in the FeatureManager. Insert and dimension a hole on a cylindrical face.

Copy the Hole Wizard 3-14 part from the Chapter 3 Homework folder to your hard drive.

Open the Hole Wizard 3-14 part.

Note: With a 3D sketch, press the Tab key to move between planes.

Click the Hole Wizard Features tool. The Hole Specification PropertyManager is displayed.

Select the Counterbore Hole Type. Select ANSI Inch for Standard. Select Socket Head Cap Screw for fastener Type.

Select 1/4 for Size. Select Normal for Fit.

Select Through All for End Condition.

Enter .100 for Head clearance in the Options box.

Click the Positions Tab. The Hole Position PropertyManager is displayed.

Click the 3D Sketch button. SOLIDWORKS displays a 3D interface with the Point ×Y × tool active.

When the Point tool is active, wherever you click, you will create a point.

Click the cylindrical face of the model as illustrated. The selected face is displayed in blue.

Insert a dimension between the top face and the Sketch point. Click the Smart Dimension ⟋ Sketch tool.

Click the top flat face of the model and the sketch point.

Enter .25in.

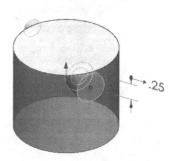

Locate the point angularly around the cylinder. Apply construction geometry.

Activate the Temporary Axes. Click View, Hide/show, check the Temporary Axes box from the Main menu toolbar.

Click the Line ∕ Sketch tool. Note: 3D sketch is still activated.

Ctrl+click the top flat face of the model. This moves the red space handle origin to the selected face. This also constrains any new sketch entities to the top flat face. Note the mouse pointer icon.

Move the mouse pointer near the center of the activated top flat face as illustrated. View the small black circle. The circle indicates that the end point of the line will pick up a Coincident relation.

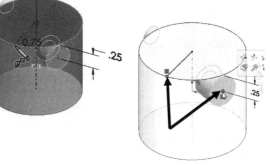

Click the center point of the circle.

Sketch a line so it picks up the **AlongZ sketch relation**. The cursor displays the relation to be applied. **This is a very important step**. Insert an AlongZ relation is needed.

Create an AlongY sketch relation between the centerpoint of the hole on the cylindrical face and the endpoint of the sketched line as illustrated. The sketch is fully defined.

Click OK ✔ from the Properties PropertyManager. Click OK ✔ from the Hole Position PropertyManager.

Expand the FeatureManager and view the results. The two sketches are fully defined. One sketch is the hole profile, the other sketch is to define the position of the feature.

Close the model.

Exercise 3.15: Counter Weight Assembly

Create the Counter Weight assembly as illustrated using SmartMates and Standard mates.

Copy the Chapter 3 Homework\Counter-Weight folder to your local hard drive.

Create a new assembly. The Counter Weight assembly consists of the following items:

- Weight-Hook sub-assembly (provided).

- Weight (provided).

- Eye Hook (provided).

- Axle component (f). Fixed to the origin. Provided.

- Flat component (provided).

- Flat Washer Type A (from the SOLIDWORKS toolbox), provided.

- Pan Cross Head Screw (from the SOLIDWORKS toolbox), provided.

- Flat Washer Type A (from the SOLIDWORKS toolbox), provided.

- Machine Screw Nut Hex (from the SOLIDWORKS toolbox), provided.

Apply SmartMates with the Flat Washer Type A Narrow_AI, Machine Screw Nut Hex_AI and the Pan Cross Head_AI components.

Use a Distance mate to fit the Axle in the middle of the Flat. Note a Symmetric mate could replace the Distance mate. Think about the design of the assembly. Apply all needed Lock mates.

The symbol (f) represents a fixed component. A fixed component cannot move and is locked to the assembly Origin.

Exercise 3.16: Gear Mate

Insert a Mechanical Gear mate between two gears. Gear mates force two components to rotate relative to one another about selected axes. Valid selections for the axis of rotation for gear mates include cylindrical and conical faces, axes, and linear edges.

1. **Copy** the Chapter 3 Homework folder to your hard drive.

2. Open **Mechanical Gear Mate** from the Chapter 3 Homework\Gear Mate folder.

3. Click and drag the **right gear**.

4. Click and drag the **left gear**. The gears move independently.

5. Click the **Mate** ✎ tool from the Assembly tab. The Mate PropertyManager is displayed.

6. Click the **Mechanical** tab.

7. **Clear** all selections.

8. **Align the components** before adding the Mechanical mate. To align the two components, select a Front view and move them manually.

9. Select the **inside cylindrical face** of the left hole.

10. Select the **inside cylindrical face** of the right hole.

11. Click **Gear** 🔩 mate. The GearMate1 PropertyManager is displayed. Accept the default settings.

12. Click **OK** ✔ from the GearMate1 PropertyManager.

13. **Expand** the Mates folder. View the created GearMate1.

14. **Rotate** the Gears in the Graphics window.

15. **View** the results.

Exercise 3.17: Cam Mate

Create a Tangent Cam follower mate and a Coincident Cam follower mate. A cam-follower mate is a type of Tangent or Coincident mate. It allows you to mate a cylinder, plane or point to a series of tangent extruded faces, such as you would find on a cam.

Create the profile of the cam from lines, arcs and splines as long as they are tangent and form a closed loop.

1. **Copy** the Chapter 3 Homework folder to your hard drive.

2. Open **Cam Mate** from the Chapter 3 Homework\Cam Mate folder. View the model. The model contains two cams.

Insert a tangent Cam follower mate.

3. Click the **Mate** ✎ tool from the Assembly tab.

4. **Pin** the Mate PropertyManager.

5. Click the **Mechanical** tab.

6. Click **Cam Follower** ⬭ mate.

7. Click inside of the **Cam Path** box.

8. Click the **outside face** of cam-s<1> as illustrated.

9. Click the **face of Link-c<1>** as illustrated. Face<2>@Link-c-1 is displayed.

10. Click **OK** ✔ from the CamMateTangent1 PropertyManager.

Insert a Coincident Cam follower mate.

11. Click **Cam Follower** ⟋ mate from the Mechanical Mates box.

12. Click the **outside face** of cam-s2<1>.

13. Click the vertex of **riser<1>** as illustrated. Vertex<1>@riser-1 is displayed.

14. Click **OK** ✔ from the CamMateCoincident1 PropertyManager.

15. **Un-pin** the Mate PropertyManager.

16. Click **OK** ✔ from the Mate PropertyManager.

17. **Expand** the Mates folder. View the new created mates.

18. Display an **Isometric** view.

19. **Rotate** the cams.

20. **View** the results in the Graphics window.

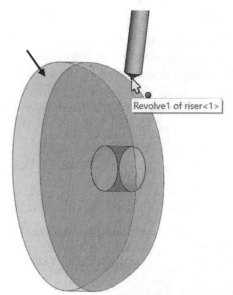

When creating a cam follower mate, make sure that your Spline or Extruded Boss/Base feature, which you use to form the cam contact face, does nothing but form the face.

In the next exercise of the spur gear and rack, the gear's pitch circle must be tangent to the rack's pitch line, which is the construction line in the middle of the tooth cut.

Exercise 3.18: Rack Pinion Mate

The Spur gear and rack components for this exercise were obtained from the SOLIDWORKS Toolbox folder marked Power Transmission. Insert a Mechanical Rack Pinion mate.

1. **Copy** the Chapter 3 Homework folder to your hard drive.

2. Open **Rack Pinion Mate** from the Chapter 3 Homework\Rack Pinion Mate folder.

3. **View** the displayed sketches; (TooCutSkeSim and ToothCutSim). If needed, display the illustrated sketches to mate the assembly.

The Diametral Pitch, Pressure Angle and Face Width values should be the same for the spur gear as the rack.

💡 Tangency is required between the spur gear and rack.

4. **Expand** the Mates folder to view the existing mates. Four Coincident mates and a Distance mate were created in the assembly.

Insert a Rack Pinion mate.

5. Click the **Mate** ✎ tool from the Assembly tab. Click the **Mechanical** tab.

6. **Expand** the Mechanical Mates box from the Mate PropertyManager.

7. Click **Rack Pinion** ⚖ mate.

8. Mate the rack. Make sure the teeth of the spur gear and rack are **meshing properly**; they are not interfering with each other. It is important to create a starting point where the gear and rack have no interference.

9. Click the **bottom linear edge of the rack** in the direction of travel. Any linear edge that runs in the direction of travel is acceptable.

10. Click the **spur gear pitch circle** as illustrated. The Pinion pitch diameter, 1in is taken from the sketch geometry. Accept the default settings.

11. Click **OK** ✔ from the RackPinionMate1 PropertyManager.

12. **Rotate** the spur gear on the rack. View the results.

13. **Hide** the sketches in the model. **Close** the model.

The Pitch Diameter is the diameter of an imaginary pitch circle on which a gear tooth is designed. Pitch circles of the two spur gears are required to be tangent for the gears to move and work properly.

The distance between the center point of the gear and the pitch line of the rack represents the theoretical distance required.

🔅 Diametral Pitch is the ratio equal to the number of teeth on a gear per inch of pitch diameter.

🔅 Before applying the Rack Pinion mate, make sure the teeth of the spur gear and rack are meshing properly.

Pressure Angle is the angle of direction of pressure between contacting teeth. Pressure angle determines the size of the base circle and the shape of the involute teeth. It is common for the pressure angle to be 20 or 14 1/2 degrees.

Exercise 3.19: Hinge Mate

Create a Mechanical Hinge mate. The two components in this tutorial were obtained from the SOLIDWORKS What's New section.

1. **Copy** the Chapter 3 Homework folder to your hard drive.

2. Open **Hinge Mate** from the Chapter 3 Homework\Hinge Mate folder. Two components are displayed.

A Hinge mate has the same effect as adding a Concentric mate plus a Coincident mate. You can also limit the angular movement between the two components.

3. Click the **Mate** ✎ tool from the Assembly tab.

4. Click the **Mechanical** tab.

5. Click **Hinge** ⊞ mate.

Set two Concentric faces.

6. Click the **inside cylindrical face** of the first component as illustrated.

7. Click the **outside cylindrical face** of the second component as illustrated. The two selected faces are displayed.

Set two Coincident faces.

8. Click the **front flat face** of the first component as illustrated.

9. Click the **bottom flat face** of the second component as illustrated. View the results in the Graphics window.

You can specify a limit angle for rotation by checking the **Specify angle limits** box and selecting the required faces.

10. Click **OK** ✔ from the Hinge1 PropertyManager.

11. **Rotate** the flap component about the pin.

12. **Close** the model.

A Hinge mate limits the movement between two components to one rotational degree of freedom.

Similar to other mate types, Hinge mates do not prevent interference or collisions between components. To prevent interference, use Collision Detection or Interference Detection.

Exercise 3.20: AIR RESERVOIR SUPPORT AND PLATE Assembly

The project team developed a concept sketch of the PNEUMATIC TEST MODULE assembly. Develop the AIR RESERVOIR SUPPORT AND PLATE assembly.

Create three new parts:

PNEUMATIC TEST MODULE Assembly Layout

- FLAT-PLATE

- IM15-MOUNT

- ANGLE-BRACKET

The Reservoir is a purchased part. The assembly file is located in the Chapter 3 Homework folder.

SMC AIR RESERVOIR

AIR RESERVOIR SUPPORT Assembly
Courtesy of Gears Educational Systems & SMC
Corporation of America

- Create a new assembly named AIR RESERVOIR SUPPORT AND PLATE.

- Two M15-MOUNT parts and two ANGLE-BRACKET parts hold the SMC AIR RESERVOIR.

- The ANGLE-BRACKET parts are fastened to the FLAT-PLATE.

Exercise 3.21a: FLAT-PLATE Part

Create the FLAT-PLATE Part on the Top Plane. The FLAT-PLATE is machined from .090, [2.3] 6061 Alloy flat stock. The default units are inches.

Utilize the Top Plane for the Sketch plane.

Locate the Origin at the Midpoint of the left vertical line.

The 8.690, [220.68mm] x 5.688, [144.48mm] FLAT PLATE contains a Linear Pattern of ∅.190, [4.83mm] Thru holes.

The Holes are equally spaced, .500, [12.7mm] apart.

Determine the maximum
number of holes contained in
the FLAT-PLATE.

Maximum # of
holes_____.

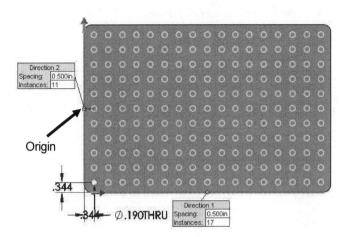

- Utilize a Linear Pattern in
  two Directions to create the
  holes.

- Utilize the Geometric
  Pattern Option.

Exercise 3.21b: IM15-MOUNT Part

- Create the IM15-MOUNT part on the Right plane.

- Center the part on the Origin. Utilize the features in the FeatureManager.

- The IM15-MOUNT Part is machined from 0.060, [1.5mm] 6061 Alloy flat stock. The
  default units are inches.

Exercise 3.21c: ANGLE BRACKET Part

- Create the ANGLE BRACKET part.

- The Extruded Base (Boss-Extrude1) feature is sketched with an L-Shaped profile on the Right Plane. The ANGLE BRACKET Part is machined from 0.060, [1.5mm] 6061 Alloy flat stock. The default units are inches.

Exercise 3.21d: Reservoir Assembly

The Reservoir stores compressed air. Air is
pumped through a Schrader Valve into the
Reservoir.

A Quick Connect Straight Fitting is utilized to
supply air to the Pneumatic Test Module
Assembly. Quick Connect Fittings allow air
tubing to be assembled and disassembled without
removing the fitting.

Copy the Pneumatic Components folder to your
hard drive.

Open the part Reservoir from the Pneumatics
Components folder. The Reservoir default units
are in millimeters (MMGS).

Schrader Valve

Quick Connect
Straight Fitting

Reservoir and Fittings
Courtesy of SMC Corporation of America and
Gears Educational Systems

Engineers and designers work in metric and english units. Always verify your units for
parts and other engineering data. In pneumatic systems, common units for volume,
pressure and temperature are defined in the below table.

Magnitude	Metric Unit (m)	English (e)
Mass	kg	pound
	g	ounce
Length	m	foot
	m	yard
	mm	inch
Temperature	°C	°F
Area, Section	m ^{2}	sq.ft
	cm ^{2}	sq.inch
Volume	m ^{3}	cu.yard
	cm ^{3}	cu.inch
	dm ^{3}	cu.ft.
Volume Flow	m ^{3}n / min	scfm
	dm ^{3}n /min (ℓ/min)	scfm
Force	N	pound force (ℓbf.)
Pressure	bar	ℓbf./sq.inch (psi)

Common Metric and English Units

The ISO unit of pressure is the Pa (Pascal). $1Pa = 1N/m$.

Exercise 3.21e: AIR RESERVOIR SUPPORT AND PLATE Assembly

Create the AIR RESERVOIR SUPPORT AND PLATE assembly. Note: There is more than one solution for the mate types illustrated below.

The FLAT-PLATE is the first component in the AIR RESERVOIR SUPPORT AND PLATE assembly. Insert the FLAT-PLATE. The FLAT-PLATE is fixed to the Origin.

- Insert the ANGLE BRACKET.

- Mate the ANGLE BRACKET to the FLAT-PLATE. The bottom flat face of the ANGLE BRACKET is coincident to the top face of the FLAT-PLATE.

- The center hole of the ANGLE BRACKET is concentric to the upper left hole of the FLAT-PLATE.

- The first hole of the ANGLE bracket is concentric with the hole in the 8th row, 1st column of the FLAT-PLATE.

- Insert the IM15-MOUNT.

- Mate the IM15-MOUNT. The IM15-MOUNT flat back face is coincident to the flat inside front face of the ANGLE BRACKET.

- The bottom right hole of the IM15-MOUNT is concentric with the right hole of the ANGLE BRACKET.

- The bottom edge of the IM15-MOUNT is parallel to the bottom edge of the ANGLE BRACKET.

AIR RESERVOIR SUPPORT AND PLATE Assembly
Courtesy of SMC Corporation of America

ANGLE BRACKET
Mate Holes

FLAT PLATE HOLE,
8th row, 1st column.

FLAT-PLATE

IM15-MOUNT
Mate Holes
Concentric

Mate Hole Edges
Parallel

ANGLE BRACKET
Inside face

- Insert the Reservoir Assembly.

- Mate the Reservoir Assembly. The conical face of the Reservoir is concentric to the IM15-MOUNT center hole.

- The left end cap of the Reservoir Assembly is coincident to the front face of the IM15-MOUNT.

- The Hex Nut flat face is parallel to the top face of the FLAT-PLATE.

- Insert the second ANGLE BRACKET.

- Mate the ANGLE BRACKET to the FLAT-PLATE. The bottom flat face of the ANGLE BRACKET is coincident to the top face of the FLAT-PLATE.

- The center hole of the ANGLE BRACKET is concentric with the hole in the 11th row, 13th column of the FLAT-PLATE.

- The first hole of the ANGLE bracket is concentric with the hole in the 8th row, 13th column of the FLAT-PLATE.

- Insert the second IM15-MOUNT.

- Mate the IM15-MOUNT to the outside face of the ANGLE BRACKET. The bottom right hole of the IM15-MOUNT is concentric with the right hole of the ANGLE BRACKET.

- The top edge of the IM15-MOUNT is parallel to the top edge of the ANGLE BRACKET.

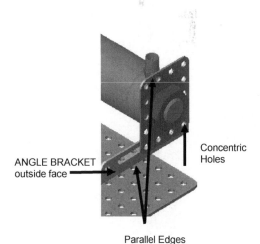

- Save the assembly. Insert the required SCREWS. The AIR RESERVOIR SUPPORT AND PLATE assembly is complete.

Notes:

Chapter 4
Fundamentals of Drawing

Below are the desired outcomes and usage competencies based on the completion of Chapter 4.

Desired Outcomes:	Usage Competencies:
• CUSTOM-A Sheet Format. • A-ANSI-MM Drawing Template. • FLATBAR Configurations. • FLATBAR Part Drawing. • LINKAGE Assembly Drawing. • FLATBAR-SHAFTCOLLAR Assembly.	• Ability to create a Custom Sheet Format, Drawing Template, Company logo and Title block.
	• Understand Standard, Isometric, Detail, Section and Exploded views. • Knowledge of the View Palette. • Ability to incorporate a Bill of Materials with Custom Properties.
	• Proficiency to create and edit drawing dimensions and annotations. • Aptitude to create a Design Table.

Notes:

Chapter 4 - Fundamentals of Drawing

Chapter Objective

Create a FLATBAR drawing with a customized Sheet Format and a Drawing Template containing a Company logo and Title block.

Obtain an understanding to display the following views with the ability to insert, add, and edit dimensions and annotations:

- Standard: Top, Front, and Right.

- Isometric, Detail, Section, and Exploded.

Create a LINKAGE assembly drawing with a Bill of Materials. Obtain knowledge to develop and incorporate a Bill of Materials with Custom Properties. Create a FLATBAR-SHAFTCOLLAR assembly.

On the completion of this chapter, you will be able to:

- Create a customized Sheet Format.

- Generate a custom Drawing Template.

- Produce a Bill of Materials with Custom Properties.

- Develop various drawing views.

- Reposition views on a drawing.

- Move dimensions in the same view.

- Apply Edit Sheet Format mode and Edit Sheet mode.

- Modify the dimension scheme.

- Create a Parametric drawing note.

- Link notes in the Title block to SOLIDWORKS properties.

- Generate an Exploded view.

- Create and edit a Design Table.

- Insert a Center of Mass Point.

Chapter Overview

Generate the LINKAGE assembly drawing with two sheets and the FLATBAR part drawing using configurations from a design table.

The LINKAGE assembly drawing contains two sheets:

- Sheet1 contains the LINKAGE assembly in an Exploded view with a Bill of Materials.

- Sheet2 contains the AirCylinder assembly with a Section view, Detail view and a Scale view.

The FLATBAR drawing utilizes a custom Drawing Template and a custom Sheet Format. The FLATBAR drawing contains two sheets:

- Sheet1 contains a Front, Top and Isometric view with dimensions and a linked Parametric note.

- Sheet2 contains the 3HOLE configuration of the FLATBAR. Configurations are created with a Design Table.

Create the FLATBAR-SHAFTCOLLAR
assembly. Utilize a Design Table to create four
new configurations of the assembly.

There are two major design modes used to
develop a drawing:

- Edit Sheet Format and Edit Sheet.

The Edit Sheet Format mode provides the ability
to:

- Change the Title block size and text headings.

- Incorporate a Company logo.

- Define the Zone Tag Editor.

- Add Custom Properties, text and more.

The Edit Sheet mode provides the ability to:

- Add or modify views.

- Add or modify dimensions.

- Add or modify notes and more.

Drawing Template and Sheet Format

The foundation of a SOLIDWORKS drawing is the
Drawing Template. Drawing size, drawing
standards, company information, manufacturing
and/or assembly requirements, units and other
properties are defined in the Drawing Template.

The Sheet Format is incorporated into the Drawing
Template. The Sheet Format contains the border,
Title block information, Revision block
information, Company name and or Logo
information, Custom Properties, and
SOLIDWORKS Properties.

Custom Properties and SOLIDWORKS Properties
are shared values between documents. Utilize an
A-size Drawing Template with Sheet format for the
FLATBAR drawing and LINKAGE assembly
drawing.

During the initial SOLIDWORKS installation, you are requested to select either the ISO or ANSI drafting standard. ISO is typically a European drafting standard and uses First Angle Projection. The book is written using the ANSI (US) overall drafting standard and Third Angle Projection for drawings.

Views from the part or assembly are inserted into the SOLIDWORKS drawing.

Create Sheet Formats for different parts types. Example: sheet metal parts, plastic parts, and high precision machined parts.

Create Sheet Formats for each category of parts that are manufactured with unique sets of title block notes.

Note: The Third Angle Projection scheme is illustrated in the chapter.

Third Angle
Projection icon

Third Angle Projection

For non-ANSI dimension standards, the dimensioning techniques are the same, even if the displayed arrows and text size are different. For printers supporting millimeter paper sizes, select A4 (ANSI) Landscape (297mm × 210mm).

The default Drawing Templates contain predefined Title block Notes linked to Custom Properties and SOLIDWORKS Properties.

Activity: Create a new Drawing

Create a new drawing. Close all parts and drawings.

1) Click **Windows**, **Close All** from the Main menu.

2) Click **New** ⬜ from the Main menu.

3) Double-click **Drawing** from the Templates tab.

4) Select **A (ANSI) Landscape**.

5) Click **OK** from the Sheet Format/Size box. The Model View PropertyManager is displayed.

6) Click **Cancel** ✕ from the Model View PropertyManager. The Draw1 FeatureManager is displayed.

The A (ANSI) Landscape paper is displayed in a new Graphics window. The sheet border defines the drawing size, 11″ × 8.5″ or (279.4mm × 215.9mm).

The Model View PropertyManager is displayed if the Start command when creating new drawing is checked.

🔅 To view the Sheet properties, right-click Properties in the drawing sheet. Expand the drop-down menu. View the Sheet Properties dialog box.
Click OK to return to the drawing sheet.

Draw1 is the default drawing name. Sheet1 is the default first Sheet name. CommandManager, FeatureManager and Task Pane tabs will vary depending on system setup and Add-ins.

Utilize the CommandManager tabs or individual toolbars to access the tools in this chapter.

Set Sheet Properties.

7)   Right-click **Sheet1** in the Draw FeatureManager.

8)   Click **Properties**. The Sheet Properties are displayed.

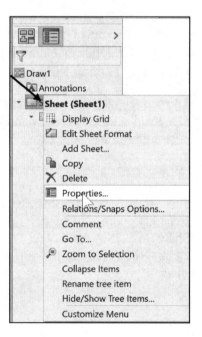

9) **View** your options.

10) Enter **Sheet Scale 1:1**.

11) Check the **Third angle** box for Type of projection.

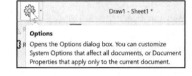

12) Click **Apply Changes** from the Sheet Properties dialog box. Apply Changes is active if you changed any of the default settings.

Activity: Drawing-Document Properties

Set Document properties. Set drafting standard, units, and precision.

13) Click **Options** ⚙ ⋅ from the Main menu.

14) Click the **Document Properties** tab from the dialog box.

15) Select **ANSI** for Overall drafting standard from the drop-down menu.

16) Click **Units**. Click **MMGS** (millimeter, gram, second) for Unit system.

17) Select **.12** for Length decimal place.

18) Select **None** for Angular decimal place.

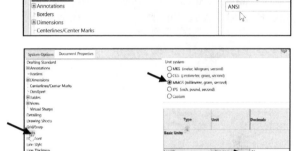

💡 Available Document Properties are document dependent.

Companies develop drawing format standards and use specific text height for Metric and English drawings. Numerous engineering drawings use the following format:

- *Font*: Century Gothic - All capitals.

- *Text height*: 3mm for drawings up to B Size, 17in. × 22in.

- *Text height*: 5mm for drawings larger than B Size, 17in × 22in.

- *Arrowheads*: Solid filled with a 1:3 ratio - width to height.

Set the dimension font.
19) Click the **Annotations** folder as illustrated.

20) Click the **Font** button. The Choose Font dialog box is displayed.

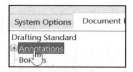

Set the dimension text height.
21) Click the **Units** box from the Choose Font dialog box.

22) Enter **3.0**mm for Height.

23) Click **OK** from the Choose Font dialog box.

Set the arrow size.
24) Click the **Dimensions** folder as illustrated.

25) Enter **1**mm for arrow Height. Enter **3**mm for arrow Width.

26) Enter **6**mm for arrow Length.

27) Click **OK** from the Document Properties dialog box.

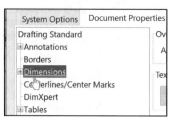

There are three dimension style type buttons: Outside, Inside, and Smart. Smart is the default option.

Title Block

The Title block contains text fields linked to System Properties and Custom Properties. System Properties are determined from the SOLIDWORKS documents. Custom Property values are assigned to named variables.

The Automatic Border command provides the ability to select columns and then select annotation notes that represent column labels for the sheet zones. Click rows and then select annotation notes that represent row labels for the sheet zones.

Save time. Utilize System Properties and define Custom Properties in your Sheet Formats.

System Properties Linked to fields in default Sheet Formats:	Custom Properties of drawings linked to fields in default Sheet Formats:		Custom Properties of parts and assemblies linked to fields in default Sheet Formats:
SW-File Name (in DWG. NO. field):	CompanyName:	EngineeringApproval:	Description (in TITLE field):
SW-Sheet Scale:	CheckedBy:	EngAppDate:	Weight:
SW-Current Sheet:	CheckedDate:	ManufacturingApproval:	Material:
SW-Total Sheets:	DrawnBy:	MfgAppDate:	Finish:
	DrawnDate:	QAApproval:	Revision:
	EngineeringApproval:	QAAppDate:	

The Title block is located in the lower right-hand corner of Sheet1. The Drawing contains two modes:

1. Edit Sheet Format

2. Edit Sheet

Insert views and dimensions in the Edit Sheet mode. Modify the Sheet Format text, lines or title block information in the Edit Sheet Format mode.

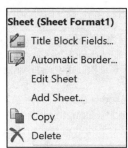

The CompanyName Custom Property is located in the title block above the TITLE box. There is no value defined for CompanyName. A small text box indicates an empty field.

Define a value for the Custom Property CompanyName. Example: D&M ENGINEERING.

Activity: Create a Title Box

Activate the Edit Sheet Format Mode.
28) Right-click in **Sheet1**. Do not click in a drawing view.

29) Click **Edit Sheet Format**. The Title block lines turn blue.

View the right side of the Title block.

30) Click the **Zoom to Area** tool from the Heads-up View toolbar.

31) **Zoom in** on the Title block.

32) Click the **Zoom to Area** tool to deactivate.

Define CompanyName Custom Property.
33) Position the **mouse pointer** in the middle of the box above the TITLE box as illustrated.

34) Click **File**, **Properties** from the Main menu. The Summary Information dialog box is displayed.

35) Click the **Custom** tab.

36) Click inside the **Property Name** box.

37) Click the **drop-down arrow** in the Property Name box.

38) Select **CompanyName** from the Property menu.

39) Enter **D&M ENGINEERING** (or your company/school name) in the Value/Text Expression box.

40) Click inside the **Evaluated Value** box. The CompanyName is displayed in the Evaluated Value box.

41) Click **OK** from the dialog box.

42) Move your **mouse pointer** in the center of the block as illustrated. The Custom Property, $PRP: "COMPANYNAME", is displayed in the Title block.

Modify the font size.
43) Double-click the **D&M ENGINEERING** (or your company/school name) text. The Formatting dialog box is displayed.

44) Click the **drop-down arrows** to set the Text Font and Height from the Formatting toolbar.

45) Click the **Style buttons** and **Justification buttons** to modify the selected text.

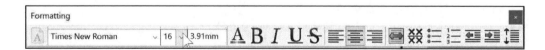

46) **Close** the Formatting dialog box.

47) Click **OK** from the Note PropertyManager.

Click a position outside the selected text box to save and exit the text.

The Tolerance block is located in the Title block. The Tolerance block provides information to the manufacturer on the minimum and maximum variation for each dimension on the drawing. If a specific tolerance or note is provided on the drawing, the specific tolerance or note will override the information in the Tolerance block.

General tolerance values are based on the design requirements and the manufacturing process.

Create Sheet Formats for different part types; examples: sheet metal parts, plastic parts and high precision machined parts. Create Sheet Formats for each category of parts that are manufactured with unique sets of Title block notes.

Modify the Tolerance block in the Sheet Format for ASME Y14.5 machined millimeter parts. Delete unnecessary text. The FRACTIONAL text refers to inches. The BEND text refers to sheet metal parts. The Three Decimal Place text is not required for this millimeter part in the chapter.

Modify the Tolerance Note.

48) Double-click the text **INTERPRET GEOMETRIC TOLERANCING PER:**

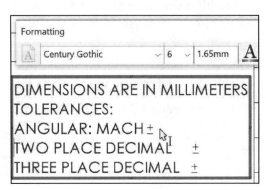

49) Enter **ASME Y14.5**.

50) Click **OK** ✔ from the Note PropertyManager.

51) Double-click inside the **Tolerance block** text. The Formatting dialog box and the Note PropertyManager is displayed.

52) Delete the text **INCHES**.

53) Enter **MILLIMETERS**.

54) Delete the line **FRACTIONAL** ±.

55) Delete the text **BEND** ±.

56) Click a **position** at the end of the
ANGULAR: MACH± line.

57) Enter **0**.

58) Click the **Add Symbol** button from the Text
Format box. The Symbols dialog box is
displayed.

59) Select **Degree** ° from the Symbols dialog box.

60) Enter **30'** for minutes of a degree.

Modify the TWO and THREE PLACE DECIMAL
LINES.
61) Delete the **TWO** and **THREE PLACE
DECIMAL lines**.

62) Enter **ONE PLACE DECIMAL ± 0.5**.

63) Enter **TWO PLACE DECIMAL ± 0.15**.

64) Click **OK** ✔ from the Note
PropertyManager.

Fit the drawing to the Graphics window.
65) Press the **f** key.

Save Draw1.

66) Click **Save** 💾. Accept the default name. Draw1 is the first default drawing file name.

UNLESS OTHERWISE SPECIFIED:

DIMENSIONS ARE IN MILLIMETERS
TOLERANCES:
ANGULAR: MACH± 0°30'
ONE PLACE DECIMAL ±0.5
TWO PLACE DECIMAL ±0.15

Various symbols are available through the Symbol dialog box. The ± symbol is located in the Modify Symbols list. The ± symbol is sometimes displayed as <MOD-PM>. The degree symbol ° is sometimes displayed as <MOD-DEG>.

Interpretation of tolerances is as follows:

- The angular dimension 110° is machined between 109.5° and 110.5°.

- The dimension 2.5 is machined between 2.0 and 3.0.

- The dimension 2.05 is machined between 1.90 and 2.20.

Company Logo

A Company logo is normally located in the Title block of the drawing. Create your own Company or School logo or copy and paste an existing picture.

If you have your own logo, skip the process of copying and applying the Logo.jpg file below. Insert the logo in the Edit Sheet Format mode.

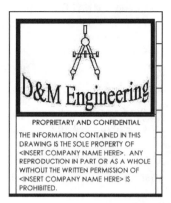

Activity: Insert a Company or School Logo

Insert a Company or School Logo.

67) Click **Insert**, **Picture** from the Main menu. The Open dialog box is displayed.

68) Select the **SW-TUTORIAL 2023\LOGO** folder.

69) Double-click the **Logo.jpg**. The Sketch Picture PropertyManager is displayed.

70) Uncheck the **Lock aspect ratio** box.

71) Uncheck the **Enable scale tool** box.

72) Drag the picture handles to size the **picture** to the left side of the Title block. Note: Text was added to the picture.

73) Click **OK** ✅ from the Sketch Picture PropertyManager.

🔆 Text can be added to create a custom logo. You can insert a picture or an object.

Return to the Edit Sheet mode.
74) Right-click in the **Graphics window**.

75) Click **Edit Sheet**. The Title block is displayed in black.

Fit the Sheet Format to the Graphics window.
76) Press the **f** key.

Draw1 displays Editing Sheet1 in the Status bar. The Title block is displayed in black when in Edit Sheet mode.

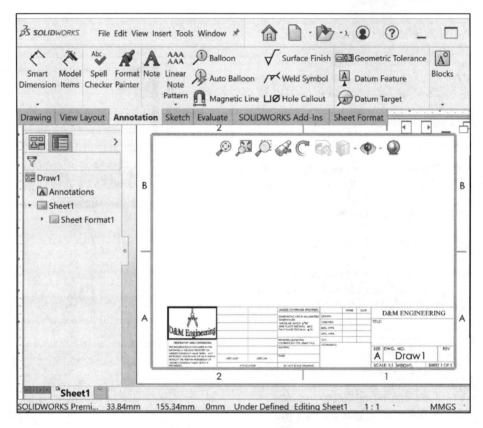

Save a Custom Sheet Format. Create a Custom Template. Create the MY-TEMPLATES Tab.

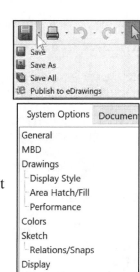

Save the drawing document in the Graphics window in two forms: Custom Sheet Format and a Custom Drawing Template. Save the Custom Sheet Format named CUSTOM-A. Use the CUSTOM-A Sheet Format for the drawings in this chapter. The Sheet Format file extension is .slddrt. Use the Save As option.

The Drawing Template can be displayed with or without the Sheet Format. Combine the Custom Sheet Format to create the Custom Drawing Template A-ANSI-MM. Utilize the Save As option to save a Drawing Template. The Drawing Template file extension is .drwdot.

Always select the Save as type option first, then select the Save in folder to avoid saving in default SOLIDWORKS installation directories.

Set the File Locations option in order to view the SW-TUTORIAL 2023\MY-TEMPLATE tab in the New Document dialog box.

If you did not create the MY-TEMPLATE tab or the drawing template, use the standard SOLIDWORKS drawing template and apply all of the needed document properties and custom properties.

Activity: Save the Custom Sheet Format. Save the Custom Drawing Template. Create the MY-TEMPLATES Tab.

Save the Custom Sheet Format.

77) Click **File**, **Save Sheet Format** from the Main menu. The Save Sheet Format dialog box is displayed.

78) Select the **SW-TUTORIAL 2023\MY-SHEETFORMATS** folder.

79) Enter **CUSTOM-A** for File name.

80) Click **Save** from the Save Sheet Format dialog box.

Save the Drawing Template.

81) Click **Save As** .

82) Click **Drawing Templates (*.drwdot)** from the Save as type box.

83) Select **SW-TUTORIAL 2023\MY-TEMPLATES** for the Save in folder.

84) Enter **A-ANSI-MM** for File name.

85) Click **Save**.

The A-ANSI-MM drawing template is displayed in the Graphics window. Create and add the MY-TEMPLATES folder, under the SW-TUTORIAL 2023 folder to the File Locations Document Template System Option.

Add the MY-TEMPLATES tab. Set System Options - File Locations. Only perform this procedure if you are working on a non-network, private system.

86) Click **Options** 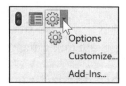 from the Main menu. The System Options General dialog box is displayed.

87) Click **File Locations** under the System Options tab.

88) Select **Document Templates** from Show folders for.

Add a new tab.

89) Click the **Add** button.

90) Select the **MY-TEMPLATES** folder.

91) Click **Select Folder**. The new folder added to the Document Templates.

92) Click the **Move Down** button.

93) Click **OK** from the System Options, File Locations dialog box.

94) Click **Yes** to the SOLIDWORKS dialog box. The SOLIDWORKS dialog box displays the new Folder location.

95) Click **Yes**. Click **Yes**. Return back to the drawing.

Close all files.

96) Click **Windows**, **Close All** from the Main menu.

Create a new drawing.

97) Click **New** ⬜ from the Main menu.

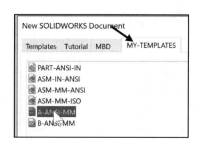

98) Click the **MY-TEMPLATES** tab. View the created drawing template. Note: Additional templates are displayed.

99) Double-click the **A-ANSI-MM** Drawing Template.

100) Click **Cancel** ✕ from the Model View PropertyManager. The Draw2 FeatureManager is displayed in the Graphics window.

💡 The Draw2-Sheet1 drawing is displayed in the Graphics window. You have successfully created a new drawing Template with a Custom sheet format.

Close all files.
101) Click **Windows**, **Close All** from the Main menu.

⛯ Combine customized Drawing Templates and Sheet Formats to match your company's drawing standards. Save the empty Drawing Template and Sheet Format separately to reuse information.

🔍 Additional details on Drawing Templates, Sheet Format and Custom Properties are available in SOLIDWORKS Help Topics. Keywords: Documents (templates, properties); Sheet Formats (new, new drawings, note text); Properties (drawing sheets); Customize Drawing Sheet Formats.

Review Drawing Templates.

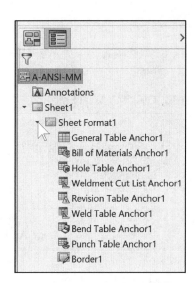

The Custom Sheet Format and Custom Drawing Template was created from the SW default Sheet and Drawing Template. You modified Sheet Properties and Document Properties.

The Sheet Format contained a Title block and Custom Property information. You inserted a Company Logo and modified the Title block.

The Save Sheet Format option was utilized to save the CUSTOM-A.slddrt Sheet Format.

The Save As option was utilized to save the A-ANSI-MM.drwdot template.

FLATBAR Drawing

A drawing contains part views, geometric dimensioning and tolerances, notes and other related design information. When a part is modified, the drawing automatically updates. When a dimension in the drawing is modified, the part is automatically updated.

Create the FLATBAR drawing from the FLATBAR part. Display the Front, Top, Right and Isometric views. Utilize the Model View tool from the View Layout toolbar.

Insert dimensions from the part. Utilize the Insert Model Items tool from the Annotation toolbar. Insert and modify dimensions and notes.

Insert a Parametric note that links the dimension text to the part depth. Utilize a user defined Part Number. Define the part material with the Material Editor. Add Custom Properties for Material and Number.

Activity: FLATBAR Drawing. Open the FLATBAR Part

Open the FLATBAR part. The FLATBAR part was created in Chapter 2.

102) Click **Open** from the Main menu.

103) Select the **SW-TUTORIAL 2023** folder.

104) Double-click **FLATBAR**. The FLATBAR FeatureManager is displayed.

Create a new drawing.

105) Click **New** ⬜ from the Main menu.

106) Click the **MY-TEMPLATES** tab.

107) Double-click **A-ANSI-MM**. The Model View PropertyManager is displayed. Additional templates are displayed. Note: if you did not create the MY-TEMPLATES tab, create a drawing document, A-(ANSI) Landscape, Third Angle projection using the MMGS unit system. Add all needed document properties and custom properties.

The Model View PropertyManager is displayed if the Start command when creating new drawing box is checked. If the Model View PropertyManager is not displayed, click the Model View tool from the View Layout toolbar.

The FLATBAR part icon is displayed in the Open documents box. Drawing view names are based on the part view orientation. The Front view is the first view inserted into the drawing. The Top view and Right view are projected from the Front view.

Insert the Front, Top and Right view.

108) Click **Next** 🔵 from the Model View PropertyManager.

109) Check the **Create multiple views** box.

110) Click ***Front**, ***Top** and ***Right** view from the Standard views box.

De-activate the Isometric view.
111) Click the ***Isometric** icon from the Standard views box.

112) Click **OK** ✔ from the Model View PropertyManager. View the three views on the drawing.

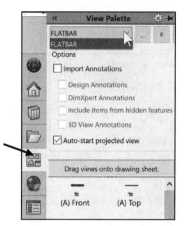

A part cannot be inserted into a drawing when the *Edit Sheet Format* mode is selected. You are required to be in the *Edit Sheet* mode.

Insert an Isometric view using the View Palette.

113) Click the **View Palette** tab on the right side of the Graphics window.

114) Select **FLATBAR** from the View Palette drop-down menu as illustrated. View the available views.

115) Click and drag the **Isometric view** in the top right corner as illustrated.

116) Click **OK** ✔ from the Drawing View PropertyManager. Your Title block may look different if you did not use the **A-ANSI-MM** Drawing Template.

Click the View Palette tab in the Task Pane. Click the drop-down arrow to view any active documents or click the Browse button to locate a document. Click and drag the desired view/views into the active drawing sheet.

Check Sheet Scale.

117) Right-click Sheet1 in the FLATBAR Drawing FeatureManager as illustrated.

118) Click **Properties**. The Sheet Properties dialog box is displayed.

119) If needed, enter **1:1** for Sheet Scale.

120) Click **Cancel** from the Sheet Properties dialog box. Apply Changes is active if you changed any of the default settings.

If needed, hide the Origins.

121) Click **View**, **Hide/Show,** un-check **Origins** from the Main menu.

Save the drawing.

122) Click **Save As** 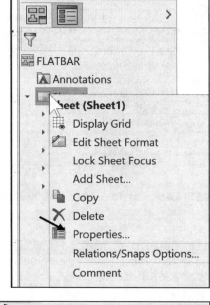.

123) Enter **FLATBAR** in the SW-TUTORIAL 2023 folder. Click **Save**.

Text in the Title block is linked to the Filename and Description created in the part. The DWG. NO. text box utilizes the Property, $PRP:"SW-File Name" passed from the FLATBAR part to the FLATBAR drawing.

The Title text box utilizes the Property, $PRPSHEET: "Description."

The filename FLATBAR is displayed in the DWG. NO. box. The Description FLATBAR 9 HOLE is displayed in the Title box. The Description was added in Chapter 2.

The FLATBAR drawing contains three Principle views (Standard views): Front, Top, Right and Isometric view.

Insert drawing views as follows:

- Utilize the Model View tool.

  o Drag a part into the drawing to create three Standard views.

  o Predefine views in a custom Drawing Template.

  o Drag a hyperlink through Internet Explorer.

- Drag an active part view from the View Palette. The View Palette is located in the Task Pane. With an open part, drag the selected view into the active drawing sheet.

The View Palette populates when you:

- Click Make Drawing from Part/Assembly.

- Browse to a document from the View Palette.

- Select from a list of open documents in the View Palette.

Move Views and Properties of the Sheet

Move Views on Sheet1 to create space for additional Drawing View placement.

The mouse pointer provides feedback in both the Drawing Sheet and Drawing View modes.

The mouse pointer displays the Drawing Sheet icon when the Sheet properties and commands are executed.

The mouse pointer displays the Drawing View icon when the View properties and commands are executed.

View the mouse pointer icon for feedback to select Sheet, View, and Component and Edge properties in the Drawing.

Use the Pack and Go option to save an assembly or drawing with references.

Sheet Properties

- Sheet Properties display properties of the selected sheet.

  Right-click in the sheet boundary to view the available commands.

View Properties

- View Properties display properties of the selected view.

  Right-click inside the view boundary. Modify the View Properties in the Display Style box or the View Toolbar.

Component Properties

- Component Properties display properties of the selected component. Right-click on the face of the component. View the available options.

Edge Properties

- Edge Properties display properties of the selected geometry. Right-click on an edge inside the view boundary. View the available options.

Reposition the views on the drawing. Provide approximately 25mm - 50mm between each view for dimension placement.

Activity: FLATBAR Drawing-Position Views

Position the views.

124) Click inside the view boundary of **Drawing View1** (Front). The mouse pointer displays the Drawing View icon.

125) Position the **mouse pointer** on the edge of the view boundary until the Drawing View icon is displayed.

126) Drag **Drawing View1** in an upward vertical direction. The Top and Right views move aligned to Drawing View1 (Front).

127) Press **Shift + Z** key to Zoom in on Sheet1.

128) Click the **Right view** boundary.

129) Position the **mouse pointer** on the edge of the view until the Drawing Move View icon is displayed.

130) Drag the **Right view** in a right to left direction towards the Front view.

Tangent edges are displayed for educational purposes.

Move the Top view in a downward vertical direction.

131) Click the **Top view**, "Drawing View3" boundary.

132) Drag the **Top view** in a downward direction towards Drawing View1.

Fit Sheet1 to the Graphics window.

133) Press the **f** key.

Detail Drawing

The design intent of this project is to work with dimensions inserted from parts and to incorporate them into the drawings. Explore methods to move, hide and recreate dimensions to adhere to a drawing standard.

There are other solutions to the dimensioning schemes illustrated in this project. Detail drawings require dimensions, annotations, tolerance, materials, Engineering Change Orders, authorization, etc. to release the part to manufacturing and other notes prior to production.

You just inserted dimensions into a drawing. The dimensions, extension lines and arrows are not in the correct locations or in the correct view. How can you address the position of these details? Answer: Dimension to an ASME Y14.5 standard.

Adds a detail view to show a portion of a view, usually at an enlarged scale.

The ASME Y14.5 standard defines an engineering drawing standard.

Apply these dimension practices to the FLATBAR and other drawings in the chapter.

A Detailed drawing is used to manufacture a part. A mistake on a drawing can cost your company substantial loss in revenue. The mistake could result in a customer liability lawsuit. Dimension and annotate your parts clearly to avoid common problems and mistakes.

Dimensions and Annotations

Dimensions and annotations are inserted from the part. The annotations are not in the correct location. Additional dimensions and annotations are required.

Dimensions and annotations are inserted by selecting individual features, views or the entire sheet. Select the entire sheet. Insert Model Items command from the Annotations toolbar.

Activity: FLATBAR Drawing - Dimensions and Annotations

Insert dimensions.

134) Click **Sheet1** in the center of the drawing. The mouse

pointer displays the Sheet ⬚ icon.

135) Click the **Annotation** tab from the CommandManager.

136) Click the **Model Items** tool from the Annotation toolbar. The Model Items PropertyManager is displayed. The Import items into all views option is checked.

137) Select **Entire model** from the Source box.

138) Click **OK** ✔ from the Model Items PropertyManager. Dimensions are inserted into the drawing. View the results.

Dimensions are inserted into the drawing. The dimensions MAY NOT BE in the correct location with respect to the feature lines per the ANSI Y-14.5 standard. Move them later in the chapter and address extension line gaps (5mm).

The dimensions and text in the next section have been enlarged for visibility. Drawing dimension location is dependent on *Feature dimension creation* and *Selected drawing views.*

Move dimensions within the same view. Use the mouse pointer to drag dimensions and leader lines to a new location. Leader lines reference the size of the profile. A gap must exist between the profile lines and the leader lines. Shorten the leader lines to maintain a drawing standard. Use the blue Arrow buttons to flip the dimension arrows.

Plan ahead for general drawing notes. Notes provide relative part or assembly information. Example: Material type, material finish, special manufacturing procedure or considerations, preferred supplier, etc.

Below are a few helpful guidelines to create general drawing notes:

- Use capital letters.

- Use left text justification.

- Font size should be the same as the dimension text.

Create parametric notes by selecting dimensions in the drawing. Example: Specify the material thickness of the FLATBAR as a note in the drawing. If the thickness is modified, the corresponding note is also modified.

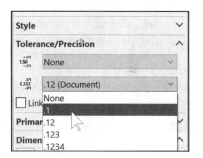

Move the linear dimensions in Drawing View1, (Front).

139) Click the dimension text **101.60**. The dimension text turns blue. **Remove** the trailing zero. Select **Precision** of 1.

140) Drag the **dimension text** downward.

141) Click the dimension text **12.70**. The dimension text turns blue. **Remove** the trailing zero. Select **Precision** of 1. Drag the **text** approximately 10mm from the profile. The smallest linear dimensions are closest to the profile.

142) Click the radial dimension **R6.35**.

143) Drag the **text** diagonally off the profile if required.

Modify dimension text.

144) Click the diameter dimension **4.83**. The Dimension PropertyManager is displayed.

145) Click inside the **Dimension Text** box.

146) Enter **9X** before <MOD-DIAM>.

147) Enter **EQ SP** after <DIM>.

148) Click **OK** ✔ from the Dimension PropertyManager.

Inserted dimensions can be moved from one drawing view to another. Hold the Shift key down. Click and drag the dimension text from one view into the other view boundary. Release the Shift key.

Modify the precision of the material thickness.

149) Click the **depth dimension** text in the Top view. Note: The part was created in Chapter 2.

150) Select **.1** from the Unit Precision drop-down menu. Click **OK** ✔ from the Dimension PropertyManager. The text displays the dimension.

Insert a Parametric note.

151) Click the **Annotation** tab from the CommandManager.

152) Click the **Note** A tool from the Annotation toolbar. The Note icon is displayed.

153) Click a **position** above the Front view.

154) Enter **MATERIAL THICKNESS**.

155) Click the **depth dimension text (2.3)** in the Top view. The variable name for the dimension is displayed in the text box. Note: Your material thickness may vary if you modified the FLATBAR model.

156) Enter **MM**.

157) Click **OK** ✔ from the Note PropertyManager.

Hide superfluous dimensions.

158) Right-click the **dimension (2.3)** in the Top view. Note: Your material thickness may vary if you modified the FLATBAR model.

159) Click **Hide**.

Hide the Right view.

160) Right-click the **Right view** boundary.

161) Click **Hide**. Note: If required, expand the drop-down menu. The Right view is not displayed in the Graphics window.

162) Click **OK** ✔ from the Drawing View PropertyManager.

Fit the model to the drawing. Address all needed extension line gaps.

163) Press the **f** key.

Save the drawing.

164) Click **Save** 💾. Click **Save All**.

Locate the Top view off of Sheet1.

165) Click and drag the **Top view boundary** off of the Sheet1 boundary as illustrated.

Views and notes outside the sheet boundary do not print. The Parametric note is controlled through the FLATBAR Boss-Extrude1 feature depth. Modify the depth to update the note.

🔅 Click the dimension Palette rollover button 🔲 to display the dimension palette. Use the dimension palette in the Graphics window to save mouse travel to the Dimension PropertyManager. Click on a dimension in a drawing view, and modify it directly from the dimension palette.

🔅 To show a hidden dimension, click **View**, **Hide/Show**, **Annotations** from the Main menu.

Open the FLATBAR part.
166) Right-click inside the **Front view** boundary.

167) Click **Open Part**. The FLATBAR part is displayed.

Modify the Boss-Extrude1 depth dimension.
168) Click **Boss-Extrude1** from the FeatureManager.

169) Click **.090**in, [2.29] in the Graphics window.

170) Enter **2.5mm** as illustrated. Note: You need to enter mm.

171) Rebuild the model.

Save the model.

172) Click **Save** .

Return to the drawing.
173) Click **Window**, **FLATBAR - Sheet1** from the Main menu. The Parametric note is updated to reflect the dimension change in the part.

174) Return to the FLATBAR part and **undo** the dimension change from 2.5mm to 2.3mm.

175) Return to the drawing. View the update.

The FLATBAR drawing references the FLATBAR part. Do not delete the part or move the part location. Work between multiple documents:

- Press Ctrl-Tab to toggle between open SOLIDWORKS documents.

- Right-click inside the Drawing view boundary. Select Open Part.

- Right-click the part icon in the FeatureManager. Select Open Drawing.

Commands are accessed through the toolbars and drop-down menus. Commands are also accessed with a right-click in the Graphics window and FeatureManager.

A majority of FLATBAR drawing dimensions are inserted from the FLATBAR part. An overall dimension is required to dimension the slot shape profile. Add a dimension in the drawing.

Add a dimension to Drawing View1 (Front View). Address precision.

176) Click the **Smart Dimension** tool from the Annotation toolbar. The Autodimension tab is selected by default.

177) Click the **Smart dimensioning** box. Click the **top horizontal line** of the FLATBAR in the Front view. Click the **bottom horizontal line** of the FLATBAR in the Front view.

178) Click a **position** as illustrated.

179) Select **.1** from the Unit Precision drop-down menu.

180) Click **OK** from the Dimension PropertyManager.

Modify the Radius text. R will be a stated reference dimension.
181) Click the **R6.35** dimension text.

182) Delete **R<DIM>** in the Dimension Text box.

183) Click **Yes** to confirm dimension override.

184) Enter **2X R** for Dimension text. Do not enter the radius value.

185) Click **OK** from the Dimension PropertyManager.

186) Insert all needed **extension line gaps (5mm)** as illustrated.

Save the FLATBAR drawing.

187) Click **Save** .

Part Number and Document Properties

Engineers manage the parts they create and modify. Each part requires a Part Number and Part Name. A part number is a numeric representation of the part. Each part has a unique number. Each drawing has a unique number. Drawings incorporate numerous part numbers or assembly numbers.

There are software applications that incorporate unique part numbers to create and perform: Bill of Materials, Manufacturing procedures, Cost analysis and Inventory control/Just in Time, JIT.

You are required to procure the part and drawing numbers from the documentation control manager. Utilize the following prefix codes to categorize created parts and drawings. The part name, part number and drawing numbers are as follows:

Category:	Prefix:	Part Name:	Part Number:	Drawing Number:
Machined Parts	56-	FLATBAR	GIDS-SC-10001-9	56-10222
		AXLE	GIDS-SC-10017	56-10223
		SHAFT-COLLAR	GIDS-SC-10012-3-16	56-10224
Purchased Parts	99-	AIRCYLINDER	99-FBM8x1.25	999-101-8
Assemblies	10-	LINKAGE ASM	GIDS-SC-1000	10-10123

Link notes in the Title block to SOLIDWORKS Properties. Properties are variables shared between documents and applications.

The machined parts are manufactured from Aluminum. Specify the Material Property in the part. Link the Material Property to the drawing title block. Create a part number that is utilized in the Bill of Materials. Create additional notes in the title block to complete the drawing.

Activity: FLATBAR Drawing - Part Number and Document Properties

---

Return to the FLATBAR part. Apply Material.
188) Right-click in the **Front view** boundary.

189) Click **Open Part**. Material was not added when you created the FLATBAR in Chapter 2. Right-click **Material** in the FLATBAR FeatureManager.

190) Click **Edit Material**. The Material dialog box is displayed.

191) **Expand** the Aluminum Alloys folder. Select **2014 Alloy**. View the material properties.

192) Click the **Apply** button.

193) Click **Close**. The 2014 Alloy is displayed in the FeatureManager.

Save the FLATBAR.

194) Click **Save**.

Define the part number property for the BOM.

195) Click the FLATBAR **ConfigurationManager** tab.

196) Right-click **Default**.

197) Click **Properties**. The Configuration Properties PropertyManager is displayed.

198) Click **User Specified Name** from the drop-down box under Document Name.

199) Enter **GIDS-SC-10001-9** in the Part number text box.

Define a material property.
200) Click the **Custom Properties** button.

201) Click inside the **Property Name** box.

202) Click the **down arrow**. Select **Material** from the Property Name list.

203) Click inside the **Value/Text Expression** box.

204) Click the **down arrow**. Select **Material**. Click inside the **Evaluated Value** box. **View** the results.

Define the Number Property.
205) Click inside the **second Property Name** box.

206) Click the **down arrow**. Select **Number** from the Name list.

207) Click inside the **Value/Text Expression** box.

208) Enter **56-10222** for Drawing Number.

209) Click **OK** from the Summary Information dialog box.

210) Click **OK** ✔ from the Configuration Properties PropertyManager.

Return to the FeatureManager.

211) Click the FLATBAR **FeatureManager** 🗋 tab.

Save the FLATBAR part.

212) Click **Save** 💾.

Return to the drawing.

213) Click **Windows**, **FLATBAR - Sheet1** from the Main menu.

The Material Property is inserted into the Title block.

Activity: FLATBAR Drawing - Linked Note

Create a Linked Note. Set the Drawing Number.

214) Click **File**, **Properties** from the Main menu. Click the **Custom** tab.

215) Click inside the **Property Name** box. Note: The CompanyName was set the created drawing template.

216) Click the **down arrow**. Select **Number** from the Property Name list.

217) Click inside the **Value/Text Expression** box.

218) Enter **56-10222**.

219) Click **OK** from the Summary Information dialog box.

220) **Right-click** in Sheet1.

	Property Name	Type	Value / Text Expression	Evaluated Value
1	SWFormatSize	Text	8.5in*11in	8.5in*11in
2	CompanyName	Text	D&M ENGINEERING	D&M ENGINEERING
3	Number	Text	56-10222	56-10222

221) Click **Edit Sheet Format**.

222) **Zoom in** on the lower right corner of the drawing.

223) Double-click on the DWG. NO. text **FLATBAR**. The Note PropertyManager is displayed.

224) Delete **FLATBAR**. Note: The Formatting dialog is still active.

225) Click **Link to Property** from the Text Format box.

226) Select **Current** document.

227) Select **Number** from the Link to Property drop-down menu.

228) Click **OK** from the Link to Property box.

229) Click **OK** ✔ from the Note PropertyManager.

Return to the drawing sheet.
230) Right-click a **position** in the Graphics window.

231) Click **Edit Sheet**.

232) If needed, **remove all Tangent Edges** from the Top and Isometric view.

233) Address **all extension line gaps (5mm) and any additional needed information** as illustrated below.

Save the FLATBAR drawing.

234) Click **Save** 💾.

As an exercise, insert the Third Angle projection icon located in the LOGO folder.

Add drawing Custom properties: DrawnBy and Drawndate. Add DWG. NO.

Insert the reference dimension symbol for the second 12.7mm dimension as illustrated.

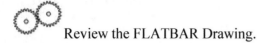   Additional details on Drawing Views, New Drawing, Details, Dimensions, Dimensions and Annotations are available in SOLIDWORKS Help.

Review the FLATBAR Drawing.

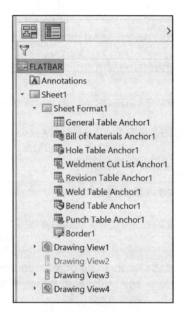

You created the FLATBAR drawing with the A-ANSI-MM Drawing Template. The FLATBAR drawing utilized the FLATBAR part with the Model View tool and the View Palette tool.

The Model View PropertyManager provided the ability to insert new views of a document. You selected the Front, Top and Right view. You applied the View Palette to insert an Isometric view.

You moved the views by dragging the blue view boundary. You inserted part dimensions and annotations into the drawing with the Insert Model Items tool. Dimensions were moved to new positions. Leader lines and dimension text were repositioned. Annotations were edited to reflect the dimension standard.

You created a Parametric note that referenced part dimensions in the drawing text. Material was assigned in the FLATBAR part. The Material Custom Property and Number Custom Property were assigned in the FLATBAR part and referenced in the drawing Title block.

Know inch/mm decimal display. The ASME Y14.5 standard states:

- For millimeter dimensions < 1, display the leading zero. Remove trailing zeroes.

- For inch dimensions < 1, delete the leading zero. The dimension is displayed with the same number of decimal places as its tolerance.

Note: The FLATBAR drawing linked Title block notes to Custom Properties in the drawing and in the part. The additional drawings in this project utilize drawing numbers linked to the model file name. The Title of the drawing utilizes a Note.

LINKAGE Assembly Drawing - Sheet1

The LINKAGE assembly drawing Sheet 1 utilizes the LINKAGE assembly. Add an Exploded view and a Bill of Materials to the drawing.

Create an Exploded view in the LINKAGE assembly. The Bill of Materials reflects the components of the LINKAGE assembly. Create a drawing with a Bill of Materials. Perform the following steps:

- Create a new drawing with the custom A-ANSI-MM size Drawing Template with the CUSTOM-A sheet format.

- Create and display the Exploded view of the LINKAGE assembly.

- Insert the Exploded view of the assembly into the drawing.

- Insert a Bill of Materials.

- Label each component with Balloon text.

Activity: LINKAGE Assembly Drawing - Sheet1

Close all parts and drawings.
235) Click **Windows**, **Close All** from the Main menu.

Create a new drawing.
236) Click **New** ⬜ from the Main menu.

237) Click the **MY-TEMPLATES** tab.

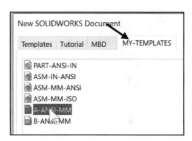

238) Double-click **A-ANSI-MM**. Additional templates are displayed. Note: if you did not create the MY-TEMPLATES tab, create a drawing document, A-(ANSI) Landscape, Third Angle projection using the MMGS unit system. Add all needed document properties and custom properties.

Insert the LINKAGE assembly using Model View feature.
239) **Browse** to locate the LINKAGE assembly. There is a LINKAGE assembly located in the Chapter 4 homework section.

240) Double-click **LINKAGE**. Single view is selected by default.

241) Click **Isometric view** from the Standard views box.

242) Select **Shaded With Edges** from the Display Style box.

243) Check the **Use custom scale** box.

244) Enter **1:2** for Scale.

245) Click a **position** in the middle of Sheet1.

246) Click **OK** from the Drawing View1 PropertyManager.

Deactivate the Origins if needed.
247) Click **View**, **Hide/Show**, un-check **Origin** from the Main menu.

Save the LINKAGE assembly drawing.

248) Click **Save**. Accept the default file name.

249) Click **Save All**.

Display modes for a Drawing view are similar to a part document. The 3D Drawing View tool provides the ability to manipulate the model view in 3D, to select a difficult face, edge, or point.

Wireframe and Shaded Display modes provide the best graphic performance. Mechanical details require Hidden Lines Visible display and Hidden Lines Removed display. Select Shaded/Hidden Lines Removed to display Auxiliary Views to avoid confusion.

Tangent Edges Visible provides clarity for the start of a Fillet edge. Tangent Edges Removed provides the best graphic performance.

Right-click in the view boundary to access the Tangent Edge options.

🔆 Utilize the Lightweight Drawing option to improve performance for large assemblies.

Hidden Lines Removed

Shaded

Shaded With Edges

Wired Frame

Hidden Lines Visible

To address Tangent lines views:

- Right-click in a Drawing view.

- Click Tangent Edge.

- Click a Tangent Edge view option.

Return to the LINKAGE assembly. Set Configuration Properties.
250) Right-click inside the **Isometric view** boundary.

251) Click **Open Assembly**. The LINKAGE assembly is
displayed.

252) Click the **ConfigurationManager** tab.

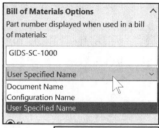

253) Right-click **Default [LINKAGE]**. Click **Properties**. The
Configuration Properties PropertyManager is displayed.

254) Select **User Specified Name** from the Part number
displayed when used in Bill of Materials.

255) Enter **GIDS-SC-1000** in the Part number text box.

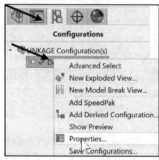

256) Click **OK** from the Configuration Properties
PropertyManager.

Save the model.

257) Click **Save**.

Exploded View

The Exploded View illustrates how to assemble the components in
an assembly. Create an Exploded View with four steps. Click and
drag components in the Graphics window. The Manipulator icon
indicates the direction to explode. Select an alternate component
edge for the Explode direction. Drag the component in the Graphics
window or enter an exact value in the Explode distance box.

Manipulate the top-level components in the assembly. Access the
Explode view option as follows:

- Right-click the configuration name in the ConfigurationManager.

- Select the Exploded View tool in the Assembly toolbar.

- Select Insert, Exploded View from the Main menu.

Activity: LINKAGE Assembly Drawing - Exploded View

Insert an Exploded view.

258) Right-click **Default [GIDS-SC-1000]**.

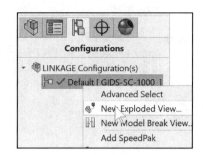

259) Click **New Exploded view** . The Explode PropertyManager is displayed.

260) Click the **Regular step (translate and rotate)** icon.

Create Chain1. Use the distance box option.

261) Click the back **SHAFT-COLLAR** as illustrated.

262) Enter **50**mm (1.97in) in the Explode Distance box.

263) Click the **Reverse Direction** box.

264) Click **Add Step**. The SHAFT-COLLAR moves 50mm to the back of the model. Chain 1 is created.

Create Chain2. Use the Manipulator icon.

265) Click the front **SHAFT-COLLAR** in the Graphics window as illustrated.

266) Drag the **Manipulator icon** to the front of the assembly approximately 50mm (1.97in).

267) Click **Done**. Chain 2 is created.

Create Chain3.

268) Click the **back FLATBAR** in the Graphics window as illustrated.

269) Drag the **Manipulator icon** approximately 32mm (1.26in) to the back of the assembly. Chain 3 is created.

270) Click **Done**.

Create Chain4.

271) Click the **front FLATBAR** in the Graphics window.

272) Drag the **Manipulator icon** approximately 32mm (1.26in) to the front of the assembly. Chain 4 is created.

273) Click **Done**.

274) Click **OK** from the Explode PropertyManager.

Save the LINKAGE part in the Exploded State.

275) Click **Save** 💾.

🔅 Create exploded view steps that rotate a component with or without linear translation. Use the rotation and translation handles of the triad. Include rotation and translation in the same explode step. You can also edit the explode step translation distance and rotation angle values in the PropertyManager.

Activity: LINKAGE Assembly Drawing - Animation

Animate the Exploded view.
276) Expand the Default [GIDS-SC-1000] folder.

277) Right-click **Exploded View1** in the ConfigurationManager.

278) Click **Animate collapse** to play the animation. View the Animation.

279) Click **Pause** .

280) Click **End** .

Return the Exploded view in its collapsed state.
281) Close the Animation Controller dialog box.

Return to the Assembly FeatureManager.
282) Click the LINKAGE **FeatureManager** tab.

Save the assembly.
283) Click **Save** .

Open the LINKAGE drawing.
284) Click **Window, LINKAGE - SHEET1** from the Main menu.

Display the Exploded view in the drawing.
285) Right-click inside the **Isometric view**.

286) Click **Show in Exploded State**. Note: You can also click **Properties,** check the **Show in Exploded State** box and then click **OK** from the Drawing Views Properties box.

287) Click **OK** from the Drawing View1 PropertyManager. View the exploded state in the drawing on Sheet1.

Bill of Materials

A Bill of Materials (BOM) is a table inserted into a drawing to keep a record of the parts used in an assembly. The default BOM template contains the Item Number, Quantity, Part No. and Description. The default Item number is determined by the order in which the component is inserted into the assembly. Quantity is the number of instances of a part or assembly.

Part No. is determined by the following: file name, default and the User Defined option, Part number used by the Bill of Materials. Description is determined by the description entered when the document is saved.

Activity: LINKAGE Assembly Drawing - Bill of Materials

Create a Bill of Materials.

288) Click inside the **Isometric view** boundary.

289) Click the **Annotation** tab from the CommandManager.

290) Click the **Bill of Materials** tool from the Consolidated Tables toolbar. The Bill of Materials PropertyManager is displayed.

291) Select **bom-material** for Table Template.

292) Select **Top-level only** for BOM Type.

293) Click **OK** from the Bill of Materials PropertyManager.

294) Double-click a position in the **upper left corner** of the Sheet1.

295) Click a **position** in Sheet1. The Descriptions were created in Chapter 2. Additional information was addressed in this Chapter.

ITEM NO.	PART NUMBER	DESCRIPTION	MATERIAL	QTY.
1	GIDS-PC-10001	LINEAR ACTUATOR		1
2	AXLE	AXLE ROD		1
3	GIDS-SC-10001-9	FLAT BAR 9-HOLES	2014 Alloy	2
4	SHAFT-COLLAR	SHAFT-COLLAR		2

The Bill of Materials requires some editing.

The current part description determines the DESCRIPTION values. Redefine the PART NUMBER for the Bill of Materials. Note: You will also scale the LINKAGE assembly.

Modify the AXLE Part number.
296) Right-click the **AXLE** part in the LINKAGE drawing.

297) Click **Open Part**. The AXLE FeatureManager is displayed.

298) Click the AXLE **ConfigurationManager** tab.

299) Right-click **Default [AXLE]** in the ConfigurationManager.

300) Click **Properties**.

301) Select **User Specified Name** from the Configuration Properties dialog box.

302) Enter **GIDS-SC-10017** for the Part Number to be utilized in the Bill of Materials. Click **OK** from the Configuration Properties PropertyManager.

303) **Return** to the FeatureManager. Click **Save** .

Return to the LINKAGE drawing. Save the drawing.
304) Click **Window**, **LINKAGE - Sheet1** from the Main menu.

Modify the SHAFT-COLLAR PART NUMBER.
305) Right-click the front **SHAFT-COLLAR** part in the LINKAGE drawing.

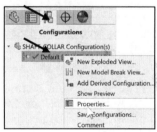

306) Click **Open Part**. The SHAFT-COLLAR FeatureManager is displayed.

307) Click the SHAFT-COLLAR **ConfigurationManager** tab.

308) Right-click **Default [SHAFT-COLLAR]** from the ConfigurationManager. Click **Properties**.

309) Select the **User Specified Name** from the Configuration Properties box.

310) Enter **GIDS-SC-10012-3-16** for the Part Number in the Bill of Materials.

311) Click **OK** from the Configuration Properties PropertyManager.

312) **Return** to the FeatureManager. Click **Save** .

Return to the LINKAGE assembly drawing.
313) Click **Window, LINKAGE - Sheet1** from the Main menu.

314) Rebuild the drawing and update the BOM.

Save the drawing.
315) Click **Save** 💾 .

ITEM NO.	PART NUMBER	DESCRIPTION	MATERIAL	QTY.
1	GIDS-PC-10001	LINEAR ACTUATOR		1
2	GIDS-SC-10017	AXLE ROD	6061 Alloy	1
3	GIDS-SC-10001-9	FLAT BAR 9-HOLES	2014 Alloy	2
4	GIDS-SC-10012-3-16	SHAFT-COLLAR	6061 Alloy	2

Note: As an exercise, complete the Bill of Materials. Add material to each component (AXLE-ROD, SHAFT-COLLAR). Add material in the **Part FeatureManager** and **Custom Properties**. Label each component with a unique item number. The item number is placed inside a circle. The circle is called Balloon text. List each item in a Bill of Materials table. Utilize Auto Balloon to apply Balloon text to all BOM components.

The Circle Split Line option contains the Item Number and Quantity. Item number is determined by the order listed in the assembly FeatureManager. Quantity lists the number of instances in the assembly.

Activity: LINKAGE Assembly Drawing - Automatic Balloons

Insert the Automatic Balloons.
316) Click inside the **Isometric view** boundary of the LINKAGE.

317) Click the **Auto Balloons** tool from the Annotation toolbar. The Auto Balloon PropertyManager is displayed. Accept the defaults.

318) Click **OK** ✔ from the Auto Balloon PropertyManager.

Reposition the Balloon text.
319) Click and drag each **Balloon** to the desired position.

320) Click and drag the **Balloon arrowhead** to reposition the arrow on a component.

321) Click **OK** ✔ from the Balloon PropertyManager.

Display Item Number/Quantity.

322) Ctrl-Select the **four Balloon text** in the Graphics window. The Balloon PropertyManager is displayed. Select **Circular Split Line** for Style.

323) Click **OK** ✔ from the Balloon PropertyManager.

Split the Component lines.

324) Click the **SHAFT Circular Split Line**. The Circular Split Line is displayed in blue. **Zoom in** on the arrow point.

325) Hold the **Ctrl** key down. **Click and drag** the arrow point to the second SHAFT-COLLAR. Release the **Ctrl** key. **Release** the mouse button.

326) **Perform** the above procedure for the FLAT BAR.

Save the LINKAGE assembly drawing.

327) Click **Save** .

LINKAGE Assembly Drawing - Sheet2

A drawing consists of one or more sheets. Utilize the Model View tool in the View Layout toolbar to insert the AirCylinder assembly. The LINKAGE drawing Sheet2 displays the Front view and Top view of the AirCylinder assembly.

Activity: LINKAGE Assembly - Drawing-Sheet2

Add Sheet2.

328) **Right-click** in the Graphics window.

329) Click **Add Sheet**. Sheet2 is displayed.

330) Right-click **Sheet2** in the Drawing FeatureManager.

331) Click **Properties**.

Select the CUSTOM-A Sheet Format.

332) Click **Browse** from the Sheet Properties box.

333) Select the **SW-TUTORIAL 2023\MY-SHEETFORMAT** folder.

334) Double-click **CUSTOM-A**.

335) Click **Apply Changes** from the Sheet Properties box. Sheet 2 of 2 is displayed in the Graphics window. CUSTOM-A was created at the beginning of the chapter. If you did not create CUSTOM A, accept the default sheet format and fill in all of the needed icons and Custom Properties.

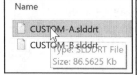

Insert the AirCylinder assembly.
336) Click the **View Layout** tab from the CommandManager.

337) Click the **Model View** tool from the View Layout toolbar. The Model View PropertyManager is displayed.

338) Click the **Browse** button.

339) Double-click the **AirCylinder** assembly from the SW-TUTORIAL 2023 folder.

340) Check the **Create multiple views** box. Create a Top and Front view drawing.

341) Click **Top view** as illustrated in the Standard views box.

342) Click **OK** from the Model View PropertyManager.

Save Sheet2.
343) Click **Save**.

Enter Title Name and Modify font size.
344) **Right-click** in the Graphics window.

345) Click **Edit Sheet Format**.

346) **Double-click** inside the Title box.

347) Enter **LINEAR ACTUATOR**.

348) Enter **5**mm for text height.

349) Click inside the **Graphics window**.

350) **Right-click** in the Graphics window.

351) Click **Edit Sheet**.

Section views display the interior features. Define a cutting plane with a sketched line in a view perpendicular to the Section view. Create a full Section view by sketching a section line in the Top view. Detailed views enlarge an area of an existing view. Specify location, shape and scale. Create a Detail view from a Section view at a 2:1 scale.

Activity: LINKAGE Assembly Drawing - Sheet2 Section View

Add a Section View to the drawing.
352) Click inside the **Drawing View3** boundary.

353) Click the **Section View** ↵ tool from the View Layout toolbar. The Section View PropertyManager is displayed.

354) Click the **Section** tab from the PropertyManager.

355) Click the **Horizontal** Cutting Line button.

356) Check the **Auto-start section view** box.

357) Click the **midpoint** of the model as illustrated.

Position Section View A-A and Modify the Drawing Scale.
358) Click **OK** from the Section View dialog box.

359) Click a **location** above the Top view. The section arrows point upward.

360) If needed, check the **Flip direction** box.

361) Click **OK** ✔ from the Section View PropertyManager.

362) **Modify** the Drawing Scale to 1:2.

Fit the views to the drawing.
363) Press the **f** key.

Activity: LINKAGE Assembly Drawing - Sheet2 Detail view

Add a Detail view to the drawing.
364) Click inside the **Section View** boundary. The Section View A-A PropertyManager is displayed.

365) **Zoom in** to enlarge the view.

366) Click the **Detail View** tool from the View Layout toolbar. The Circle Sketch tool is selected.

367) Click the **center** of the air fitting on the left side in the Section View as illustrated.

368) Sketch a **Circle** to encompass the air fitting. If required, enter **B** for Detail View Name in the Label text box.

Position Detail View B.
369) Press the **f** key.

370) Click a **location** on Sheet2 to the right of the SECTION View.

371) Enter **2:1** for Scale.

372) Click **OK** ✔ from the Detail View B PropertyManager.

Move views if required.
373) Click and drag the **view boundary** to allow for approximately 1 inch, [25mm] spacing between views.

Save the LINKAGE assembly drawing.

374) Click **Save** .

Close all parts and assemblies.
375) Click **Window**, **Close All** from the Main menu.

As an exercise, add additional drawing Custom Properties: Number, Description, DrawnBy, DrawnDate, Revision, Third Angle icon, etc.

Review the LINKAGE Assembly Drawing

The LINKAGE Assembly drawing consisted of two sheets. Sheet1 contained an Exploded view. The Exploded view was created in the LINKAGE assembly.

The Bill of Materials listed the Item Number, Part Number, Description, Material and Quantity of components in the assembly. Balloons were inserted to label top level components in the LINKAGE assembly. You developed Custom Properties in the part and utilized the Properties in the drawing and Bill of Materials.

Sheet2 contained the Front view, Top view, Section view and Detail view of the AirCylinder assembly.

Design Tables

A Design Table is a spreadsheet used to create multiple configurations in a part or assembly. The Design Table controls the dimensions and parameters in the part. Utilize the Design Table to modify the overall length and number of holes in each FLATBAR.

Create three configurations of the FLATBAR: 3HOLE, 5HOLE and 7HOLE

Utilize the Design Table to control the Part Number and Description in the Bill of Materials. Insert the custom parameter $PRP@DESCRIPTION into the Design Table. Insert the system parameter $PARTNUMBER into the Design Table.

Activity: FLATBAR Part - Design Table

Open the FLATBAR part.

376) Click **Open** from the Main menu.

377) Double-click the **FLATBAR** part from the SW-TUTORIAL 2023 folder. You created the FLATBAR in Chapter 2.

Insert a Design Table.
378) Click **Insert**, **Tables**, **Excel Design Table** from the Main menu. The Auto-create option is selected. Accept the default settings.

379) Click **OK** from the Excel Design Table PropertyManager.

Select the input dimension.
380) Hold the **Ctrl key** down.

381) Click the **D1@Sketch1, D2@Sketch1, D1@Boss-Extude1, D1@Sketch2, D3@LPattern1** and **D1@LPattern1** from the Dimensions box.

382) Release the **Ctrl key**.

383) Click **OK** from the Dimensions dialog box.

The illustrated dimension variable names in the Dimensions dialog box will be different if sketches or features were deleted when creating the FLATBAR part.

The input dimension names and default values are automatically entered into the Design Table. The Design Table displays the Primary Units of the Part. Example: Inches. The value Default is entered in Cell A3.

The values for the FLATBAR are entered in Cells B3 through G9. The FLATBAR length is controlled in Column B. The Number of Holes is controlled in Column G.

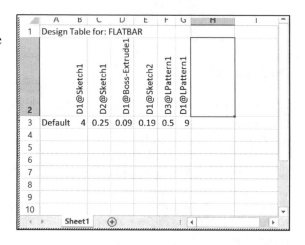

Enter the three configuration names.
384) Click **Cell A4**.

385) Enter **3HOLE**.

386) Click **Cell A5**.

387) Enter **5HOLE**.

388) Click **Cell A6**.

389) Enter **7HOLE**.

390) Click Cell D3.

Enter the dimension values for the 3HOLE configuration.
391) Click **Cell B4**. Enter **1**.

392) Click **Cell G4**. Enter **3**.

Enter the dimension values for the 5HOLE configuration.
393) Click **Cell B5**.

394) Enter **2**.

395) Click **Cell G5**.

396) Enter **5**.

Enter the dimension values for the 7HOLE configuration.
397) Click **Cell B6**.

398) Enter **3**.

399) Click **Cell G6**.

400) Enter **7**.

Build the three configurations.
401) Click a **position** outside the EXCEL Design Table in the Graphics window.

402) Click **OK** to generate the configurations. The Design Table icon is displayed in the FLATBAR FeatureManager.

Display the configurations.
403) Double-click **3HOLE**.

404) Double-click **5HOLE**.

405) Double-click **7HOLE**.

406) Double-click **Default**.

Save the model.
407) Click **Save** 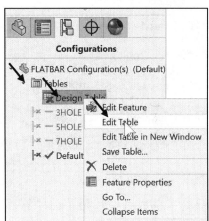.

Edit the Design Table.
408) Right-click **Design Table** in the ConfigurationManager.

409) Click **Edit Table**. The Add Rows and Columns dialog box is displayed.

410) Click **Cancel** from the Add Rows and Columns dialog box.

	A	B	C	D	E	F	G
1	Design Table for: FLATBAR						
2		D1@Sketch1	D2@Sketch1	D1@Boss-Extrude1	D1@Sketch2	D3@LPattern1	D1@LPattern1
3	Default	4	0.25	0.09	0.19	0.5	9
4	3HOLE	1					3
5	5HOLE	2					5
6	7HOLE	3					7
7							
8							
9							
10							

Sheet1 ⊕

SOLIDWORKS   ×

ⓘ The design table generated the following configurations:

  3HOLE
  5HOLE
  7HOLE

                                        OK

Configurations

FLATBAR Configuration(s) (Default)
  Tables
    Design Table   Edit Feature
  3HOLE   Edit Table
  5HOLE   Edit Table in New Window
  7HOLE   Save Table...
  Default   ✕ Delete
                      Feature Properties
                      Go To...
                      Collapse Items

Parameters
$COLOR
$DESCRIPTION
$PARTNUMBER
$PRP@Material
$PRP@Number

☐ Show unselected items again

    OK   Cancel   Help

Columns C through G are filled with the
default FLATBAR values.

Enter parameters for DESCRIPTION and
PARTNUMBER. Custom Properties
begin with the prefix "$PRP@".
SOLIDWORKS Properties begin with
the prefix "$".

Enter DESCRIPTION custom Property.
411) Double-click **Cell H2**.

412) Enter **$PRP@DESCRIPTION**.

413) Click **Cell H3**. Enter **9HOLES**.

414) Click **Cell H4**. Enter **3HOLES**.

415) Click **Cell H5**. Enter **5HOLES**.

416) Click **Cell H6**. Enter **7HOLES**.

Enter the PARTNUMBER Property.
417) Double-click **Cell I2**.

418) Enter **$PARTNUMBER**.

419) Click **Cell I3**.

420) Enter **GIDS-SC-10009-9**.

421) Click **Cell I4**.

422) Enter **GIDS-SC-10009-3**.

423) Click **Cell I5**.

424) Enter **GIDS-SC-10009-5**.

425) Click **Cell I6**.

426) Enter **GIDS-SC-10009-7**.

427) Click a **position** in the Graphics window to update
the Design Table.

428) View the updated ConfigurationManager.

Activity: FLATBAR Drawing - Sheet2

Select configurations in the drawing. The Properties option in the Drawing view displays a list of configuration names.

Open the FLATBAR drawing.

429) Click **Open** from the Main menu.

430) Double-click the **FLATBAR drawing** from the SW-TUTORIAL 2023 folder. The FLATBAR drawing is displayed. The dimensions tied to the Design table are displayed in a different color. The dimension color is controlled from the System Options, Colors section.

Copy the Front view.

431) Click inside the FLATBAR **Front view** boundary.

432) Press **Ctrl C**.

433) **Right-click** in the Graphics window.

434) Click **Add Sheet**. Sheet2 is displayed.

435) Right-click **Properties**. Note: Click the down arrow if needed.

436) Click **Browse** from the Sheet Properties dialog box.

437) Double-click **CUSTOM-A** from the SW-TUTORIAL 2023/MY-SHEETFORMAT folder.

438) Click **Apply Changes** from the Sheet Properties box. CUSTOM-A was created at the beginning of the chapter. If you did not create CUSTOM-A, accept the default sheet format and fill in all of the needed document properties and custom properties.

Paste the Front view from Sheet1.

439) Click a **position** inside the Sheet2 boundary. Press **Ctrl V**. The Front view is displayed.

Display the 3HOLE FLATBAR configuration on Sheet2.

440) Right-click inside the **Front view** boundary.

441) Click **Properties**. Click the down arrow if needed. Select **3HOLE** from the Use named configuration list.

442) Click **OK** from the Drawing View Properties dialog box. The 3HOLE FLATBAR configuration is displayed.

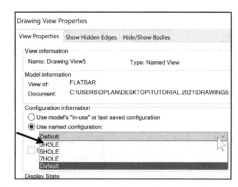

443) Right-click inside the **Drawing View** boundary.

444) Click **Open Part**.

445) View the **model configuration**. 3 HOLE is selected by the drawing.

Return to the FLATBAR Drawing Sheet2.
446) Click **Window**, **FLATBAR - Sheet2** from the Main menu.

447) Click on the **9X** dimension text in the Graphics window. The Dimension PropertyManager is displayed. Note: You can also use the Pop-up menu in this procedure.

448) Replace the 9X dimension text with **3X** in the Dimension Text box as illustrated.

449) Click **OK** ✔ from the Dimension PropertyManager.

450) **Align** all dimension text as illustrated below.

451) Address **all extension line gaps and any other needed information**.

Save the FLATBAR Sheet2 drawing.

452) Click **Save** 💾.

As an exercise, add additional drawing Custom Properties: Number, Description, DrawnBy, DrawnDate, Revision, Third Angle icon, etc.

The 5HOLE and 7HOLE configurations are explored as an exercise. Combine the FLATBAR configurations with the SHAFT-COLLAR part to create three different assemblies. Select configuration in the assembly. The Properties in the FeatureManager option displays a list of configuration names.

The FLATBAR-SHAFTCOLLAR assembly contains a FLATBAR fixed to the assembly Origin and a SHAFTCOLLAR mated to the FLATBAR left hole. The default configuration utilizes the FLATBAR-9HOLE part.

Design Tables exist in the assembly. Utilize a Design Table to control part configurations 3HOLE, 5HOLE and 7HOLE. Utilize the Design Table to Control Suppress/Resolve state of a component in an assembly. Insert the parameter $STATE into the Design Table.

Activity: FLATBAR - SHAFTCOLLAR Assembly

Return to the Default FLATBAR configuration.
453) Right-click in the **3HOLE FLATBAR** view boundary.

454) Click **Open Part**.

455) Click the **ConfigurationManager** 🔖 tab.

456) Double-click the **Default** configuration.

457) Click the **FeatureManager** 🔖 tab. The FLATBAR (Default) FeatureManager is displayed.

Save the model.
458) Click **Save** 💾.

Create the FLATBAR-SHAFTCOLLAR assembly.

459) Click **New** 🗋 from the Main menu.

460) Double-click **Assembly** from the Templates tab. The Begin Assembly PropertyManager is displayed. FLATBAR is an active document. FLATBAR is displayed in the Open documents box.

461) Click **FLATBAR** from the Open documents box.

462) Click **OK** ✔ from the Begin Assembly PropertyManager. The FLATBAR is fixed to the Origin.

Save the FLATBAR-SHAFTCOLLAR assembly.

463) Click **Save** 💾 .

464) Enter **FLATBAR-SHAFTCOLLAR** for Assembly name.

465) Click **Save**.

Insert the SHAFTCOLLAR part.

466) Click the **Insert Components** 🖼 tool from the Assembly toolbar.

467) Double-click the **SHAFT-COLLAR** part from the SW-TUTORIAL 2023 folder.

468) Click a **position** to the front left of the FLATBAR as illustrated.

Fit the model to the Graphics window. Save the model.
469) Press the **f** key.

470) Click **Save** 💾 .

Mate the SHAFTCOLLAR. Insert a Concentric mate.
471) Click the **left inside hole face** of the FLATBAR.

472) Hold the **Ctrl** key down.

473) Click the inside **cylindrical face** of the SHAFT-COLLAR.

474) Release the **Ctrl** key. The Mate pop-up menu is displayed.

475) Click **Concentric** from the Mate pop-up menu.

Insert a Coincident mate.
476) Click the **back flat face** of the SHAFT-COLLAR.

477) **Rotate** to view the front face of the FLATBAR.

478) Hold the **Ctrl** key down.

479) Click the **front face** of the FLATBAR.

480) Release the **Ctrl** key.

481) Click **Coincident** from the Mate pop-up menu.

Insert and mate the second SHAFT-COLLAR to the right hole.

482) Click the **Insert Components** tool from the Assembly toolbar.

483) Double-click the **SHAFT-COLLAR** part from the SW-TUTORIAL 2023 folder.

484) Click a **position** to the front right of the FLATBAR as illustrated.

Mate the second SHAFTCOLLAR. Insert a Concentric mate.

485) Click the **right inside hole face** of the FLATBAR.

486) Hold the **Ctrl** key down.

487) Click the **inside cylindrical face** of the SHAFT-COLLAR.

488) Release the **Ctrl** key.

489) Click **Concentric** from the Mate pop-up menu.

Insert a Coincident mate.

490) Click the **back flat face** of the SHAFT-COLLAR.

491) Rotate to view the front face of the FLATBAR.

492) Hold the **Ctrl** key down. Click the **front face** of the FLATBAR.

493) Release the **Ctrl** key.

494) Click **Coincident** from the Mate pop-up menu.

Save the FLATBAR-SHAFTCOLLAR assembly.

495) Click **Save** .

The FLATBAR-SHAFTCOLLAR FeatureManager displays the Default configuration of the FLATBAR in parentheses, FLATBAR<1> (Default).

The instance number <1> indicates the first instance of the FLATBAR. Note: Your instance number will be different if you delete the FLATBAR and then reinsert into the assembly. The exact instance number is required for the Design Table.

Create a Design Table that contains three new configurations. Each configuration utilizes a different FLATBAR configuration. Control the Suppress/Resolve State of the second SHAFT-COLLAR.

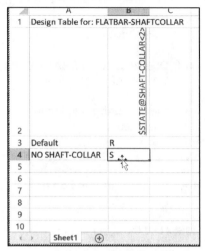

Insert a Design Table.
496) Click **Insert**, **Tables**, **Design Table** from the Main menu. The Auto-create option is selected by default.

497) Click **OK** ✔ from the Design Table PropertyManager.

Enter the Design Table values.
498) Default is displayed in Cell A3. Click **Cell A4**.

499) Enter **NO SHAFT-COLLAR**.

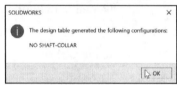

500) Double-click **CELL B2**.

501) Enter **$STATE@SHAFT-COLLAR<2>**.

502) Click **Cell B3**. Enter **R** for Resolved.

503) Click **Cell B4**.

504) Enter **S** for Suppressed.

505) Click a **position** outside the Design Table in the Graphics window.

506) Click **OK** to display the NO SHAFT-COLLAR configuration.

Display the configurations.
507) Click the **ConfigurationManager** 🔖 tab.

508) Double-click the **NO SHAFT-COLLAR** configuration. The second SHAFT-COLLAR is suppressed in the Graphics window.

509) Double-click the **Default** configuration. The second SHAFT-COLLAR is resolved. Both SHAFT-COLLARs are displayed in the Graphics window.

Insert FLATBAR configurations.
510) Right-click **Design Table**.

511) Click **Edit Table**.

512) Click **Cancel** from the Add Rows and Columns dialog box.

Enter Configuration names.
513) Click **Cell A5**. Enter **3HOLE FLATBAR**.

514) Click **Cell A6**. Enter **5HOLE FLATBAR**.

515) Click **Cell A7**. Enter **7HOLE FLATBAR**.

Enter STATE values.
516) Click **Cell B5**. Enter **S** for Suppress.

517) Click Cell **B6**. Enter **S** for Suppress.

518) Click **Cell B7**. Enter **S** for Suppress.

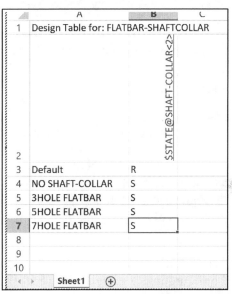

Enter Design Table values.
519) Double-click **Cell C2**.

520) Enter **$CONFIGURATION@FLATBAR<1>**.

521) Click **Cell C5**.

522) Enter **3HOLE**.

523) Click **Cell C6**.

524) Enter **5HOLE**.

525) Click **Cell C7**.

526) Enter **7HOLE**.

527) Click a **position** in the Graphics window to exit.

528) Click **OK** to create the three configurations.

Display the configurations.
529) Double-click the **3HOLE FLATBAR** configuration.

530) Double-click the **5HOLE FLATBAR** configuration.

531) Double-click the **7HOLE FLATBAR** configuration.

532) Double-click the **Default** configuration.

533) Click the **Assembly FeatureManager** tab.

Display an Isometric view. Save the assembly.
534) Display an **Isometric view** .

535) Click **Save** .

Close all documents.
536) Click **Windows**, **Close All** from the Main menu.

Always return to the default configuration in the assembly. Control the individual configuration through properties of a view in a drawing and properties of a component in the assembly.

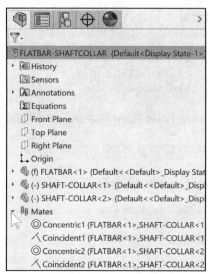

 Additional details on Design Tables and Configurations are available in SOLIDWORKS Help.

Insert a Center of Mass Point

Add a center of mass (COM) point to parts, assemblies or drawings. COM points added in component documents also appear in the assembly document. In a drawing document of parts or assemblies that contain a COM point, you can display and reference the COM point.

Add a COM to a part or assembly by clicking **Center of Mass** (Reference Geometry toolbar) or **Insert, Reference Geometry, Center of Mass** or checking the **Create Center of Mass feature** box in the Mass Properties dialog box.

The center of mass of the model is displayed in the Graphics window and in the FeatureManager design tree just below the origin.

The position of the **COM** point updates when the model's center of mass changes. The COM point can be suppressed and unsuppressed for configurations.

You can measure distances and add reference dimensions between the COM point and entities such as vertices, edges, and faces.

Add a center of mass (COM) point to a drawing view. The center of mass is a selectable entity in drawings, and you can reference it to create dimensions.

In a drawing document, click **Insert, Model Items**. The Model Items PropertyManager is displayed. Under Reference Geometry, click the **Center of Mass** icon. Enter any needed additional information. Click **OK** from the Model Items PropertyManager. View the results in the drawing.

The part or assembly **needs to have a COM before** you can view the COM in the drawing. To view the center of mass in a drawing, click **View, Hide/Show, Center of mass**.

Chapter Summary

You created two drawings: the FLATBAR drawing and the LINKAGE assembly drawing. The drawings contained Standard views, Detail view, Section view and an Isometric view.

The drawings utilized a Custom Sheet Format and a Custom Drawing Template. The Sheet Format contained the Company logo and Title block information.

The FLATBAR drawing consisted of two Sheets: Sheet1 and Sheet2. You obtained an understanding of displaying views with the ability to insert, add, and modify dimensions. You used two major design modes in the drawings: Edit Sheet Format and Edit Sheet.

The LINKAGE assembly drawing contained two sheets. Sheet1 contained an Exploded view and a Bill of Materials. The Properties for the Bill of Materials were developed in each part and assembly. Sheet2 utilized a Detail view and a Section view of the AirCylinder assembly.

You created three configurations of the FLATBAR part with a Design Table. The Design Table controlled parameters and dimensions of the FLATBAR part. You utilized these three configurations in the FLATBAR-SHAFTCOLLAR assembly.

Drawings are an integral part of the design process. Part, assemblies and drawings all work together. From your initial design concepts, you created parts and drawings that fulfilled the design requirements of your customer.

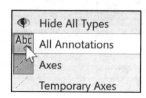

To show a hidden dimension, click **View**, **Hide/Show**, **All Annotations** from the Main menu.

Questions

1. Describe a Bill of Materials and its contents.

2. Name the two major design modes used to develop a drawing in SOLIDWORKS.

3. True or False. Units, Dimensioning Standards, Arrow size, Font size are modified in the Options, Document Properties section.

4. How do you save a Sheet Format?

5. Identify four components that are commonly found in a title block.

6. Describe the procedure to insert an Isometric view to the drawing.

7. In SOLIDWORKS, a drawing file name ends with a _____ suffix. A part file name ends with a _____ suffix.

8. True or False. In SOLIDWORKS, if a part is modified, the drawing is updated with a Rebuild command.

9. True or False. In SOLIDWORKS, when a dimension in the drawing is modified, the part is updated with a Rebuild command.

10. Name three guidelines to create General Notes on a drawing.

11. True or False. Most engineering drawings use the following font: Times New Roman - all small letters.

12. What are Leader lines? Provide an example.

13. Describe the key differences between a Detail view and a Section view on a drawing.

14. Identify the procedure to create an Exploded view.

15. Describe the purpose of a Design Table for a part and for an assembly.

16. Review the Design Intent section in the Introduction. Identify how you incorporated design intent into the drawing.

17. Identify how you incorporate design intent (FLATBAR) into configurations with a Design Table.

Exercises

Exercise 4.1: FLATBAR - 3 HOLE Drawing

Create the A (ANSI) Landscape - IPS - Third Angle 3HOLES drawing as illustrated below. Do not display Tangent Edges. Do not dimension to Hidden Lines.

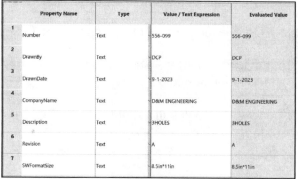

- First create the part from the drawing, then create the drawing. Use the default A (ANSI) Landscape Sheet Format/Size.

	Property Name	Type	Value / Text Expression	Evaluated Va
1	Material	Text	"SW-Material@3HOLE.SLDPRT"	1060 Alloy
2	<Type a new property>			

- Insert a Front, Top and a Shaded Isometric view as illustrated. Insert dimensions.

	Property Name	Type	Value / Text Expression	Evaluated Value
1	Number	Text	556-099	556-099
2	DrawnBy	Text	DCP	DCP
3	DrawnDate	Text	9-1-2023	9-1-2023
4	CompanyName	Text	D&M ENGINEERING	D&M ENGINEERING
5	Description	Text	3HOLES	3HOLES
6	Revision	Text	A	A
7	SWFormatSize	Text	8.5in*11in	8.5in*11in

- Address extension line gap between the feature lines.

- Address proper display mode.

- Add a Smart (Linked) Parametric note for MATERIAL THICKNESS in the drawing as illustrated.

- Hide the dimension in the Top view. Insert Centerlines.

- Modify the Hole dimension text to include 3X THRU EQ. SP. and 2X as illustrated.

- Insert Company and Third Angle projection icons. The icons are available in the homework folder.

- Insert Custom Properties: Material, Number, Description, DrawnBy, DrawnDate, CompanyName, Revision, etc.

- Material is 1060 Alloy.

Exercise 4.2: CYLINDER Drawing

Create the A (ANSI) Landscape - IPS - Third Angle
CYLINDER drawing as illustrated below. Do not display
Tangent Edges. Do not dimension to Hidden lines.

- First create the part from the drawing - then create the
  drawing. Use the default A (ANSI) Landscape Sheet
  Format/Size.

- Insert the Front, Right and Isometric view as illustrated.

- Insert dimensions.

- Address extension line gap between the feature lines.

- Insert proper display modes.

- Insert Company and Third Angle
  projection icons. The icons are
  available in the homework folder.

	Property Name	Type	Value / Text Expression	Evaluated Value
1	Material	Text	"SW-Material@CylinderPart1.SLDPRT"	AISI 1020
2	Description	Text	CYLINDER	CYLINDER

- Insert Centerlines, Center Marks, and
  Annotations.

	Property Name	Type	Value / Text Expression	Evaluated Value
1	DrawnDate	Text	9-1-2023	9-1-2023
2	DrawnBy	Text	DCP	DCP
3	CompanyName	Text	D&M ENGINEERING	D&M ENGINEERING
4	Revision	Text	A	A
5	Number	Text	667-888	667-888
6	SWFormatSize	Text	8.5in*11in	8.5in*11in
7	<Type a new property>			

- Insert Custom Properties: Material,
  Revision, Description, Number,
  DrawnBy, DrawnDate,
  CompanyName.

- Material is AISI 1020.

Exercise 4.3: PRESSURE PLATE Drawing

Create the A (ANSI) Landscape - IPS - Third Angle
PRESSURE PLATE drawing. Do not display Tangent edges.
Do not dimension to hidden lines.

- First create the part from the drawing,
  then create the drawing. Use the default
  A (ANSI) Landscape Sheet Format/Size.

- Insert the Front and Right view as
  illustrated.

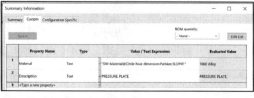

- Insert dimensions.

- Address extension line gap between
  the feature lines.

- Address proper display modes.

- Insert Company and Third Angle
  projection icons. The icons are
  available in the homework folder.

- Insert Centerlines, Center Marks, and
  Annotations.

- Insert Custom Properties:
  Material, Revision,
  Description, Number,
  DrawnBy, DrawnDate,
  CompanyName.

- Material is 1060 Alloy.

Exercise 4.3B TUBE Drawing

Create the A (ANSI) Landscape - IPS - Third Angle TUBE-2 drawing as illustrated below. Do not display Tangent edges. Phantom lines are fine.

- First create the part from the drawing, then create the drawing. Use the default A (ANSI) Landscape Sheet Format/Size.

- Insert views as illustrated.

- Insert dimensions.

- Address extension line gap between the feature lines.

- Insert Company and Third Angle projection icons. The icons are available in the homework folder.

- Insert Centerlines, Center Marks, and Annotations.

- Address proper display modes.

- Insert Custom Properties: Material, Revision, Description, Number, DrawnBy, DrawnDate, CompanyName.

- Material is 6061 Alloy.

	Property Name	Type	Value / Text Expression	Evaluated Value
1	CompanyName	Text	D&M ENGINEERING	D&M ENGINEERING
2	DrawnDate	Text	9-1-2023	9-1-2023
3	DrawnBy	Text	DCP	DCP
4	Revision	Text	A	A
5	Number	Text	3334-667A	3334-667A
6	SWFormatSize	Text	8.5in*11in	8.5in*11in
7	<Type a new property>			

Exercise 4.4: LINKS Assembly Drawing

- Create the LINK assembly. Utilize three different FLATBAR configurations and a SHAFT-COLLAR. The parts are located in the chapter folder.

- Create the LINK assembly drawing as illustrated. Use the default A (ANSI) Landscape Sheet Format/Size.

- Insert Company name, logo and Third Angle projection icon. The icon is available in the homework folder.

- Remove all Tangent Edges.

- Insert Custom Properties: Description, Number, Revision, DrawnBy, DrawnDate, and CompanyName.

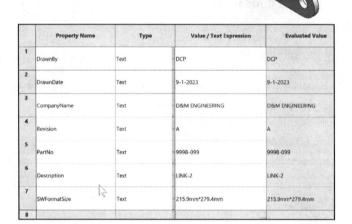

	Property Name	Type	Value / Text Expression	Evaluated Value
1	DrawnBy	Text	DCP	DCP
2	DrawnDate	Text	9-1-2023	9-1-2023
3	CompanyName	Text	D&M ENGINEERING	D&M ENGINEERING
4	Revision	Text	A	A
5	PartNo	Text	9998-099	9998-099
6	Description	Text	LINK-2	LINK-2
7	SWFormatSize	Text	215.9mm*279.4mm	215.9mm*279.4mm
8				

- Insert a Bill of Materials as illustrated with Balloons.

Exercise 4.5: PLATE-1 Drawing

Create the A (ANSI) Landscape - MMGS - Third Angle
PLATE-1 drawing as illustrated below. Do not display
Tangent edges. Do not dimension to Hidden lines.

- First create the part from the drawing, then create the
  drawing. Use the default A (ANSI) Landscape Sheet
  Format/Size.

- Insert the Front and Right view as
  illustrated.

- Insert dimensions.

- Address extension line gap between the
  feature lines.

- Address display modes.

- Insert Company and Third
  Angle projection icons. The
  icons are available in the
  homework folder.

- Insert Centerlines, Center
  Marks, and Annotations.

- Insert Custom Properties:
  Material, Description,
  Number, Revision,
  DrawnBy,
  DrawnDate, and
  CompanyName.

- Material is 1060
  Alloy.

	Property Name	Type	Value / Text Expression	Evaluated Value
1	Description	Text	PLATE-1	PLATE-1
2	DrawnBy	Text	DCP	DCP
3	DrawnDate	Text	9-1-2023	9-1-2023
4	CompanyName	Text	D&M ENGINEERING	D&M ENGINEERING
5	Revision	Text	A	A
6	Number	Text	54 -999	5444-999
7	SWFormatSize	Text	215.9mm*279.4mm	215.9mm*279.4mm
8	<Type a new property>			

Exercise 4.6: PLATE Drawing

Create the A (ANSI) Landscape - IPS - Third Angle FLAT-PLATE drawing. Do not display Tangent edges. Phantom lines are fine.

- First create the part from the drawing, then create the drawing. Use the default A (ANSI) Landscape Sheet Format/Size.

- Insert the Front, Top, Right and Isometric views as illustrated.

- Insert dimensions.

- Address extension line gap between the feature lines.

- Address proper display modes.

- Insert Company and Third Angle projection icons. The icons are available in the homework folder.

- Insert Centerlines, Center Marks, and Annotations.

- Insert Custom Properties: Description, Number, Material, Revision, DrawnBy, DrawnDate, and CompanyName.

- Material is 1060 Alloy.

	Property Name	Type	Value / Text Expression	Evaluated Value
1	DrawnBy	Text	DCP	DCP
2	DrawnDate	Text	9-1-2023	9-1-2023
3	CompanyName	Text	D&M ENGINEERING	D&M ENGINEERING
4	Description	Text	PLATE	PLATE
5	Revision	Text	A	A
6	Number	Text	2445-990	2445-990
7	SWFormatSize	Text	8.5in*11in	8.5in*11in
8				

Exercise 4.7: LINKAGE-2 Drawing

- Create the LINKAGE-2 assembly drawing. Utilize the default A (ANSI) Landscape Sheet Format/Size. Note: If needed, open the LINKAGE-2 assembly from the Chapter 4 Homework/AirCylinder 2 folder.

- Insert an Isometric shaded view of the LINKAGE-2 Assembly created in the Chapter 2 exercises.

- Define the PART NO. Property and the DESCRIPTION Property for the AXLE, FLATBAR- 9HOLE, FLATBAR - 3HOLE and SHAFT COLLAR.

	Property Name	Type	Value / Text Expression	Evaluated Value	
1	SWFormatSize	Text	8.5in*11in	8.5in*11in	
2	Revision	Text	A	A	
3	DrawnBy	Text	DCP	DCP	
4	DrawnDate	Text	9-1-2023	9-1-2023	
5	Description	Text	LINKAGE-2	LINKAGE-2	
6	CompanyName	Text	D&M ENGINEERING	D&M ENGINEERING	
7	PartNo	Text	333-223-22	333-223-22	
8	<Type a new property>				

- Save the LINKAGE-2 assembly to update the properties. Return to the LINKAGE-2 Drawing. Insert a Bill of Materials with Auto Balloons as illustrated.

- Insert the Company and Third Angle Projection icon.

- Insert Custom Properties.

 Use the Pack and Go option to save an assembly or drawing with references.

Exercise 4.8: Vertical Section View Drawing

Create the A (ANSI) Landscape - IPS - Third Angle Vertical
Section View drawing as illustrated below. Do not display Tangent
edges.

Insert Company and Third Angle projection icons. The icons are
available in the homework folder.

Insert Custom Properties: Material,
Description, DrawnBy, DrawnDate,
CompanyName, Revision, etc.

Property Name	Type	Value / Text Expression	Evaluated Valu
PartNo	Text	445-777	445-777
Revision	Text	A	A
DrawnBy	Text	DCP	DCP
DrawnDate	Text	9-1-2023	9-1-2023
CompanyName	Text	D&M ENGINEERING	D&M ENGINEERING
Description	Text	SECTION VIEW	SECTION VIEW
SWFormatSize	Text	8.5in*11in	8.5in*11in

1. Open **Section View** from the Chapter 4
    Homework\Section folder.

2. Create the **drawing document**.

3. Insert the **Front view (Drawing View1)**.

4. Click inside the **Drawing View1** view
    boundary. The Drawing View1
    PropertyManager is displayed.

Display the origin on Sheet1.

5. Click **View**, **Hide/Show**,
    Origins from the Main menu.

6. Click the **Section View**
    drawing tool. The Section View
    PropertyManager is displayed.

7. Click the **Section** tab. Click the
    Vertical Cutting Line button.

8. Click the **origin**. Note: You can
    select the midpoint vs. the
    origin.

Place the Section view.

9. Click a **position** to the right of Drawing View1. The Section arrows point to the right.
    If required, click Flip direction. Check **Auto hatching** from the Section View box.

10. Check the **Shaded With Edges** option from the Display Style box. Click **OK**
    from the Section View A-A PropertyManager. Section View A-A is created and is
    displayed in the Drawing FeatureManager.

11. Click inside the **Drawing View1** view boundary. The Drawing View1
    PropertyManager is displayed. Modify the Scale to **1.5:1**. Click **OK** from the
    Drawing View1 PropertyManager. Both drawing views are modified (parent and
    child).

Exercise 4.9: Aligned Section View Drawing

Create an Aligned Section view drawing.

Create the A (ANSI) Landscape - IPS - Third Angle drawing as illustrated below.

Insert Company and Third Angle projection icons. The icons are available in the homework folder.

Insert Custom Properties: Material, Description, DrawnBy, DrawnDate, CompanyName, etc.

1. Open **Aligned Section** from the Chapter 4 Homework\Aligned Section folder.

2. Create the **drawing document**. Insert the **Front (Drawing View1)** and **Right** view.

3. Click inside the **Drawing View1** view boundary. The Drawing View1 PropertyManager is displayed. Click the **Section View** ⬍ drawing tool. The Section View PropertyManager is displayed.

4. Click the **Half Section** tab.

5. Click the **Leftside Down** Half Section button.

6. Select the **center** of the drawing view as illustrated. The Section View dialog box is displayed.

7. Click the **LENSCAP** in the Graphics window. The LENSCAP component is displayed in the Excluded components box. Check the **Auto hatching** box.

8. Click **OK** from the Section View dialog box.

Place the Section View.

9. Click a **position** above Drawing View1. Click **OK** ✔ from the Section View A-A PropertyManager.

10. Modify the **scale** to **1:4**. **Rebuild** 🔘 the drawing.

11. **Move** the drawing views to fit the Sheet. View the drawing FeatureManager.

12. **Close** the drawing.

💡 Use Section View Assist, previously called the Section View User Interface, to add offsets to existing section views. To edit existing section views with Section View Assist: Right-click the section view or its cutting line and click **Edit Cutting Line**.

Exercise 4.10: Auxiliary View Drawing

Create an Auxiliary View drawing as illustrated from an existing Front view.

Create the A (ANSI) Landscape - IPS - Third Angle drawing as illustrated.

Insert Company and Third Angle projection icons.

The icons are available in the homework folder.

Insert Custom Properties: Material, Description, DrawnBy, DrawnDate, CompanyName, etc.

	Property Name	Type	Value / Text Expression	Evaluated Value
1	Material	Text	"SW-Material@Auxiliary View.SLDPRT"	1060 Alloy
2	<Type a new property>			

1. **Copy** the Chapter 4 Homework folder to your hard drive.

2. Open **Auxiliary view** from the Chapter 4 Homework\Auxiliary folder.

3. Create the **drawing** document.

4. Insert the **Front (Drawing View1)**, **Top**, **Right** and **Isometric** view.

5. Click the **angled edge** in Drawing View1.

6. Click the **Auxiliary View** drawing tool.

7. Click a **position** up and to the right.

8. Drag the **A-A arrow** as illustrated.

9. Drag the **text** off the view.

10. Click **OK** ✔ from the Auxiliary View PropertyManager.

11. **Close** the drawing.

	Property Name	Type	Value / Text Expression	Evaluated Value
1	SWFormatSize	Text	215.9mm*, .4mm	215.9mm*279.4mm
2	Revision	Text	A	A
3	PartNo	Text	3345-88	3345-88
4	DrawnBy	Text	DCP	DCP
5	Description	Text	AUXILIARY	AUXILIARY
6	CompanyName	Text	D&M ENGINEERING	D&M ENGINEERING
7	DrawnDate	Text	9-1-2023	9-1-2023

Chapter 5

Advanced Features

Below are the desired outcomes and usage competencies based on the completion of Chapter 5.

Desired Outcomes:	Usage Competencies:

• Six parts for the PNEUMATIC-TEST-MODULE assembly:
   ○ WEIGHT
   ○ HOOK
   ○ WHEEL
   ○ HEX-ADAPTER
   ○ AXLE-3000
   ○ SHAFTCOLLAR-500
	• Apply the following Advanced modeling features: Plane, Lofted Base, Extruded Cut, Swept Base, Dome, Thread, Extruded Boss/Base, Revolved Cut, Extruded Cut, Circular Pattern, Axis, Instant3D, Hole Wizard, Split Line, and Thread Wizard.
	
	• Reuse geometry.
	
	• Modify existing parts to create new parts with the Save as copy command.

Notes:

Chapter 5 - Advanced Features

Chapter Objective

Obtain the ability to reuse geometry by modifying existing parts and to create new parts. Knowledge of the following SOLIDWORKS features: *Plane, Lofted Base, Extruded Cut, Swept Base, Dome, Thread, Extruded Boss/Base, Revolved Cut, Extruded Cut, Circular Pattern, Axis, Instant3D, Hole Wizard, Split Line and Thread Wizard.*

Create six individual parts:

- WEIGHT

- HOOK

- WHEEL

- HEX-ADAPTER

- AXLE-3000

- SHAFTCOLLAR-500

Develop a working understanding with multiple documents in an assembly. Build on sound assembly modeling techniques that utilize symmetry, component patterns, and mirrored components.

Create five assemblies:

- 3HOLE-SHAFTCOLLAR assembly

- 5HOLE-SHAFTCOLLAR assembly

- WHEEL-FLATBAR assembly

- WHEEL-AND-AXLE assembly

- PNEUMATIC-TEST-MODULE assembly

On the completion of this chapter, you will be able to:

- Create new parts and copy parts with the Save As command to reuse similar geometry.

- Utilize Construction geometry in a sketch.

- Apply the following SOLIDWORKS features:

  o Extruded Boss/Base

  o Extruded Cut

  o Dome

  o Plane

  o Lofted Boss/Base

  o Swept Boss/Base

  o Thread

  o Revolved Cut

  o Hole Wizard

  o Axis

  o Circular Pattern

  o Instant3D

  o Split Line

Chapter Overview

Six additional parts are required for the final PNEUMATIC-TEST-MODULE assembly. Each part explores various modeling techniques.

Create three new parts in this chapter:

- WEIGHT

- HOOK

- WHEEL

The WEIGHT and HOOK parts were supplied in the Chapter 3 Homework exercises. View the Chapter 3 Homework folder for the parts if needed.

Utilize the Save As command and modify existing parts that were created in the previous chapters to create three additional parts for the PNEUMATIC-TEST-MODULE assembly.

- HEX-ADAPTER

- AXLE-3000

- SHAFTCOLLAR-500

The HEX-ADAPTER part utilizes modified geometry from the HEX-STANDOFF part.

The AXLE-3000 part utilizes modified geometry from the AXLE part.

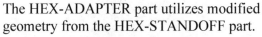

The SHAFTCOLLAR-500 part utilizes modified geometry from the SHAFT-COLLAR part.

Press the **g** key to activate the Magnifying glass tool. Use the Magnifying glass tool to inspect a model and make selections without changing the overall view of your model in the Graphics window.

Download all needed model files (SW-TUTORIAL 2023 folder) from the SDC Publication website (www.SDCpublications.com/downloads/978-1-63057-556-4) to a local hard drive.

WEIGHT Part

The WEIGHT part is a machined part. Utilize the

Lotted feature. Create a Loft by blending two or more profiles. Each profile is sketched on a separate plane.

Create Plane1. Offset Plane1 from the Top Plane.

Sketch a rectangle for the first profile on the Top Plane.

Sketch a square for the second profile on Plane1.

Select the corner of each profile to create the Lofted feature.

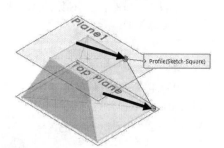

Utilize the Instant3D tool to create an

Extruded Cut feature to create a Through All hole centered on the top face of the Loft feature.

Reference geometry defines the shape or form of a surface or a solid. Reference geometry includes planes, axes, coordinate systems, and points.

When using the Instant3D tool, you lose the ability to select various End Conditions to maintain design intent.

All parts in this chapter utilize a custom part template. Create the custom part template from the default part template. Save the Custom Part template in the SW-TUTORIAL 2023\MY-TEMPLATE folder.

Activity: Create the WEIGHT Part

Create a new part template.

1) Click **New** from the Main menu.

2) Double-click **Part** from the Templates tab. The Part FeatureManager is displayed.

Set Document Properties. Set drafting standard.

3) Click **Options** from the Main menu.

4) Click the **Document Properties** tab from the dialog box.

5) Select **ANSI** from the Overall drafting standard drop-down menu.

Set document units and precision.
6) Click **Units**.

7) Select **IPS**, [**MMGS**] for Unit system.

8) Select **.123**, [**.12**] for linear unit decimal place.

9) Select **None** for angular unit decimal place.

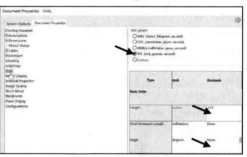

Set Leader arrow direction.
10) Click **Dimensions**. Check the **Smart** box.

11) Click **OK** from the Document Properties - Detailing - Dimensions dialog box.

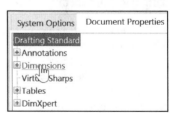

Save the Part template. Enter name.
12) Click **Save As** .

13) Select **Part Templates (*.prtdot)** for Save as type.

14) Select **SW-TUTORIAL 2023\MY-TEMPLATES** for Save in folder.

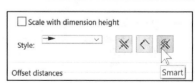

15) Enter **PART-ANSI-IN** for File name.

16) Click **Save**.

Close the Part template.

17) Click **File**, **Close** from the Main menu.

Create a new part.

18) Click **New** from the Main menu.

19) Click the **SW-TUTORIAL 2023\MY-TEMPLATES** tab. Note: Additional templates are displayed.

20) Double-click **PART-ANSI-IN**. The Part FeatureManager is displayed.

Save the part. Enter part name. Enter part description.

21) Click **Save As** .

22) Select the **SW-TUTORIAL 2023** folder.

23) Enter **WEIGHT** for File name.

24) Enter **WEIGHT** for Description.

25) Click **Save**.

Insert Plane1. Display and Isometric view.

26) Click the **Isometric view** icon.

27) Right-click **Top Plane** from the FeatureManager.

28) Click **Show**. The Top Plane is displayed in the Graphics window.

29) Hold the **Ctrl** key down.

30) Click the **boundary** of the Top Plane as illustrated.

31) Drag the **mouse pointer** upward.

32) Release the **mouse pointer**.

33) Release the **Ctrl** key. The Plane PropertyManager is displayed. Top Plane is displayed in the First Reference box.

Add relations, then dimensions. This keeps the user from having too many unnecessary dimensions. This also helps to show the design intent of the model. Dimension what geometry you intend to modify or adjust.

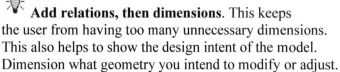

34) Enter **.500**in, **[12.70]** for Distance.

35) Click **OK** from the Plane PropertyManager.

Plane1 is displayed in the Graphics window and is listed in the FeatureManager. Plane1 is offset from the Top Plane.

A Lofted feature requires two sketches. The first sketch, Sketch1, is a rectangle sketched on the Top Plane centered about the Origin. The second sketch, Sketch2, is a square sketched on Plane1 centered about the Origin.

Create Sketch1 in the Top Plane.
36) Right-click **Top Plane** from the FeatureManager.

37) Click **Sketch** from the Context toolbar. The Sketch toolbar is displayed.

38) Click **Center Rectangle** from the Consolidated Sketch tool. The Center Rectangle icon is displayed.

39) Click the **Origin**.

40) Click a **position** to the top right as illustrated.

The Center Rectangle tool provides the ability to sketch a rectangle located at a center point, in this case the Origin. This eliminates the need for centerlines to the Origin with a Midpoint geometric relation.

CommandManager, FeatureManager tabs and tree folders will vary depending on system setup and Add-ins.

Add dimensions.

41) Click the **Smart Dimension** Sketch tool.

42) Click the **top horizontal** line.

43) Click a **position** above the line.

44) Enter **1.000**in, [25.40].

45) Click the **right vertical** line.

46) Click a **position** to the right.

47) Enter **.750**in, [19.05].

48) Click the **Green Check mark** ✔.

Close Sketch1.
49) Click **Exit Sketch** from the Sketch toolbar. The sketch is fully defined and is displayed in black.

Rename Sketch1.
50) Rename **Sketch1** to **Sketch-Rectangle**.

Save the part.
51) Click **Save** 💾.

Display an Isometric view.
52) Click **Isometric view** 🔲 from the Heads-up View toolbar.

Create Sketch2 on Plane1. Plane1 is your Sketch plane.
53) Right-click **Plane1** from the FeatureManager. Plane1 is your Sketch plane.

54) Click **Sketch** 🔲 from the Context toolbar. The Sketch toolbar is displayed.

55) Click the **Center Rectangle** 🔲 Consolidated Sketch tool. The Center Rectangle icon is displayed.

56) Click **Top View**.

57) Click the **Origin** ↳.

58) Click a **position** as illustrated.

59) Right-click **Select** to de-select the Center Rectangle tool.

If needed, add an Equal relation between the left vertical line and the top horizontal line.

60) Click the **left vertical line** of the rectangle.

61) Hold the **Ctrl** key down.

62) Click the **top horizontal line** of the rectangle.

63) Release the **Ctrl** key.

64) Click **Equal** = from the Add Relations box.

65) Click **OK** ✔ from the Properties PropertyManager.

Add a dimension.

66) Click the **Smart Dimension** Sketch tool.

67) Click the **top horizontal** line.

68) Click a **position** above the line.

69) Enter **.500**in, [**12.70**].

70) Click the **Green Check mark** ✔. The sketch is displayed in black.

Close Sketch2. Display an Isometric view.

71) Click **Exit Sketch** from the Sketch toolbar. Sketch2 is fully defined.

72) Click **Isometric view** ⬛ from the Heads-up View toolbar. View the results in the Graphics window.

If you did not select the Origin, insert a Coincident relation between the rectangle and the Origin to fully define Sketch2.

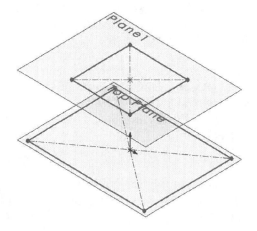

Rename Sketch2.

73) Rename **Sketch2** to **Sketch-Square**.

Save the WEIGHT part.

74) Click **Save** 📳.

💡 Lofted features are comprised of multiple sketches. Name sketches for clarity.

Activity: WEIGHT Part - Lofted Feature

Insert a Lofted feature.

75) Click the **Features** tab from the CommandManager.

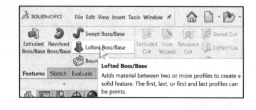

76) Click the **Lofted Boss/Base** 🔻 Feature tool. The Loft PropertyManager is displayed.

77) Clear the **Profiles** box.

78) Click the **back right corner** of Sketch-Rectangle as illustrated.

79) Click the **back right corner** of Sketch-Square. Sketch-Rectangle and Sketch-Square are displayed in the Profiles box.

80) Click **OK** ✔ from the Loft PropertyManager. Loft1 is displayed in the FeatureManager.

81) **Hide** all planes.

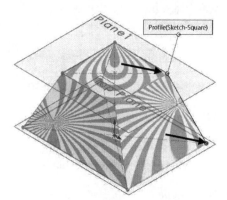

💡 To display the Selection Filter toolbar, right-click in the Graphics window, and click the **Selection Filters** drop-down menu icon. The Selection Filter is displayed.

💡 To clear a Filter icon 🔻, click **Clear All Filters** from the Selection Filter toolbar.

82) **Expand** Loft1 in the FeatureManager. Sketch-Rectangle and Sketch-Square are the two sketches that contain the Loft feature.

83) **Zoom in** on the Loft1 feature.

Activity: WEIGHT Part - Instant3D - Extruded Cut Feature

Insert a New sketch for the Extruded Cut feature.

84) Right-click the **top square face** of the Loft1 feature for the Sketch plane.

85) Click **Sketch** from the Context toolbar. The Sketch toolbar is displayed.

86) Click the **Circle** Sketch tool. The circle icon is displayed.

87) Click the **center** as illustrated.

88) Click a **position** to the right as illustrated.

Add a dimension.

89) Click the **Smart Dimension** Sketch tool.

90) Click the **circumference** of the circle.

91) Click a **position** in the Graphics window above the circle to locate the dimension.

92) Enter **.150**in, **[3.81]** in the Modify box.

Insert an Extruded feature using the Instant3D tool.

93) **Exit** the Sketch. By default, Instant3D is active.

94) Click the **diameter** of the circle, Sketch3, as illustrated.

95) Click the **Arrowhead** and drag it below the model.

96) Click a **position** on the Instant3D ruler. The Extrude feature is displayed in the FeatureManager.

Display Wireframe style.

97) Click **Wireframe** from the Heads-up View toolbar. View the Extrude feature.

Rename the Extrude feature.
98)   Rename the **Extrude** feature to **Hole-for-Hook**.

Display an Isometric view. Display Shaded With Edges. Save the WEIGHT part.

99)   Click **Isometric view** from the Heads-up View toolbar.

100) Click **Shaded With Edges** from the Heads-up View toolbar.

101) Click **Save** . The WEIGHT part is complete. Later, apply material to the part.

 Review the WEIGHT Part.

The WEIGHT part was created with the Loft feature. The Loft feature required two planes: Top Plane and Plane1. Profiles were sketched on each plane. Profiles were selected to create the Loft feature.

An Extruded Cut feature was created using the Instant3D tool to create a Through All center hole in the WEIGHT.

HOOK Part

The HOOK part fastens to the WEIGHT. The HOOK is created with a Swept Base feature. The Swept Base feature adds material by moving a profile along a path.

The Swept Base feature requires two sketches (path and profile) or a sketch path and a circular profile diameter. If the sketch profile is a circle, enter the circular profile diameter in the Swept PropertyManager.

For non-circular sketch profiles, create the sketch on a perpendicular plane to the path and use the pierce relation to locate the profile on the path.

Create the HOOK part with a Swept Base feature.

The Swept Base feature uses:

- A path sketched on the Right Plane.

- A circular profile diameter.

Utilize the Dome feature tool to create a
spherical feature on a circular face.

Utilize the Thread feature tool to create a
right-hand #8-36 thread for the HOOK. The
Thread tool can add or removes material.

Reference geometry defines the shape or
form of a surface or a solid. Reference geometry
includes planes, axes, coordinate systems and points.

Activity: Create the HOOK Part

Create the new part.

102) Click **New** from the Main menu.

103) Select the **SW-TUTORIAL 2023\MY-TEMPLATES** tab.
Additional templates are displayed.

104) Double-click **PART-ANSI-IN.**

Save the part. Enter part name. Enter part description.

105) Click **Save As** .

106) Select the **SW-TUTORIAL 2023** folder.

107) Enter **HOOK** for File name.

108) Enter **HOOK** for Description.

109) Click **Save**. The HOOK FeatureManager is displayed.

If the sketch profile is a circle, enter the diameter in the
Swept PropertyManager. For non-circular sketch profiles,
create the sketch on a perpendicular plane to the path and
use the pierce relation to locate the profile on the path.

Sketch1 is the Sweep Path sketched on the Right Plane.

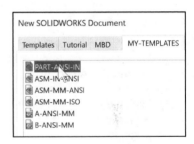

Sketch the Sweep Path.
110) Right-click **Right Plane** from the FeatureManager.

111) Click **Sketch** from the Context toolbar.

112) Click the **Line** Sketch tool. The Insert Line PropertyManager is displayed.

113) Sketch a **vertical line** from the Origin as illustrated.

Add a dimension.
114) Click the **Smart Dimension** Sketch tool.

115) Click the **vertical line**.

116) Click a **position** to the right.

117) Enter **.250**in, [**6.35**].

118) Click the **Green Check mark** .

Fit the model to the Graphics window.
119) Press the **f** key.

Create the Centerpoint arc.

120) Click the **Centerpoint Arc** Sketch tool from the Consolidated Sketch toolbar. The Centerpoint Arc icon is displayed.

121) Click the **arc center point** vertically aligned to the Origin as illustrated.

122) Click the **arc start point** as illustrated.

Origin

Start point

R = 0.213

123) Move (do not drag) the **mouse pointer** clockwise approximately 270°.

124) Click a point **horizontally aligned** to the arc start point. If needed add a horizontal relationship.

125) Click the **3 Point Arc** Sketch tool from the Consolidated Sketch toolbar. The Arc PropertyManager is displayed.

126) Click the **vertical line** endpoint.

127) Click the **center point arc** endpoint.

128) Drag and pull the center of the **3 Point Arc downwards**.

129) Click the center of the **center point arc line** as illustrated.

130) Click **OK** ✔ from the Arc PropertyManager.

It is important to draw the correct shape with the 3 Point Arc tool as illustrated.

Add a Vertical relation between the Origin and the center point of the arc.

131) Click the **Origin** .

132) Hold the **Ctrl** key down.

133) Click the **center point** of the Center point arc.

134) Release the **Ctrl** key.

135) Click **Vertical** from the Add Relations box.

Correct shape   **Incorrect shape**

If needed, add a Horizontal relation.
136) Click the **start point** of the Center point arc.

137) Hold the **Ctrl** key down.

138) Click the **end point** of the Center point arc.

139) Release the **Ctrl** key.

140) Click **Horizontal** ━ from the Add Relations box.

Add a Tangent relation.
141) Click the **vertical line**.

142) Hold the **Ctrl** key down.

143) Click the **3 Point Arc**.

144) Release the **Ctrl** key.

145) Click **Tangent** ⟋ from the Add Relations box.

Add a second Tangent relation.

146) Click the **3 Point Arc**.

147) Hold the **Ctrl** key down.

148) Click the **Center point** arc.

149) Release the **Ctrl** key.

150) Click **Tangent** ⟋ from the Add Relations box.

Add dimensions.
151) Click the **Smart Dimension** ⟋ Sketch tool.

152) Click the **3 Point Arc**.

153) Click a **position** to the left.

154) Enter .**500**in, [**12.70**].

155) Click the **Green Check mark** ✔.

Dimension the overall length of the sketch.
156) Click the **top of the arc**.

157) Click the **Origin** .

158) Click a **position** to the right of the profile. Accept the default dimension.

159) Click the **Green Check mark** .

Modify the overall length.
160) Double-click the default **dimension**.

161) Enter **1.000**in, [**25.40**]. By default, the Dimension tool utilizes the center point of an arc or circle. Select the circle profile during dimensioning.

162) Click the **Leaders** tab in the Dimension PropertyManager

163) Modify the arc condition. Click **Maximum**.

164) Click the **Green Check mark** .

Fit the model to the Graphics window.
165) Press the **f** key.

166) Move the **dimensions** as illustrated.

Close the sketch.
167) Click **Exit Sketch** from the Sketch toolbar.

Rename Sketch1.
168) Rename **Sketch1** to **Sketch-Path**.

Save the HOOK.

169) Click **Save** .

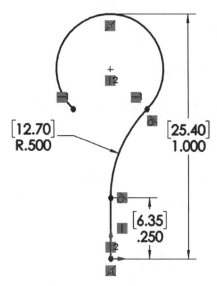

Activity: HOOK Part - Swept Base Feature

Insert the Swept feature.

170) Click the **Swept Boss/Base** Features tool. The Sweep PropertyManager is displayed.

171) Select **Circular Profile**.

172) Click **Sketch-Path** from the fly-out FeatureManager. Sketch-Profile is displayed in the Profile box.

173) Enter **0.150**in diameter.

174) Click **OK** from the Sweep PropertyManager. Sweep1 is displayed in the FeatureManager.

Save the HOOK part.

175) Click **Save** .

Activity: HOOK Part - Dome Feature

Insert a Dome feature.

176) **Rotate** the model with the middle mouse button.

177) Click the **flat face** of the Sweep1 feature in the Graphics window as illustrated.

178) Click the **Dome** Features tool (Insert, Features, Dome). The Dome PropertyManager is displayed. Face<1> is displayed in the Parameters box.

179) Enter **.050**in, [**1.27**] for Distance.

180) Click **OK** from the Dome PropertyManager. Dome1 is displayed in the FeatureManager.

The HOOK requires a simplified right-hand #8-36 thread from the bottom of the Sweep1 feature. Utilize the Thread feature. The Thread feature provides the ability to create helical threads on cylindrical faces using profile sketches. Store custom thread profiles as library features.

☀ The Thread tool Type and Size profiles are nominal thread profiles only. Do not use them for production-quality threads. To create production-quality threads, modify the nominal profiles to meet your design requirements.

Define the start thread location, specify an offset, set end conditions, specify the type, size, diameter, pitch and rotation angle, and choose options such as right-hand or left-hand thread.

☀ Use the Stud Wizard to create external threaded stud features. This tool works similar to the Hole Wizard. You define the stud parameters then position the studs on the model. You can also apply thread parameters to existing circular studs.

Activity: HOOK Part - Thread Feature

Create a right-hand #8-36 thread from the bottom of the Sweep1 feature.

181) Click **Thread** 🪛 from the Features toolbar.

182) Click **OK**. The Thread PropertyManager is displayed.

183) Click the **bottom circular edge** of Sweep1. Edge <1> is displayed in the Edge of cylinder box.

184) Click inside the **Optional start location** box.

185) Click the **bottom face**. Face<1> is displayed.

Select End Condition.

186) Select **Up To Selection** for End Condition.

187) Click inside the **End Condition** box.

188) Click the **ending edge** as illustrated. Edge<2> is displayed in the End Condition box.

Set Thread Specification.

189) Click **Inch Die** from the drop-down menu.

Set Thread Size.

190) Click **#8-36** for size from the drop-down menu. View the Override diameter and Override pitch number.

Set the Thread method.

191) Click the **Cut thread** box.

Create a Right-hand thread.

192) Click **Right-hand thread** as illustrated.

193) Click the **Trim with start face** box.

194) Click **OK** ✔ from the Thread PropertyManager. Thread1 is displayed in the FeatureManager.

Apply material.

195) Right-click the **Material** folder in the FeatureManager.

196) Click **Edit Material**. The Material dialog box is displayed.

197) Click **Plain Carbon Steel** under the Steel category. Click **Apply**. Click **Close**.

Display an Isometric view. Save the model.

198) Click **Isometric view** 🔲 from the Heads-up View toolbar.

199) Click **Save** 💾. The HOOK part is finished.

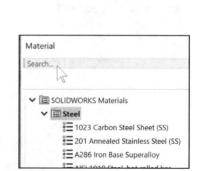

💡 Utilize the new Search feature in the Material dialog box to quickly locate the desired material.

Review the HOOK Part.

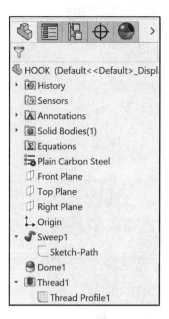

The HOOK part was created with a Swept, Dome, and Thread feature. A Swept Base feature added material by moving a profile along a path. The Swept Base feature requires two sketches (path and profile) or a sketch path and a circular profile diameter.

If the sketch profile is circular, enter the diameter in the Swept PropertyManager. For non-circular sketch profiles, create the sketch on a perpendicular plane to the path and use the pierce relation to locate the profile on the path.

The Dome feature created a spherical face on the end of the Swept Base feature.

The Thread feature provides the ability to create helical threads on cylindrical faces using profile sketches. Store custom thread profiles as library features.

The Thread tool Type and Size profiles are nominal thread profiles only. Do not use them for production quality threads. To create production quality threads, modify the nominal profiles to meet your design requirements.

Use the Stud Wizard to create external threaded stud features. This tool works similar to the Hole Wizard. You define the stud parameters then position the studs on the model. You can also apply thread parameters to existing circular studs.

If you modify a document property from an Overall drafting standard, a modify message is displayed as illustrated.

Overall drafting standard
ANSI-MODIFIED
Derived from: ANSI

When you create a new part or assembly, the three default Planes (Front, Right and Top) are aligned with specific views. The Plane you select for the Base sketch determines the orientation of the part.

WHEEL Part

The WHEEL part is a machined part.

Create the WHEEL part with the Extruded Boss/Base feature tool. Utilize the Mid Plane option to center the WHEEL on the Front Plane.

Utilize the Revolved Cut feature tool to remove material from the WHEEL and to create a groove for a belt.

The WHEEL contains a complex pattern of holes. Apply the Extruded Cut feature tool.

Simplify the geometry by dividing the four holes into two Extruded Cut features.

The first Extruded Cut feature contains two small circles sketched on two bolt circles. The bolt circles utilize Construction geometry.

The second Extruded Cut feature utilizes two small circles sketched on two bolt circles. The bolt circles utilize Construction geometry.

Utilize the Circular Pattern Feature tool. The two Extruded Cut features are contained in the Circular Pattern. Revolve the Extruded Cut features about the Temporary Axis located at the center of the Hexagon.

Create a Reference Axis. The Reference Axis is utilized in the WHEEL-AXLE assembly.

Construction geometry is used only to assist in creating the sketch entities and geometry that are ultimately incorporated into the part. Construction geometry is ignored when the sketch is used to create a feature. Construction geometry uses the same line style as centerlines.

💡 You can utilize the Hole Wizard feature tool instead of the Cut-Extrude feature tool, or use the Instant3D tool to create a Through All hole for any part. See SOLIDWORKS Help for additional information.

💡 Slots are available in the Hole Wizard. Create regular slots as well and counterbore and countersink slots. You also have options for position and orientation of the slot. If you have hardware already mated in place, the mates will not be broken if you switch from a hole to a slot.

Activity: WHEEL Part

Create the new part.

200) Click **New** from the Main menu.

201) Click the **SW-TUTORIAL 2023\MY-TEMPLATES** tab. Additional templates are displayed.

202) Double-click **PART-ANSI-IN**.

Save the part. Enter part name. Enter part description.

203) Click **Save As** .

204) Select the **SW-TUTORIAL 2023** folder.

205) Enter **WHEEL** for File name.

206) Enter **WHEEL** for Description.

207) Click **Save**. The WHEEL FeatureManager is displayed.

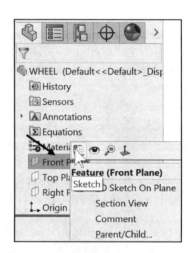

Insert the sketch for the Extruded Base feature.
208) Right-click **Front Plane** from the FeatureManager.

209) Click **Sketch** from the Context toolbar. The Sketch toolbar is displayed.

210) Click the **Circle** Sketch tool. The Circle PropertyManager is displayed.

211) Click the **Origin** as illustrated.

212) Click a **position** to the right of the Origin.

Insert a polygon.
213) Click the **Polygon** Sketch tool. The Polygon PropertyManager is displayed.

214) Click the **Origin** .

215) Drag and click the **mouse pointer** horizontally to the right of the Origin to create the hexagon as illustrated.

216) Click **OK** from the Polygon PropertyManager.

De-select the Polygon Sketch tool.
217) Right-click **Select**.

Add a Horizontal relation.

218) Click the **Origin** .

219) Hold the **Ctrl** key down.

220) Click the **right point** of the hexagon.

221) Release the **Ctrl** key.

222) Click **Horizontal** ▬ from the Add
Relations box.

223) Click **OK** ✔ from the Properties
PropertyManager.

Add dimensions.

224) Click the **Smart Dimension** Sketch
tool. Click the **circumference** of the large
circle.

225) Click a **position** above the circle. Enter
3.000in, **[76.20]**.

226) Click the **Green Check mark** ✔.

227) Click the **circumference** of the inscribed circle for
the Hexagon.

228) Click a **position** above the Hexagon.

229) Enter **.438**in, **[11.13]**.

230) Click the **Green Check mark** ✔.

Activity: WHEEL Part - Extruded Boss/Base Feature

Insert an Extruded Boss/Base feature.

231) Click **Extruded Boss/Base** from the Features toolbar.
The Boss-Extrude PropertyManager is displayed.

232) Select **Mid Plane** for End Condition in Direction 1.

233) Enter **.250**in, **[6.35]** for Depth.

234) Click **OK** ✔ from the Boss-
Extrude PropertyManager. Boss-
Extrude1 is displayed in the
FeatureManager.

Fit the model to the Graphics window.
235) Press the **f** key.

Display Hidden Lines Removed. Save the WHEEL part.
236) Display **Hidden Lines Removed**.

237) Click **Save** .

Activity: WHEEL Part - Revolved Cut Feature

Insert a new sketch for the Revolved Cut feature.
238) Right-click **Right Plane** from the FeatureManager.

239) Click **Sketch** from the Context toolbar. The Sketch toolbar is displayed.

240) Click **Right view** from the Heads-up View toolbar.

Sketch the axis of revolution.
241) Click the **Centerline** Sketch tool from the Consolidated Sketch toolbar. The Insert Line PropertyManager is displayed.

242) Click the **Origin**.

243) Click a **position** horizontally to the right of the Origin as illustrated.

De-select the sketch tool.
244) Right-click **Select**.

245) Zoom in on the top edge.

Sketch the profile.
246) Click the **Line** Sketch tool.

247) Sketch the **first vertical line** as illustrated.

248) Click the **Tangent Arc** Sketch tool. The Arc PropertyManager is displayed.

249) Click the **end point** of the vertical line.

250) Sketch a **180° arc** as illustrated.

De-select the sketch tool.
251) Right-click **Select** in the Graphics window.

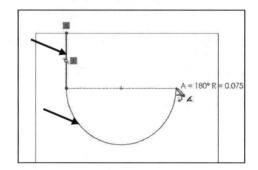

252) Click the **Line** ✎ Sketch tool.

253) Sketch the **second vertical line** as illustrated. The end point of the line is Coincident with the top horizontal edge of Extrude1.

254) Sketch a **horizontal line** collinear with the top edge to close the profile.

Add a Vertical relation.
255) Right-click **Select** in the Graphics window.

256) Click the **Origin** ↳ from the FeatureManager.

257) Hold the **Ctrl** key down.

258) Click the **center point** of the arc. Release the **Ctrl** key.

Center point

259) Click **Vertical** | from the Add Relations box.

Add an Equal relation.
260) Click the **left vertical** line.

261) Hold the **Ctrl** key down.

262) Click the **right vertical** line.

263) Release the **Ctrl** key.

264) Click **Equal** from the Add Relations box.

Add dimensions.
265) Click the **Smart Dimension** ✎ Sketch tool.

266) Click the **arc**.

267) Click a position to the **left** of the profile.

[1.60]
R.063

268) Enter **.063**in, [**1.6**]. Click the **Green Check mark** ✔.

269) Click the **right vertical** line.

270) Click a position to the **right** of the profile.

[1.98]
.078

271) Enter **.078**in, [**1.98**]. Click the **Green Check mark** ✔. The sketch should be fully defined.

[1.60]
R.063

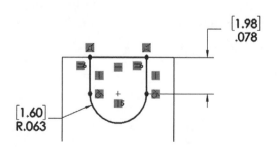

Fit the model to the Graphics window.
272) Press the **f** key.

De-select the sketch tool.
273) Right-click **Select** in the Graphics window.

Activity: WHEEL Part - Revolved Cut Feature

Insert a Revolved Cut feature.
274) Select the Axis of Revolution. Click the **centerline**
  in the Graphics window as illustrated.

275) Click **Revolved Cut** from the Features toolbar.
  The Cut-Revolve PropertyManager is displayed.
  The Cut-Revolve PropertyManager displays 360
  degrees for Direction 1 Angle.

276) Click **OK** from the Cut-Revolve
  PropertyManager. Cut-Revolve1 is displayed in the
  FeatureManager.

Save the WHEEL part.
277) Click **Save**.

Four bolt circles, spaced 0.5in, [12.7] apart, locate
the 8 - ∅.190, [4.83] holes. Simplify the situation.
Utilize two Extruded Cut features on each bolt
circle.

Position the first Extruded Cut feature hole on the
first bolt circle and third bolt circle.

Position the second Extruded Cut feature hole on
the second bolt circle and fourth bolt circle.

Activity: WHEEL Part - First Extruded Cut Feature

Display the Top Plane.
278) Right-click **Top Plane** from the FeatureManager.

279) Click **Show** 👁 from the Context toolbar.

Display a Front view - Hidden Lines Visible.
280) Click **Front view** 📦 from the Heads-up View toolbar.

281) Click **Hidden Lines Visible** ⬚ from the Heads-up View toolbar.

Insert a new sketch for the first Extruded Cut feature.
282) Right-click the **Boss-Extrude1 front face** as illustrated.

283) Click **Sketch** ▣ from the Context toolbar. The Sketch toolbar is displayed.

Create the first construction bolt circle.
284) Click the **Circle** ⊙ Sketch tool. The Circle PropertyManager is displayed.

285) Click the **Origin** ↳.

286) Click a **position** to the right of the hexagon as illustrated.

287) Check the **For construction** box.

Create the second construction bolt circle.

288) Click the **Origin** ↳.

289) Click a **position** to the right of the first construction bolt circle as illustrated.

290) Check the **For construction** box. The two bolt circles are displayed with Construction style lines.

💡 Construction geometry is used only to assist in creating the sketch entities and geometry that are ultimately incorporated into the part. Construction geometry is ignored when the sketch is used to create a feature. Construction geometry uses the same line style as centerlines.

De-select the circle Sketch tool.
291) Right-click **Select**.

Insert a centerline.

292) Click the **Centerline** Sketch tool. The Insert Line PropertyManager is displayed.

293) Sketch a **45° centerline** (approximately) from the Origin to the second bolt circle as illustrated.

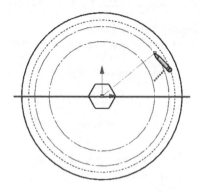

Sketch the two circle profiles.

294) Click the **Circle** ⊙ Sketch tool. The Circle PropertyManager is displayed.

295) Sketch a **circle** at the intersection of the centerline and the first bolt circle.

296) Sketch a **circle** at the intersection of the centerline and the second bolt circle.

De-select the Circle Sketch tool.

297) Right-click **Select** in the Graphics window.

Note: An Intersection relation is created between three entities: the center point of the small circle, the centerline, and the bolt circle.

Add an Equal relation.

298) Click the **first circle**.

299) Hold the **Ctrl** key down.

300) Click the **second circle**.

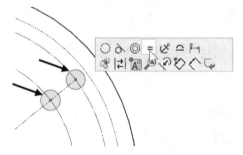

301) Release the **Ctrl** key.

302) Right-click **Make Equal** = from the Context toolbar.

Add dimensions.

303) Click the **Smart Dimension** Sketch tool.

304) Click the **first construction** circle.

305) Click a **position** above the profile.

306) Enter **1.000**in, [**25.4**].

307) Click the **Green Check mark** ✔.

[25.40]
⌀1.000

308) Click the **second construction** circle.

309) Click a **position** above the profile.

310) Enter **2.000**in, [**50.80**].

311) Click the **Green Check mark** .

312) Click the **second small** circle.

313) Click a **position** above the profile.

314) Enter **.190**in, [**4.83**].

315) Click the **Green Check mark** .

316) Click **Top Plane** from the fly-out FeatureManager.

317) Click the **45° centerline**.

318) Click a **position** between the two lines.

319) Enter **45**deg for angle.

320) Click the **Green Check mark** .

Note: If the sketch is not fully defined, you may need to add an Intersection relation between the center point of the small circle, the centerline, and the bolt circle.

Insert an Extruded Cut feature.

321) Click **Extruded Cut** from the Features toolbar. The Cut-Extrude PropertyManager is displayed.

322) Select **Through All** for the End Condition in Direction 1.

323) Click **OK** from the Cut-Extrude PropertyManager. Cut-Extrude1 is displayed in the FeatureManager.

Activity: WHEEL Part - Second Extruded Cut Feature

Insert a new sketch for the second Extruded Cut feature.
324) Right-click the **Boss-Extrude1** front face.

325) Click **Sketch** from the Context toolbar.

Sketch two additional Construction line bolt circles, 1.500in, [38.1] and 2.500in, [63.5]. Create the first Construction bolt circle.

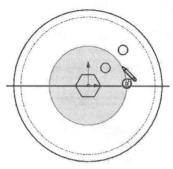

326) Click the **Circle** ⌖ Sketch tool. The Circle PropertyManager is displayed.

327) Click the **Origin** ↳.

328) Click a **position** between the two small circles.

329) Check the **For construction** box.

Create the second additional construction bolt circle.

330) Click the **Origin** ↳.

331) Click a **position** to the right of the large construction bolt circle as illustrated.

332) Check the **For construction** box from the Circle PropertyManager. The two bolt circles are displayed with the two construction lines.

Insert a centerline.

333) Click the **Centerline** Sketch tool. The Insert Line PropertyManager is displayed.

334) Sketch a **22.5° centerline** to the right from the Origin to the second bolt circle as illustrated.

335) Select **.1** from the Unit Precision box.

Display Hidden Lines Removed.

336) Click **Hidden Lines Removed** from the Heads-up View toolbar.

Sketch the two circle profiles.

337) Click the **Circle** Sketch tool. The Circle PropertyManager is displayed.

338) Sketch a **circle** at the intersection of the centerline and the first bolt circle.

339) Sketch a **circle** at the intersection of the centerline and the second bolt circle.

De-select the Circle Sketch tool.

340) Right-click **Select**.

Add an Equal relation.

341) Click the **first circle**.

342) Hold the **Ctrl** key down.

343) Click the **second circle**.

344) Release the **Ctrl** key.

345) Right-click **Make Equal** $=$ from the shortcut toolbar.

Add dimensions.

346) Click the **Smart Dimension** Sketch tool. The Smart Dimension icon is displayed.

347) Click the **first construction** circle.

348) Click a **position** above the profile.

349) Enter **1.500**in, [**38.1**].

350) Click the **second construction** circle.

351) Click a **position** above the profile.

352) Enter **2.500**in, [**63.5**].

353) Click the **small circle** as illustrated.

354) Click a **position** above the profile.

355) Enter **.190**in, [**4.83**]. The sketch should be fully defined.

Insert an Extruded Cut feature.

356) Click **Extruded Cut** 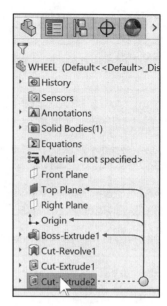 from the Features toolbar. The Cut-Extrude PropertyManager is displayed.

357) Select **Through All** for End Condition in Direction 1.

358) Click **OK** ✔ from the Cut-Extrude PropertyManager. Cut-Extrude2 is displayed in the FeatureManager.

Save the model.

359) Click **Save** 🖫.

View the Temporary Axes.

360) Click **View**, **Hide/Show**, check **Temporary Axes** from the Main menu.

💡 The book is designed to expose the new SOLIDWORKS user to many different tools, techniques and procedures. It may not always use the most direct tool or process.

Activity: WHEEL Part - Circular Pattern Feature

Insert a Circular Pattern.

361) Click **Isometric view** 🔲 from the Heads-up View toolbar.

362) Click **Circular Pattern** 🔄 from the Consolidated Features toolbar. The Circular Pattern PropertyManager is displayed.

363) Click **inside** the Pattern Axis box.

364) Click the **Temporary Axis** in the Graphics window at the center of the Hexagon. Axis<1> is displayed in the Pattern Axis box.

365) Click the **Equal spacing** box.

366) Enter **360**deg for Angle.

367) Enter **8** for Number of Instances.

368) Click inside the **Features to Pattern** box.

369) Click **Cut-Extrude1** and **Cut-Extrude2** from the fly-out FeatureManager. Cut-Extrude1 and Cut-Extrude2 are displayed in the Features to Pattern box.

370) Check the **Geometry pattern** box.

371) Click **OK** ✔ from the Circular Pattern PropertyManager. CirPattern1 is displayed in the FeatureManager.

Save the WHEEL part.

372) Click **Save** 🖫.

Utilize a Reference Axis to locate the WHEEL in the PNEUMATIC-TEST-MODULE assembly. The Reference Axis is located in the FeatureManager and Graphics window. The Reference Axis is a construction axis defined between two planes.

Insert a two Plane Reference axis.

373) Click the **Axis** 🖊 tool from the Reference Geometry Consolidated Features toolbar. The Axis PropertyManager is displayed.

374) Click **Top Plane** from the fly-out FeatureManager.

375) Click **Right Plane** from the fly-out FeatureManager. The selected planes are displayed in the Selections box.

376) Click **Two Planes**.

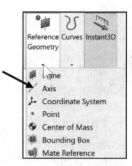

377) Click **OK** ✔ from the Axis PropertyManager. Axis1 is displayed in the FeatureManager.

Axis1 is positioned through the Hex Cut centered at the Origin ⌊.

378) Click and drag the **Axis1 handles** outward to extend the length on both sides as illustrated.

Display an Isometric view - Shaded With Edges. Clear Temporary Axes. Hide all Planes.

379) Click **Isometric view** .

380) Click **View**, **Hide/Show**, un-check **Temporary Axes** from the Main menu. **Hide** all Planes.

381) Click **Shaded With Edges** from the Heads-up View toolbar.

Save the WHEEL part.

382) Click **Save** .

Sketched lines, arcs or circles are modified from profile geometry to construction geometry. Select the geometry in the sketch. Check the For construction box option.

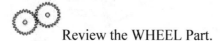 Review the WHEEL Part.

The WHEEL part was created with the Extruded Boss/Base feature. You sketched a circular sketch on the Front Plane and extruded the sketch with the Mid Plane option.

A Revolved Cut feature removed material from the WHEEL and created the groove. The Revolved Cut feature utilized an arc sketched on the Right Plane. A sketched centerline was required to create the Revolved Cut feature.

The WHEEL contained a complex pattern of holes. The first Extruded Cut feature contained two small circles sketched on two bolt circles. The bolt circles utilized construction geometry. Geometric relationships and dimensions were used in the sketch. The second Extruded Cut feature utilized two small circles sketched on two bolt circles.

The two Extruded Cut features were contained in one Circular Pattern and revolved about the Temporary Axis. The Reference Axis was created with two perpendicular planes. Utilize the Reference Axis, Axis1 in the WHEEL-AXLE assembly.

Modify a Part

Conserve design time and cost. Modify existing parts and assemblies to create new parts and assemblies. Utilize the Save as copy tool to avoid updating the existing assemblies with new file names.

The HEX-STANDOFF part was created in a previous chapter. The HEX-ADAPTER is required to fasten the WHEEL to the AXLE. Start with the HEX-STANDOFF part. Note: It is also located in the Chapter 5 Homework folder.

Utilize the Save As command and enter the HEX-ADAPTER for the new file name. Check the Save as copy and continue check box. The HEX-ADAPTER is the new part name. Open the HEX-ADAPTER. Modify the dimensions of the Extruded

Base feature. Utilize Edit Definition to modify the Hole Wizard Tap Hole to a Standard Hole. Insert an Extruded Boss/Base feature to create the head of the HEX-ADAPTER.

Insert an Extruded Cut feature. Sketch a circle on the Right Plane. Extrude the circle in Direction1 and Direction2 with the Through All End Condition option. Note: You can use the Hole Wizard feature with a 3D Sketch. Feature order determines the internal geometry of the Hole. If the Hole feature is created before the Extrude2-Head feature, the Through All End Condition will extend through the Boss-Extrude1 feature.

If the Hole feature is created after the Extrude2-Head feature, the Through All End Condition will extend through the Boss-Extrude1 feature and the Extrude2-Head feature.

Modify feature order by dragging feature names in the FeatureManager. Utilize the Save As command to create the AXLE3000 part from the AXLE part.

Utilize the Save As command to create the SHAFTCOLLAR-500 part from the SHAFT-COLLAR part. Save the HEX-STANDOFF as the HEX-ADAPTER part.

Activity: HEX-ADAPTER Part

Create the HEX-ADAPTER.

383) Click **Open** from the Main menu.

384) Double-click **HEX-STANDOFF**. The HEX-STANDOFF FeatureManager is displayed. Note: HEX-STANDOFF was created in a previous chapter. It is also located in the Chapter 5 Homework folder.

385) Click **Save As** .

386) Select the **SW-TUTORIAL 2023** folder.

387) Enter **HEX-ADAPTER** for File name.

388) Enter **HEX-ADAPTER 10-24** for Description.

389) Check the **Save as copy and continue** box.

390) Click **Save**.

391) Click **File**, **Close** from the Main menu.

Open the HEX-ADAPTER.

392) Click **Open** 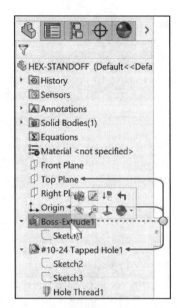 from the Main menu.

393) Double-click **HEX-ADAPTER**. The HEX-ADAPTER FeatureManager is displayed.

Modify the Boss-Extrude1 dimensions.
394) Double-click **Boss-Extrude1** from the FeatureManager.

395) Move the **dimensions** off the model.

396) Double-click **.735**in, **[18.67]**.

397) Enter **.700**in, **[17.78]** for depth.

398) Double-click **.313**in, **[7.95]**.

399) Enter **.438**in, **[11.13]** for diameter.

Modify the #10-24 Tapped Hole1 feature.
400) Right-click **#10-24 Tapped Hole1** from the FeatureManager.

401) Click **Edit Feature** from the Context toolbar. The Hole Specification PropertyManager is displayed.

The Hole Specification PropertyManager is part of the Hole Wizard tool located in the Features toolbar.

The Type tab is selected by default.

402) Select the **Hole type** from the Hole Specification box.

403) Select **ANSI Inch** for Standard.

404) Select **Tap Drills** for Type.

405) Select **#10-24** for Size.

406) Select **Through All** for End Condition.

407) Click **OK** ✔ from the Hole Specification PropertyManager. The Tap Hole is modified.

Insert a sketch for the Extruded Boss feature.
408) **Rotate** the model to view the bottom face.

409) Right-click the **bottom hexagonal face** of the Boss-Extrude1 feature as illustrated.

410) Click **Sketch** from the Context toolbar.

411) Click **Bottom view** from the Heads-up View toolbar.

412) Click the **Circle** Sketch tool. The Circle PropertyManager is displayed.

413) Click the **Origin** as illustrated.

414) Click a **position** in the Graphics window to the right of the Origin.

Add a dimension.
415) Click the **Smart Dimension** Sketch tool.

416) Click the **circumference** of the circle.

417) Click a **position** above the circle to locate the dimension.

418) Enter **.625**in, **[15.88]** in the Modify dialog box.

Fit the model to the Graphics window.
419) Press the **f** key.

$[15.88]$
$\emptyset.625$

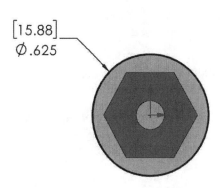

Activity: HEX-ADAPTER Part - Extruded Boss/Base Feature

Extrude the sketch to create the Extruded Boss/Base feature.

420) Click **Isometric view** from the Heads-up View toolbar.

421) Click **Extruded Boss/Base** from the Features toolbar. The Boss-Extrude PropertyManager is displayed.

422) Enter **.200**in, **[6.35]** for Depth. The Direction arrow points downward. Flip the **Direction arrow** if required.

423) Click **OK** from the Boss-Extrude PropertyManager. Boss-Extrude2 is displayed in the FeatureManager.

Rename Boss-Extrude2. Save the model.
424) Rename **Boss-Extrude2** to **Extrude2-Head**.

425) Click **Save**.

Rename a feature or sketch for clarity. Slowly click the feature or sketch name twice and enter the new name when the old one is highlighted.

Activity: HEX-ADAPTER Part - Extruded Cut Feature

Insert a new sketch for the Extruded Cut on the Right Plane.
426) Right-click **Right Plane** from the FeatureManager.

427) Click **Sketch** from the Context toolbar.

428) Click **Right view** from the Heads-up View toolbar. Note the location of the Origin.

429) Click **Hidden Lines Visible**.

430) Click the **Circle** Sketch tool. The Circle PropertyManager is displayed.

431) Sketch a **circle** below the Origin. The center point is vertically aligned to the Origin (vertical relation).

432) If required, add a **Vertical relation** between the center point of the circle and the Origin.

Add dimensions.

433) Click the **Smart Dimension** Sketch tool.

434) Click the **middle horizontal** edge.

435) Click the **center point** of the circle.

436) Click a **position** to the right of the profile.

437) Enter **.100**in, [**2.54**].

438) Click the **Green Check mark** .

439) Click the **circumference** of the circle.

440) Click a **position** below the profile.

441) Enter .**120**in, [**3.95**].

442) Click the **Green Check mark** .

Insert an Extruded Cut feature.

443) Click **Extruded Cut** from the Features toolbar. The Cut-Extrude PropertyManager is displayed.

444) Select **Through All - Both** for End Condition in Direction 1.

445) Click **OK** from the Cut-Extrude PropertyManager.

Display an Isometric view. Rename the feature.

446) Click **Isometric view** .

447) Rename the Cut-Extrude# feature to **Extrude3-SetScrew**.

Save the HEX-ADAPTER part.

448) Click **Save** .

The Through All End Condition is required to penetrate both the Boss-Extrude1 and Extrude2 features. Reorder features in the FeatureManager. Position the Extrude2 feature before the Tap Drill for # 10-24 Tap 1 feature in the FeatureManager.

Reorder the Features.
449) Click and drag **Extrude2-Head** from the FeatureManager upward as illustrated.

450) Click a **position** below Boss-Extrude1. The Through All End Condition option for the Tap Drill for # 10-24 Tap 1 feature creates a hole through both Boss-Extrude1 and Boss-Extrude2.

Display a Section view.
451) Click **Front Plane** from the FeatureManager.

452) Click **Section view** from the Heads-up View toolbar in the Graphics window. The Section View PropertyManager is displayed. View the results.

453) Click **OK** from the Section View PropertyManager.

Display the full view.
454) Click **Section view** from the Heads-up View toolbar in the Graphics window.

455) Click **Shaded With Edges** from the Heads-up View toolbar.

Save the HEX-ADAPTER.
456) Click **Save**. Note the location of the Origin in the model.

Origin

Close all documents.
457) Click **Windows**, **Close All** from the Main menu.

💡 Utilize the Save As command and work on the copied version of the document before making any changes to the original. Keep the original document intact.

💡 Tangent edges and Origins are displayed for educational purposes in this book.

Review the HEX-ADAPTER Part.

The HEX-ADAPTER part was created by utilizing the Save As and the Save as copy command with the HEX-STANDOFF part. The Boss-Extrude1 feature dimensions were modified. Edit Definition was utilized to modify the Hole type from the Hole Wizard feature.

An Extruded Boss feature added material. An Extruded Cut feature, sketched on the Right Plane with the Through All End Condition for both Direction1 and Direction2, created a hole through the Extruded Boss feature. Reordering features in the FeatureManager modified the Hole. Utilizing existing geometry saved time with the Save as copy command. The original part and its references to other assemblies are not affected with the Save as copy command.

Utilize Design Table configurations for the AXLE part and SHAFT-COLLAR part developed in the previous chapter.

The AXLE-3000 part and SHAFT-COLLAR-500 part utilize the Save As option in the next section.

Utilize the Save As components or the configurations developed with Design Tables to create the WHEEL-AXLE assembly.

Activity: AXLE-3000 Part

Create the AXLE-3000 part from the AXLE part.

458) Click **Open** from the Main menu.

459) Double-click **AXLE** from the SW-TUTORIAL 2023 folder. **Note:** You created this part in Chapter 2. If you did not create the AXLE, open the AXLE part from the SW-TUTORIAL 2023\Homework 5 folder.

460) Click **Save As** .

461) Select the **SW-TUTORIAL 2023** folder.

462) Enter **AXLE-3000** for File name.

463) Enter **AXLE-3000** for Description.

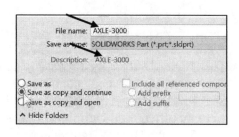

464) Check the **Save as copy and continue** check box. Note: You can use the Save as copy and open command. This will open the new copy.

465) Click **Save**.

Close the AXLE part.
466) Click **File**, **Close** from the Main menu.

Open AXLE-3000 part.

467) Click **Open** from the Main menu.

468) Double-click **AXLE-3000** from the SW-TUTORIAL 2023 folder. The AXLE-3000 FeatureManager is displayed.

Modify the depth dimension.
469) Double-click the **cylindrical face** in the Graphics window.

470) **Move** the dimensions off the model.

471) Click **1.375**in, [**34.93**].

472) Enter **3.000**in, [**76.20**].

Fit the model to the Graphics window.
473) Press the **f** key.

Save the AXLE-3000 part.

474) Click **Save** .

475) Click **inside** the Graphics window.

Activity: SHAFTCOLLAR-500 Part

Create the SHAFTCOLLAR-500 part.

476) Click **Open** from the Main menu.

477) Double-click **SHAFT-COLLAR** from the SW-TUTORIAL 2023 folder. **Note:** You created the SHAFT-COLLAR part in Chapter 2. If you did not create the SHAFT-COLLAR part, open the SHAFT-COLLAR part from the SW-TUTORIAL 2023\Homework 5 folder.

478) Click **Save As** .

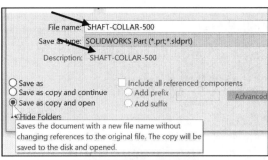

479) Enter **SHAFT-COLLAR-500** for File name.

480) Enter **SHAFT-COLLAR-500** for Description.

481) Click the **Save as copy and open** check box.

482) Click **Save**. Both models are open at this time.

Modify the diameter dimensions of the SHAFT-COLLAR-500 Part.

483) Right-click **Annotations** in the FeatureManager.

484) Check the **Show Feature Dimensions** box.

485) Press the **f** key to fit the model to the graphics area.

486) Click **.438**in, [**11.11**].

487) Enter **.750**in, [**19.05**] for outside diameter.

488) Click **.190**in, [**4.83**].

489) Enter **.500**in, [**12.70**] for inside diameter. View the results.

490) Right-click **Annotations** in the FeatureManager.

491) Un-check the **Show Feature Dimensions** box.

492) **Rebuild** the model.

Fit the model to the Graphics window.
493) Press the **f** key.

Save the SHAFT-COLLAR-500 part.

494) Click **Save** .

Close all documents.
495) Click **Windows**, **Close All** from the Main
menu.

Press the **s** key in the Graphics window.
A Context Pop-up features toolbar is
displayed. The features toolbar displays the
last few feature tools applied.

Select the types of Annotations that you want
to display and set text scale and other Annotations
options. In the FeatureManager design tree, right-
click the **Annotations** folder, and click **details**.
View the options from the Annotation Properties
dialog box.

Rename a feature or sketch for clarity.
Slowly click the feature or sketch name twice and
enter the new name when the old one is highlighted.

Add relations, then dimensions. This will keep
the user from having too many unnecessary
dimensions. This helps to show the design intent of
the model. Dimension what geometry you intend to
modify or adjust.

Chapter Summary

In this chapter, you created six parts. The WEIGHT part utilized the Plane feature, Lofted Base feature and the Extruded Cut (Instant3D tool) feature. The HOOK part utilized the Swept Base feature, Dome feature, and Thread feature. The WHEEL part utilized the Extruded Base feature, Revolved Cut feature, Extruded Cut feature, Circular Pattern feature and Axis feature.

The second three parts utilized existing parts created in an early chapter. The HEX-ADAPTER part, AXLE-3000 part and the SHAFTCOLLAR-500 part utilized existing part geometry along with the Hole Wizard feature.

Conserve design time and cost. Modify existing parts and assemblies to create new parts and assemblies. Utilize the Save As/Save as command to save the file in another file format.

Utilize the Save as copy and continue command to save the document to a new file name without replacing the active document. Utilize the Save as copy and open command to save the document to a new file name that becomes the active document. The original document remains open. References to the original document are not automatically assigned to the copy.

You applied the following Sketch tools in this chapter: Circle, Line, Centerline, Tangent Arc, Polygon, Smart Dimension, Center Rectangle, Centerpoint Arc, 3 Point Arc and Convert Entities.

The book is designed to expose the new SOLIDWORKS user to many different tools, techniques and procedures. It may not always use the most direct tool or process.

Questions

1. What is the minimum number of profiles and planes required for a Loft feature?

2. A Swept Boss/Base feature requires a _____ and a _____.

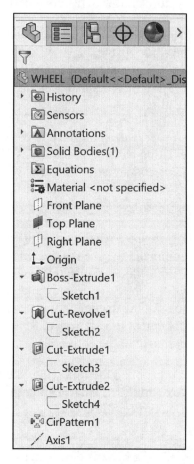

3. Describe the differences between a Loft feature and a Swept feature.

4. Identify the three default reference planes in an assembly.

5. True or False. A Revolved-Cut feature requires an axis of revolution.

6. True or False. A Circular Pattern Feature does not require a seed feature to pattern.

7. Describe the difference between a Swept Boss/Base feature and a Swept Cut feature.

8. Identify the type of Geometric relations that can be added to a sketch.

9. What function does the Save as option perform when saving a part under a new name?

10. True or False. Never reuse geometry from one part to create another part.

Exercises

Exercise 5.1: Advanced Part

Build the illustrated model. Calculate the volume of the part and locate the Center of mass with the provided information.

Three holes are displayed with an Ø1.00in.

Set the document properties for the model.

Tangent edges and Origin are displayed for educational purposes.

Given:
A = Ø1.00
All Thru Holes
Material: Brass
Density = .307 lb/in^3
Units: IPS
Decimal Place: 2

Exercise 5.2: Advanced Part

Build the illustrated model. Calculate the overall mass of the part and locate the Center of mass with the provided information. Insert a Revolved Base feature and Extruded Cut feature to build this part.

Note: Select the Front Plane as the Sketch plane. Apply the Centerline Sketch tool for the Revolve1 feature. Insert the required geometric relations and dimensions. Sketch1 is the profile for the Revolve1 feature.

Given:
A = Ø12
Material: Cast Alloy Steel
Density = .0073 g/mm^3
Units: MMGS
Decimal Place: 2

Exercise 5.3: Advanced Part

Build the illustrated model. Calculate the overall mass of the part and locate the Center of mass with the provided information. Insert the required geometric relations and dimensions. Decimal Place: 2.

Insert two features: Extruded Base and Revolved Boss.

Given:
A = 60, B = 40, C = 8
Material: Cast Alloy Steel
Density = .0073 g/mm^3
Units: MMGS

Exercise 5.4: Advanced Part

Build the illustrated model. Calculate the overall mass of the part and locate the Center of mass with the provided information.

Think about the various features that create the model. Insert the required geometric relations and dimensions.

Apply symmetry. Create the left half of the model first, and then apply the Mirror feature. Decimal Place: 2.

Thru ALL holes.

Tangent Edges and Origin are displayed for educational purposes.

Given:
A = 76, B = 127
Material: 2014 Alloy
Density: .0028 g/mm^3
Units: MMGS
ALL ROUNDS
EQUAL 6MM

Exercise 5.5: Advanced Part

Build the illustrated model.
Calculate the overall mass of
the part and locate the Center
of mass with the provided
information. Decimal
Place: 2. Thru ALL holes.

Think about the various
features that create the part.

Apply reference
construction planes to build
the circular features. Insert
the required geometric
relations and dimensions.

Given:
A = 52, B = 58
ALL-ROUNDS R 4MM
Material: 6061 Alloy
Density: .0027 g/mm^3
Units: MMGS

Exercise 5.6: Advanced Part

Build the illustrated model. Calculate the
volume of the part and locate the Center of
mass with the provided information.
Decimal Place: 2. Thru ALL holes.

Think about the various features that create
this model. Insert the required geometric
relations and dimensions.

Given:
A = Ø3.00, B = 1.00
Material: 6061 Alloy
Density: .097 lb/in^3
Units: IPS
Decimal Place: 2

Exercise 5.7: Advanced Part

Build the illustrated model.

Calculate the overall mass and locate the Center of mass of the illustrated model.

Think about the steps that you would take to build the illustrated part.

Identify the location of the part Origin.

Review the provided dimensions and annotations in the part illustration.

Decimal Place: 2. Thru ALL holes.

 Tangent edges and Origin are displayed for educational purposes.

Use Symmetry. When possible and if it makes sense, model objects symmetrically about the origin.

Given:
A = 2.00, B = Ø.35
Material: 1060 Alloy
Density: 0.097 lb/in^3
Units: IPS
Decimal Place: 2

Exercise 5.8: Traditional Ice cream cone

Create a traditional Ice Cream Cone as illustrated.

Create an ANSI - IPS model.

Think about the design features that create this model. Why does the cone use ribs?

Ribs are used for structural integrity.

View the sample FeatureManager. Your FeatureManager can (should) be different. This is just one way to create this part.

Create your own ice cream cone design.

Below are a few sample models from my Freshman Engineering class.

Below are a few sample models from my Freshman Engineering (Cont:).

Exercise 5.9: Gem® Paper clip

Create a simple paper clip. Create an ANSI - IPS model.

Apply material to the model (Plain Carbon Steel). Decimal Place: 2.

What is your Base Sketch?

The paper clip model uses (lines and arcs) as the path.

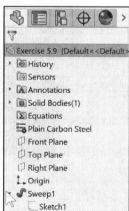

Exercise 5.10: Spring

Create a variable pitch spring (ANSI - IPS) with 6 coils; two active as illustrated.

Sketch a circle, Coincident to the Origin on the Top plane with a .235in dimension.

Create the Helix/Spiral feature.

Create a Region parameters table.

Enter the following information as illustrated. Coils 1, 2, 5, 6 & 7 are the closed ends of the spring. The pitch needs to be slightly larger than the wire.

Enter .021in for the Pitch.

Enter .080in for the free state of the two active coils.

Enter Start angle of 0deg.

Create the Swept Boss feature.

Enter .015in for Depth (Circular Profile).

Add material to the model.

View the results.

Helix/Spiral1

Defined By: Pitch and Revolution

Parameters
- ○ Constant pitch
- ◉ Variable pitch

Region parameters:

	P	Rev	H	Dia
1	0.021in	0	0in	0.235in
2	0.021in	1	0.021i	0.235in
3	0.08in	2	0.0715	0.235in
4	0.08in	3	0.1515	0.235in
5	0.021in	4	0.202i	0.235in
6	0.021in	5	0.223i	0.235in
7	0.021in	6	0.244i	0.235in
8				

☐ Reverse direction

Start angle:
0.00deg

◉ Clockwise
○ Counterclockwise

Sweep1

Profile and Path
- ○ Sketch Profile
- ◉ Circular Profile
- Helix/Spiral1
- ⊘ 0.015in

Options
- ☐ Merge tangent faces
- ☑ Show preview

Curvature Display

Path(Helix/Spiral1)

Helix/Spiral1

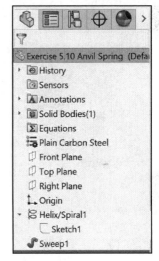

Exercise 5.10 Anvil Spring (Defa
- ▸ ⊙ History
- Sensors
- ▸ Annotations
- ▸ Solid Bodies(1)
- Σ Equations
- Plain Carbon Steel
- Front Plane
- Top Plane
- Right Plane
- Origin
- ▾ Helix/Spiral1
  - Sketch1
- Sweep1

Exercise 5.11: Bottle

Create the container as illustrated.

Create an ANSI - IPS model.

Apply material to the model.

What is your Base Sketch?

What is your Base Feature?

What are the dimensions?

View the sample FeatureManager. Your FeatureManager can (should) be different. This is just one way to create this part. Estimate any needed dimension.

Helix/Spiral1

Defined By:

Pitch and Revolution

Parameters

○ Constant pitch
● Variable pitch

Region parameters:

	P	Rev	H	Dia
1	0.1875in	0	0in	1.3in
2	0.1875in	0.25	0.0468	1.5in
3	0.1875in	2.25	0.4218	1.5in
4	0.1875in	2.5	0.4687	1.3in
5				

☑ Reverse direction

Start angle:

0.00deg

● Clockwise
○ Counterclockwise

Exercise 5.12: Explicit Equation Driven Curve tool

Create an Explicit Equation Driven Curve on the Front plane. Revolve the curve. Calculate the volume of the solid.

Create a New part. Use the default ANSI, IPS Part template.

Create a 2D Sketch on the Front Plane.

Activate the Equation Driven Curve Sketch f_x tool from the Consolidated drop-down menu.

Enter the Equation y_x as illustrated.

Enter the parameters x_1, x_2 that define the lower and upper bounds of the equation as illustrated. View the curve in the Graphics window.

Size the curve in the Graphics window. The Sketch is under defined.

Insert three lines to close the profile as illustrated. Fully define your sketch. Enter dimensions and any needed geometric relation.

Create the Revolved feature. View the results in the Graphics window. Revolve1 is displayed. Utilize the Section tool parallel with the Right plane to view how each cross section is a circle.

Apply Brass for material.

Decimal Place: 2.

Calculate the volume of the part using the Mass Properties tool. View the results.

Density = 0.31 pounds per cubic inch

Mass = 16.10 pounds

Volume = 52.43 cubic inches

Surface area = 103.20 square inches

You can create parametric (in addition to explicit) equation-driven curves in both 2D and 3D sketches.

Use regular mathematical notation and order of operations to write an equation. x_1 and x_2 are for the beginning and end of the curve. Use the transform options at the bottom of the PropertyManager to move the entire curve in x-, y- or rotation. To specify $x = f(y)$ instead of $y = f(x)$, use a 90-degree transform.

Exercise 5.13: Curve Through XYZ Points tool

The Curve Through XYZ Points $\mathcal{U}$ feature provides the ability to either type in (using the Curve File dialog box) or click Browse and import a text file with x-, y-, z-, coordinates for points on a curve.

A text file can be generated by any program which creates columns of numbers. The Curve $\mathcal{U}$ feature reacts like a default spline that is fully defined.

Create a curve using the Curve Through XYZ Points tool. Import the x-, y-, z- data.

Verify that the first and last points in the curve file are the same for a closed profile.

1. Create a new part.

2. Click the Curve Through XYZ Points $\mathcal{U}$ tool from the Features CommandManager. The Curve File dialog box is displayed.

Import the curve data.

3. Click Browse from the Curve File dialog box.

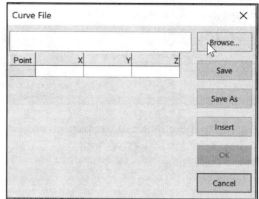

4. Browse to the SW-TUTORIAL\Chapter 5 Homework folder.

5. Set file type to Text Files.

6. Double-click cam2.text. View the data in the Curve File dialog box.

7. Click OK from the Curve File dialog box.

Fix the Curve to the Graphics window. Curve1 is displayed in the FeatureManager. You created a curve using the Curve Through XYZ Points tool with imported x-, y-, z- data from a cam program.

This curve can now be used to create a sketch (closed profile). In this case a cam.

Close the existing model.

Split Line Curve tool

The Split Line Curve tool projects an entity (sketch, solid, surface, face, plane, or surface spline) to surfaces, or curved or planar faces. It divides a selected face into multiple separate faces. You can split curves on multiple bodies with a single command. You can create the following split lines:

- **Silhouette**. Creates a split line on a cylindrical part.

- **Projection**. Project a sketch on a surface. You can pattern split line features that were created using projected curves. You can create split lines using sketched text. This is useful for creating items such as decals or applying a force or pressure when using SOLIDWORKS Simulation.

- **Intersection**. Splits faces with an intersecting solid, surface, face, plane, or surface spline.

When you create a split line with an open profile sketch, the sketch must span at least two edges of the model.

Exercise 5.15: Split Line

Create a Split Line feature to apply a distributed load to a selected face. The Split Line feature in this case does not add depth to the part.

1. Open **Split Line** from the SW-TUTORIAL\Chapter 5 Homework folder. Set units and Decimal place.

2. Create a **center rectangular sketch** on the top face as illustrated. Add all **dimensions** as illustrated. Use the sketch for the Split Line feature.

Create the Split Line feature.

3. Click the **Features** tab.

4. Click the **Split Line** tool from the Curves drop-down menu. Projection is selected by default.

5. Click the **top face** of Boss-Extrude1. Face 1 is displayed in the Faces to Split box.

6. Click **OK** ✔ from the Split Line PropertyManager. Split Line1 is created and is displayed.

7. **Save** the part.

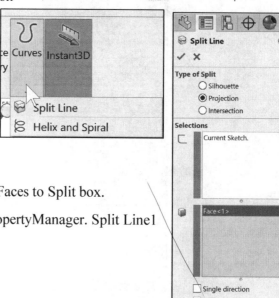

Start a SOLIDWORKS Simulation study and utilize the Split Line feature as your selected face for the applied force on the model.

8. Click **SOLIDWORKS Add-Ins tab, SOLIDWORKS Simulation**. The Simulation tab is displayed.

Create a new Static Study.

9. Click the **Simulation** tab.

10. Click **New Study** from the Study drop-down menu. Static 1 is selected by default. Accept the default study name.

11. Click **OK** ✔ from the Study PropertyManager.

Fix the model at one end and apply a distributed load using the Split Line feature.

12. Right-click the **Fixtures** folder.

13. Click **Fixed Geometry**. The Fixture PropertyManager is displayed.

14. Click the **left face of Boss-Extrude1** as illustrated. The left face is fixed.

15. Click **OK** ✔ from the Fixture PropertyManager. Fixed-1 is displayed.

Apply an External load (force) to the model using the Split Line feature.

16. Right-click the **External Load** folder.

17. Click **Force**. The Force PropertyManager is displayed. In this example, the area of the applied force is normal to the Split Line feature.

18. Click the **Split Line1** feature on the model in the Graphics window. Normal is selected by default.

19. Select **English (IPS)**.

20. Enter **1000lbf** for force.

21. Click **OK** ✔ from the Force PropertyManager. Force-1 is created and is displayed.

22. Apply **Alloy Steel** to the part.

23. **Mesh and Run** the model. Click **Yes**. Use the Large Displacement method. View the Displacement results.

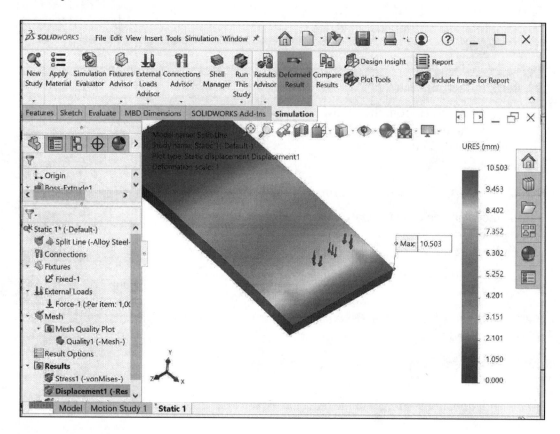

24. **Close** the part. The Split Line feature divides a selected face into multiple separate faces.

Chapter 6

PNEUMATIC-TEST-MODULE and Final ROBOT Assembly

Below are the desired outcomes and usage competencies based on the completion of Chapter 6.

Desired Outcomes:	Usage Competencies:

• Create five sub-assemblies:
   o WHEEL-AND-AXLE.
   o WHEEL-FLATBAR.
   o 3HOLE-SHAFTCOLLAR.
   o 5HOLE-SHAFTCOLLAR.
   o PNEUMATIC-TEST-MODULE.
• Create the final ROBOT assembly.
	• Reuse geometry.
	
	• Apply Standard Mate types.
	
	• Modify existing assemblies to create new assemblies.
	
	• Utilize the following Assembly tools: Mate, Linear Component Pattern, Feature Driven Component Pattern, Mirror Components, Replace Components, and Performance Evaluation.
	
	• Work with multiple documents in an assembly.

Notes:

Chapter 6 - PNEUMATIC-TEST-MODULE and ROBOT Assembly

Chapter Objective

Develop a working understanding with multiple documents in an assembly. Build on sound assembly modeling techniques that utilize symmetry, component patterns and mirrored components.

Create five sub-assemblies and the final ROBOT assembly:

- 3HOLE-SHAFTCOLLAR sub-assembly.

- 5HOLE-SHAFTCOLLAR sub-assembly.

- WHEEL-FLATBAR sub-assembly.

- WHEEL-AND-AXLE sub-assembly.

- PNEUMATIC-TEST-MODULE sub-assembly.

- ROBOT final assembly.

On the completion of this chapter, you will be able to:

- Utilize various Assembly techniques.

- Suppress and hide components.

- Create new assemblies and copy assemblies to reuse similar parts.

- Use the following Assembly tools:

  o Insert Component.

  o Standard Mates: Concentric, Coincident, and Parallel.

  o Linear Component Pattern.

  o Feature Driven Component Pattern.

  o Circular Component Pattern.

  o Mirror Components.

  o Replace Components.

  o Performance Evaluation.

Chapter Overview

Create the 3HOLE-SHAFTCOLLAR sub-assembly.

Utilize the 3HOLE-SHAFTCOLLAR sub-assembly to create the 5HOLE-SHAFTCOLLAR sub-assembly.

The WHEEL-FLATBAR sub-assembly contains the following items:

- 3HOLE-SHAFTCOLLAR assembly.

- 5HOLE-SHAFTCOLLAR assembly.

- WHEEL part.

Mate to the first component added to the assembly. If you mate to the first component or base component of the assembly and decide to change its orientation later, all the components will move with it.

Remove Tangent edges. Click Display from the Options dialog box, check the Removed box as illustrated.

🔅 Determine the static and dynamic behavior of mates in each sub-assembly before creating the top-level assembly.

The WHEEL-AND-AXLE assembly contains the following items:

- WHEEL-FLATBAR assembly.

- AXLE-3000 part.

- SHAFTCOLLAR-500 part.

- HEX-ADAPTER part.

Combine the created new assemblies and parts to develop the PNEUMATIC-TEST-MODULE assembly.

Create the final ROBOT assembly. Insert the Robot-platform assembly, PNEUMATIC-TEST-MODULE assembly, basic_integration assembly and the HEX-STANDOFF components.

All assemblies and components for the final ROBOT assembly are in the Chapter 6 Models folder. Copy the folder to your hard drive.

Tangent edges and Origins are displayed for educational purposes.

Utilize the Pack and Go option to save an assembly or drawing with references and toolbox components. The Pack and Go tool saves either to a folder or creates a zip file to e-mail. View SOLIDWORKS help for additional information.

Assembly Techniques

Assembly modeling requires practice and time. Below are a few helpful techniques to address Standard Mates. These techniques are utilized throughout the development of all assemblies.

Mating Techniques:
• Plan your assembly and sub-assemblies in an assembly layout diagram. Group components together to form smaller sub-assemblies.
• Utilize symmetry in an assembly. Utilize Mirror Component and Component Pattern to create multiple instances (copies) of components. Reuse similar components with Save as copy and configurations.
• Use the Zoom and Rotate commands to select the geometry in the Mate process. Zoom to select the correct face.
• Apply various colors to features and components to improve display.
• Activate Temporary Axes and Show Planes when required for Mates, otherwise Hide All Types from the View menu.
• Select Reference planes from the FeatureManager for complex components. Expand the FeatureManager to view the correct plane.
• Remove display complexity. Hide components when visibility is not required.
• Suppress components when Mates are not required. Group fasteners at the bottom of the FeatureManager. Suppress fasteners and their assembly patterns to save rebuild time and file size.
• Utilize Section views to select internal geometry.
• Use the Move Component and Rotate Component commands before Mating. Position the component in the correct orientation.
• Create additional flexibility in a Mate. Distance Mates are modified in configurations and animations. Rename Mates in the FeatureManager.
• Verify the position of the components. Use Top, Front, Right and Section views.

PNEUMATIC TEST MODULE Layout

The PNEUMATIC TEST MODULE assembly is comprised of four major sub-assemblies:

- LINKAGE assembly.

- RESERVOIR assembly.

- FRONT-SUPPORT assembly.

- WHEEL-AND-AXLE assembly.

There are over one hundred components in the PNEUMATIC TEST MODULE assembly. Complex assemblies require planning. The Assembly Layout diagram provides organization for a complex assembly by listing sub-assemblies and parts.

2 LINKAGE
Assembly

AIR RESERVOIR
Assembly

FRONT SUPPORT Assembly

WHEEL-AND-AXLE
Assembly

Review the Assembly Layout diagram for the PNEUMATIC TEST MODULE assembly.

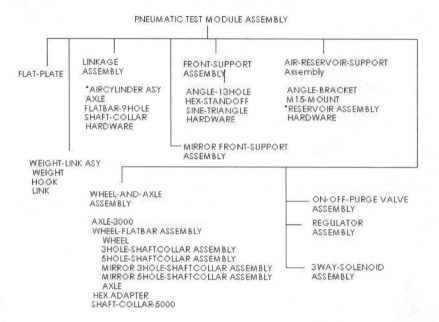

Physical space on the FLAT-PLATE is at a premium. Determine the requirements for hardware and placement after the mechanical components are assembled to the FLAT-PLATE part.

The LINKAGE assembly, FRONT-SUPPORT assembly, AIR-RESERVOIR SUPPORT assembly, and any other component or assembly are provided in the SW-TUTORIAL 2023/Chapter 6 Models folder.

The book is designed to expose the new SOLIDWORKS user to many different tools, techniques and procedures. It may not always use the most direct tool or process.

The WHEEL-FLATBAR assembly consists of the following:

- WHEEL part.

- 3HOLE-SHAFTCOLLAR assembly.

- 5HOLE-SHAFTCOLLAR assembly.

FLATBAR Sub-assembly

There are two similar sub-assemblies contained in the WHEEL-FLATBAR assembly:

- 3HOLE-SHAFTCOLLAR assembly.

- 5HOLE-SHAFTCOLLAR assembly.

Create the 3HOLE-SHAFTCOLLAR assembly. Utilize parts and mating techniques developed in Chapter 2 and 3.

Utilize the Save As command and create the 5HOLE-SHAFTCOLLAR assembly.

Combine the 3HOLE-SHAFTCOLLAR assembly, 5HOLE-SHAFTCOLLAR assembly and the WHEEL part to create the WHEEL-FLATBAR assembly.

The FLATBAR-3HOLE and FLATBAR 5HOLE parts were created in the Chapter 3 exercises. If needed, copy the models from the SW-TUTORIAL 2023/Chapter 6 Models folder.

Activity: 3HOLE - SHAFTCOLLAR Assembly

Create the 3HOLE-SHAFTCOLLAR assembly.

1)   Click **New** ⬜ from the Main menu.

2)   Double-click **Assembly** from the Templates tab.

3)   Double-click **FLATBAR-3HOLE** from the SW-TUTORIAL 2023\Chapter 6 Models folder.

4)   Click **OK** ✔ from the Begin Assembly PropertyManager. The FLATBAR-3HOLE is fixed to the Origin.

Save the assembly. Enter name. Enter description.

5)   Click **Save As** from the drop-down Main menu.

6)   Select the **Save in folder** location.

7)   Enter **3HOLE-SHAFTCOLLAR** for File name.

8)   Enter **3HOLE-SHAFTCOLLAR** for Description.

9)   Click **Save**.

Save the 3HOLE-SHAFTCOLLAR assembly.

10) Click **Save** .

Utilize a Concentric/Coincident SmartMate between the SHAFT-COLLAR and the FLATBAR-3HOLE.

Open the **SHAFT-COLLAR** part.

11) Click **Open** from the Main menu.

12) Double-click **SHAFT-COLLAR** from the SW-TUTORIAL 2023\Chapter 6 Models folder. SHAFT-COLLAR is the current document name.

13) **Rotate** the SHAFT-COLLAR to view the back circular edge.

14) Click **Window**, **Tile Horizontally** from the Main menu.

15) Drag the **back circular edge** of the SHAFT-COLLAR to the left circular hole edge of the FLATBAR-3HOLE in the Assembly Graphics window as illustrated. The mouse pointer displays the Concentric/Coincident icon.

16) **Release** the mouse button. Note: Select the back circular edge of the SHAFT-COLLAR, not the face.

Save the 3HOLE-SHAFTCOLLAR assembly.

17) **Close** the SHAFT-COLLAR window.

18) **Maximize** the 3HOLE-SHAFTCOLLAR assembly.

Fit the model to the Graphics window. Save the assembly.

19) Press the **f** key.

20) Click **Save** .

Create the 5HOLE-SHAFTCOLLAR assembly. Utilize the Save As command with the Save as copy and continue option. Recover from Mate errors.

Save the 3HOLE-SHAFTCOLLAR assembly as the 5HOLE-SHAFTCOLLAR assembly.

21)　Click **Save As** from the drop-down Main menu.

22)　Check the **Save as copy and continue** box.

23)　Enter **5HOLE-SHAFTCOLLAR** for File name.

24)　Enter **5HOLE-SHAFTCOLLAR** for Description.

25)　Click **Save**.

Close the model.

26)　Click **File**, **Close** from the Main menu.

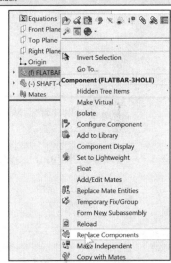

27)　Click **Save all**.

Open the **5HOLE-SHAFTCOLLAR** assembly.

28)　Click **Open** from the Main menu.

29)　Double-click **5HOLE-SHAFTCOLLAR**. The 5HOLE-SHAFTCOLLAR FeatureManager is displayed.

30)　Right-click **FLATBAR-3HOLE** from the FeatureManager.

31)　**Expand** the Pop-up menu.

32)　Click **Replace Components**. The Replace PropertyManager is displayed.

33)　Click the **Browse** button.

34)　Double-click **FLATBAR-5HOLE**. Note: The FLATBAR-5HOLE part is located in the SW-TUTORIAL 2023\Chapter 6 Models folder.

35)　Check the **Re-attach mates** box.

36)　Click **OK** from the Replace PropertyManager. The Mate Entities PropertyManager and the What's Wrong dialog box is displayed. There are two red Mate error marks displayed in the Mate Entities box.

37)   The What's Wrong dialog box is displayed. Recover from the Mate errors. Click **Close** from the What's Wrong dialog box.

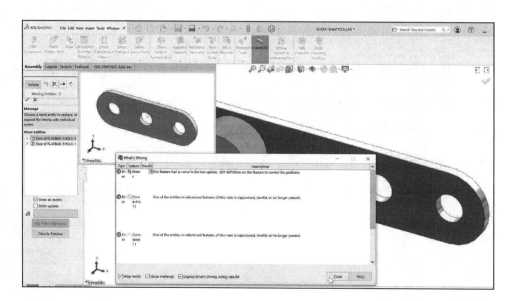

38)   Click **OK** ✔ from the Mated Entities PropertyManager.

39)   **View** the location of the SHAFT-COLLAR in the Graphics window.

Recover from the Mate errors.

40)   **Expand** the Mates folder from the FeatureManager. Right-click the first mate, **Concentric #** from the Mates folder.

41)   Click **Edit Feature** from the Context toolbar. The Mate PropertyManager is displayed.

42)   Right-click the **Mate Face error** in the Mate Selections box. Click **Delete**.

43)   Click the **inside face** of the left hole of the FLATBAR as illustrated. Concentric is selected by default.

44)   Click **OK** ✔ from the Mate PropertyManager.

45)   Right-click the second mate, **Coincident #** from the Mates folder.

46)   Click **Edit Feature** from the Context toolbar. The Mate PropertyManager is displayed.

47) Right-click the **Mate Face error** in the Mate Selections box.

48) Click **Delete**.

49) Click the **front face** of the FLATBAR as illustrated. The selected faces are displayed in the Mate Selections box. Coincident is selected by default.

50) Click **OK** ✔ from the Mate PropertyManager.

51) **Expand** the Mate folder from the FeatureManager.

52) **View** the created mates.

🔆 The Mate Entities box will list red X's if the faces, edges or planes are not valid. Expand the Mate Entities and select new references in the Graphics window to redefine the mates.

The FLATBAR-3HOLE is replaced with the FLATBAR-5HOLE part. The Mates are updated.

Fit the model to the Graphics window.
53) Press the **f** key.

Display an Isometric view. Save the 5HOLE-SHAFTCOLLAR assembly.
54) Click **Isometric view** 🧊 .

55) Click **Save** 💾 .

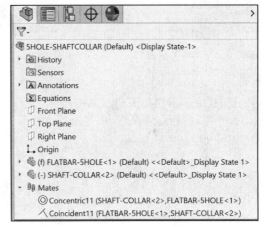

🔆 Incorporate symmetry into the assembly. Divide large assemblies into smaller sub-assemblies.

WHEEL-FLATBAR Assembly

The WHEEL-FLATBAR assembly consists of the following components:

- 3HOLE-SHAFTCOLLAR assembly.

- 5HOLE-SHAFTCOLLAR assembly.

- WHEEL part.

Create the WHEEL-FLATBAR assembly. Mate the 3HOLE-SHAFTCOLLAR assembly 67.5 degrees counterclockwise from the Top Plane.

The 3HOLE-SHAFTCOLLAR assembly is concentric with holes on the second and fourth bolt circle.

Mate the 5HOLE-SHAFTCOLLAR assembly 22.5 degrees clockwise from the Top Plane.

The 5HOLE-SHAFTCOLLAR assembly is concentric with holes on the second and fourth bolt circle.

☀ Remove the fixed state. Right-click the component name in the FeatureManager. Click **Float**. The component is free to move.

☀ CommandManager and Feature Manager tabs and folder files will vary, depending on system setup and Add-Ins.

Activity: WHEEL-FLATBAR Assembly

Create the WHEEL-FLATBAR assembly.

56)   Click **New** from the Main menu.

57)   Double-click **Assembly** from the Templates tab.

Insert the WHEEL.
58)   Double-click **WHEEL** from the SW-TUTORIAL 2023\Chapter 6 Models folder.

59)   Click **OK** ✔ from the Begin Assembly PropertyManager. The WHEEL part is fixed to the assembly Origin.

If needed, deactivate the Origins.
60)   Click **View**, **Hide/Show**, uncheck **Origins** from the Main menu menu.

Save the assembly. Enter part name. Enter part description.
61)   Click **Save As** from the drop-down Main menu.

62)   Select the **SW-TUTORIAL 2023** folder.

63)   Enter **WHEEL-FLATBAR** for File name.

64)   Enter **WHEEL-FLATBAR** for Description.

65)   Click **Save**.

Display the Top Plane in the Front view.
66)   Click **Front view** from the Heads-up View toolbar. View the WHEEL part.

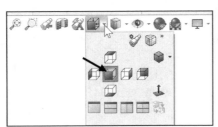

Locate the first set of holes from the Right plane (-Y-axis). Left Hole1 and Left Hole2 are positioned on the second and fourth bolt circle, 22.5° from the Right plane. Select Left Hole1. The x, y, z coordinates, -.287, -.693, .125 are displayed.

Tan^{-1} (-.287/.693) = 22.5°

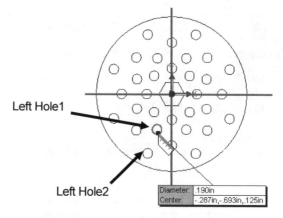

As an exercise, utilize the Measure tool to determine the center-to-center distance between the Left Hole1 and Left Hole2. The center-to-center distance is .500in.

Insert two Concentric mates between Left Hole1 and Left Hole2 and the 3HOLE-SHAFTCOLLAR assembly holes. The FLATBAR-3HOLE center-to-center distance is also .500in.

To determine tolerance issues, utilize two Concentric mates between components with mating cylindrical geometry. If the mating components center-to-center distance is not exact, a Mate error is displayed on the second Concentric mate.

Insert a Coincident mate between the back face of the 3HOLE-SHAFTCOLLAR assembly and the front face of the WHEEL.

Right Hole1 and Right Hole2 are 22.5° from the Top Plane.

Insert two Concentric mates between Right Hole1 and Right Hole2 and the 5HOLE-SHAFTCOLLAR assembly holes.

Insert a Coincident mate between the back face of the 5HOLE-SHAFTCOLLAR assembly and the front face of the WHEEL.

Activity: WHEEL-FLATBAR Assembly - Insert 3HOLE-SHAFTCOLLAR Assembly

Insert the 3HOLE-SHAFTCOLLAR assembly.

67) Click **Isometric view** from the Heads-up View toolbar.

68) Click **Insert Components** from the Assembly toolbar.

69) Double-click the **3HOLE-SHAFTCOLLAR** assembly from the SW-TUTORIAL 2023 folder or the folder that you saved it in. The assembly is also located in the SW-TUTORIAL 2023\Chapter 6 Models folder.

70) Click a **position** to the left of the WHEEL as illustrated.

Move and rotate the 3HOLE-SHAFTCOLLAR component.

71) Click the front face of the **3HOLE-SHAFTCOLLAR**.

72) Right-click **Move with Triad**.

73) Hold the **left mouse button** down on the X-axis (red).

74) Drag the **component** to the left. View the ruler.

75) Hold the **right mouse button** down on the Z-axis (blue).

76) Drag the **component** and rotate it about the Z-axis. View the ruler.

77) **Position** the component until the SHAFT-COLLAR part is approximately in front of the WHEEL Left Hole1.

78) Release the **right mouse** button.

79) Click a **position** in the Graphics window to deselect the face.

	Component Display
↳	Move with Triad
⚒	Temporary Fix/Group
	Form New Subassembly
⧉	Make Independent
⧉	Copy with Mates

Insert a Concentric mate.

80) Click the **back top inside cylindrical face** of the SHAFT-COLLAR.

81) **Rotate** the model to view the WHEEL.

82) Hold the **Ctrl** key down.

83) Click the **WHEEL Left Hole1** cylindrical face as illustrated.

84) Release the **Ctrl** key. The Mate pop-up menu is displayed.

85) Click **Concentric** from the Mate pop-up menu.

Insert the second Concentric mate.

86) Click the **back middle inside cylindrical face** of the FLATBAR.

87) **Rotate** the model to view the WHEEL.

88) Hold the **Ctrl** key down.

89) Click the **WHEEL Left Hole2** inside cylindrical face as illustrated.

90) Release the **Ctrl** key. The Mate pop-up menu is displayed.

91) Click **Concentric** from the Mate pop-up menu.

Insert a Coincident mate.

92)   Click the **FLATBAR-3HOLE back** face.

93)   **Rotate** the model to view the WHEEL.

94)   Hold the **Ctrl** key down.

95)   Click the front face of the **WHEEL**.

96)   Release the **Ctrl** key. The Mate pop-up menu is displayed.

97)   Click **Coincident** from the Mate pop-up menu.

Display a Front view.

98)   Click **Front view** .

Save the WHEEL-FLATBAR. assembly.

99)   Click **Save** .

Activity: WHEEL-FLATBAR Assembly - Insert 5HOLE-SHAFTCOLLAR Assembly

Insert the 5HOLE-SHAFTCOLLAR assembly.

100)   Click **Insert Components** from the Assembly toolbar.

101)   Double-click the **5HOLE-SHAFTCOLLAR** assembly from the SW-TUTORIAL 2023\Chapter 6 Models folder.

102)   Click a **position** to the right of the WHEEL.

Move the 5HOLE-SHAFTCOLLAR component.

103)   Click the **5HOLE-SHAFTCOLLAR** front face.

104)   Right-click **Move with Triad**.

105)   Hold the **left mouse button** down on the X-axis (red).

106)   Drag the **component** to the right. View the results.

107)   Click **Isometric view** .

108)   Click **inside** the Graphics window.

Mate the 5HOLE-
SHAFTCOLLAR assembly.
Insert a Concentric mate.

109) Click the **back inside cylindrical face** as illustrated on the FLATBAR-5HOLE assembly.

110) **Rotate** the model to view the WHEEL.

111) Hold the **Ctrl** key down.

112) Click the **WHEEL Right Hole1** cylindrical face.

Right Hole2

Right Hole1

113) Release the **Ctrl** key. The Mate pop-up menu is displayed.

114) Click **Concentric** from the Mate pop-up menu.

Insert a Concentric mate.

115) **Move** the 5HOLE-SHAFTCOLLAR to view the back side.

116) Click the **back inside cylindrical** face of the second hole on the FLATBAR-5HOLE assembly.

117) **Rotate** the model to view the WHEEL.

118) Hold the **Ctrl** key down.

119) Click the **WHEEL Right Hole2** cylindrical face.

Right Hole2

Right Hole1

120) Release the **Ctrl** key. The Mate pop-up menu is displayed.

121) Click **Concentric** from the Mate pop-up menu.

Insert a Coincident mate.

122) Click the **back face** of the FLATBAR-5HOLE.

123) **Rotate** the model to view the WHEEL.

124) Hold the **Ctrl** key down.

125) Click the **front face** of the WHEEL.

126) Release the **Ctrl** key. The Mate pop-up menu is displayed.

127) Click **Coincident** from the Mate pop-up menu.

Measure the angle between the 3HOLE-SHAFTCOLLAR assembly and the 5HOLE-SHAFTCOLLAR assembly.

128) Click **Front view** .

129) Click the **Measure** tool from the Evaluate tab in the CommandManager. The Measure dialog box is displayed.

130) Select the **two inside edges** of the FLATBAR assemblies. View the results.

131) If required, click the **Show XYZ Measurements button**. The items are perpendicular.

The Measure tool provides the ability to display dual units. Click the **Units/Precision** icon from the Measure dialog box. Set the desired units.

Utilize the Measure tool to measure distances and add reference dimensions between the COM point and entities such as vertices, edges, and faces.

132) **Close** ✕ the Measure dialog box.

Apply the Performance Evaluation tool. The Performance Evaluation tool displays statistics and checks the health of the current assembly.

133) Click the **Performance Evaluation** tool in the Evaluate toolbar. The Performance Evaluation dialog box is displayed. Review the Status and description for the assembly.

134) Click **OK** from the Performance Evaluation dialog box.

Fit the model to the Graphics window. Display an Isometric view.
135) Press the **f** key.

136) Click **Isometric view** 🔲 from the Heads-up View toolbar.

Save the WHEEL-FLATBAR assembly.
137) Click **Save** 💾.

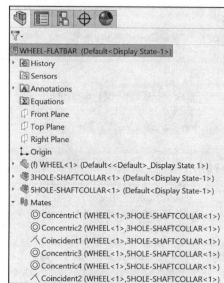

💡 To remove the fixed state, Right-click the component name in the FeatureManager. Click **Float**. The component is free to move.

💡 Mate the first component added to the assembly. If you mate to the first component or base component of the assembly and decide to change its orientation later, all the components will move with it.

WHEEL-AND-AXLE Assembly

The WHEEL-AND-AXLE assembly contains the following items:

- WHEEL-FLATBAR assembly.

- AXLE-3000 part.

- SHAFTCOLLAR-500 part.

- HEX-ADAPTER part.

Create the WHEEL-AND-AXLE assembly. The AXLE-3000 part is the first component in the assembly. A part or assembly inserted into a new assembly is called a component. The WHEEL-FLATBAR assembly rotates about the AXLE part.

Combine the created new assemblies and parts to develop the PNEUMATIC-TEST-MODULE assembly.

Activity: WHEEL-AND-AXLE Assembly

Create the WHEEL-AND-AXLE assembly.

138) Click **New** ⬜ from the Main menu.

139) Double-click **Assembly** from the Templates tab.

Insert the AXLE-3000 part.

140) Double-click **AXLE-3000** from the SW-TUTORIAL 2023 Chapter 6 Models folder.

141) Click **OK** ✔ from the Begin Assembly PropertyManager. The AXLE-3000 part is fixed to the assembly Origin.

Save the assembly. Enter name. Enter description.
142) Click **Save As** from the drop-down Main menu.

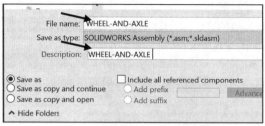

143) Enter **WHEEL-AND-AXLE** for File name in the SW-TUTORIAL 2023 folder.

144) Enter **WHEEL-AND-AXLE** for Description.

145) Click **Save**.

Insert a Coincident mate between the Axis of the AXLE-3000 and the Axis of the WHEEL. Insert a Coincident mate between the Front Plane of the AXLE-3000 and the Front Plane of the WHEEL. The WHEEL-FLATBAR assembly rotates about the AXLE-3000 axis.

Display the Temporary Axes.
146) Click **View**, **Hide/Show**, check **Temporary Axes** from the Main menu.

Insert the WHEEL-FLATBAR assembly.

147) Click **Insert Components** from the Assembly toolbar.

148) Double-click the **WHEEL-FLATBAR** assembly from the SW-TUTORIAL 2023 Chapter 6 Models folder.

149) Click a **position** to the right of AXLE-3000.

View the Reference WHEEL Axis.
150) Click **View**, **Hide/Show**, check **Axes** from the Main menu.

Reference geometry defines the shape or form of a surface or a solid. Reference geometry includes planes, axes, coordinate systems, and points.

The book is designed to expose the new SOLIDWORKS user to many different tools, techniques and procedures. It may not always use the most direct tool or process.

Insert a Coincident mate.

151) Click **Axis1** in the Graphics window.

152) Hold the **Ctrl** key down.

153) Click the **AXLE-3000 Temporary Axis**.

154) Release the **Ctrl** key. The Pop-up Mate menu is displayed.

155) Click **Coincident** from the Mate Pop-up menu.

Insert a Coincident mate.

156) **Expand** WHEEL-AND-AXLE from the fly-out FeatureManager.

157) **Expand** AXLE-3000 from the fly-out FeatureManager.

158) Click **Front Plane** of AXLE-3000<1>.

159) **Expand** WHEEL from the fly-out FeatureManager.

160) Hold the **Ctrl** key down.

161) Click **Front Plane** of the WHEEL.

162) Release the **Ctrl** key. The Mate Pop-up menu is displayed.

163) Click **Coincident** from the Mate Pop-up menu.

Rotate the WHEEL-FLATBAR assembly about AXLE-3000.

164) Click and drag the **WHEEL** around AXLE-3000.

Save the WHEEL-AND-AXLE assembly.

165) Click **Save** 🖫.

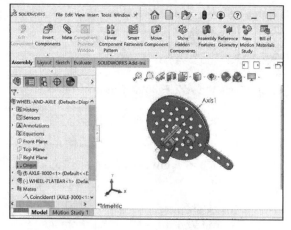

Activity: WHEEL-AND-AXLE Assembly - Insert the HEX-ADAPTER Part

Insert the HEX-ADAPTER part.

166) Click the **Insert Components** 🗗 tool from the Assembly toolbar.

167) Double-click **HEX-ADAPTER** from the SW-TUTORIAL 2023 Chapter 6 Models folder.

168) Click a **position** near the assembly as illustrated. The HEX-ADAPTER is displayed in the FeatureManager.

169) **Expand** the Mates folder.

170) **View** the created mates for the assembly. View the inserted components: AXLE-3000, WHEEL-FLATBAR and HEX-ADAPTER.

Insert a Concentric mate. Mate the HEX-ADAPTER.

171) Click the **Mate** 🖉 tool from the Assembly toolbar. The Mate PropertyManager is displayed.

172) Click the **Standard** tab.

173) Click the **HEX-ADAPTER** cylindrical face as illustrated.

174) Click the **AXLE-3000** cylindrical face as illustrated. Concentric is selected by default. The selected faces are displayed in the Mate Selections box.

175) Click **Aligned** ⬆⬆ from the Concentric1 PropertyManager to flip the HEX-ADAPTER, if required.

176) Click **OK** ✔ from the Concentric PropertyManager.

Insert a Coincident mate.
177) Click the **front face** of the WHEEL.

178) **Rotate** the assembly to view the flat back circular face of the HEX-ADAPTER as illustrated.

179) Hold the **Ctrl** key down.

180) Click the **flat back circular face** of the HEX-ADAPTER.

181) Release the **Ctrl** key. The Mate pop-up menu is displayed.

182) Click **Coincident** from the Mate pop-up menu.

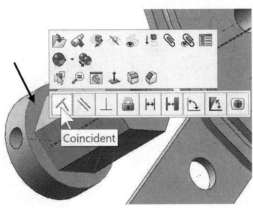

Insert a Parallel mate.

183) **Rotate** the WHEEL-AND-AXLE assembly to view the back-bottom edge of the HEX-ADAPTER.

184) **Zoom in** on the back-bottom edge of the HEX-ADAPTER.

185) Click the **back-bottom edge** of the WHEEL. Do not select the midpoint.

186) Hold the **Ctrl** key down.

187) Click the **top edge** of the HEX-ADAPTER. Do not select the midpoint.

188) Release the **Ctrl** key. The Mate pop-up menu is displayed.

189) Click **Parallel** from the Mate pop-up menu.

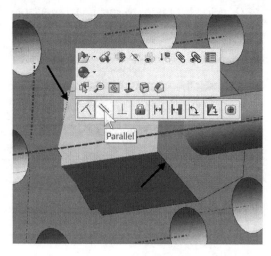

Fit the model to the Graphics window.
190) Press the **f** key.

Display an Isometric view.

191) Click **Isometric view** ⬚.

View the created mates.
192) **Expand** the Mates folder.

Save the WHEEL-AND-AXLE assembly.

193) Click **Save** 💾.

Activity: WHEEL-AND-AXLE Assembly - Insert SHAFTCOLLAR-500 Part

Insert the SHAFTCOLLAR-500 part.

194) Click the **Insert Components** tool from the Assembly toolbar.

195) Double-click **SHAFTCOLLAR-500** from the SW-TUTORIAL 2023 folder.

196) Click a **position** behind the WHEEL-AND-AXLE assembly.

Insert a Concentric mate.

197) Click **View**, **Hide/Show**, un-check **Temporary Axes** from the Main menu.

198) Click **View**, **Hide/Show**, un-check **Axes** from the Main menu.

199) Hold the **Ctrl** key down.

200) Click the **inside cylindrical face** of the SHAFTCOLLAR-500 part.

201) Click the **cylindrical face** of the AXLE-3000 part.

202) Release the **Ctrl** key. The Mate pop-up menu is displayed.

203) Click **Concentric** from the Mate pop-up menu.

Insert a Coincident mate.

204) Click the **front face** of the SHAFTCOLLAR-500 part.

205) **Rotate** the WHEEL to view the back face.

206) Hold the **Ctrl** key down.

207) Click the **back face** of the WHEEL.

208) Release the **Ctrl** key. The Mate pop-up menu is displayed.

209) Click **Coincident** from the Mate pop-up menu.

Display an Isometric view.

210) Click **Isometric view** .

View the created Mates.

211) **Expand** the Mates folder. View the created mates.

Save the WHEEL-AND-AXLE assembly.

212) Click **Save** .

Close all files.

213) Click **Windows**, **Close All** from the Main menu.

💡 Determine the static and dynamic behavior of mates in each sub-assembly before creating the top-level assembly.

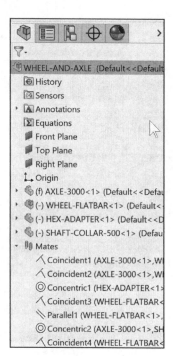

Review the WHEEL-AND-AXLE Assembly.

You combined the WHEEL-FLATBAR sub assembly, the AXLE-3000 part, the HEX-ADAPTER part and the SHAFTCOLLAR-500 part to create the WHEEL-AND-AXLE assembly.

The WHEEL-FLATBAR sub-assembly rotated about the AXLE-3000 part. The WHEEL-FLATBAR assembly combined the 3HOLE-SHAFTCOLLAR assembly and the 5HOLE-SHAFTCOLLAR assembly. The 5HOLE-SHAFTCOLLAR assembly was created from the 3HOLE-SHAFTCOLLAR assembly by replacing the FLATBAR component and recovering from two Mate errors. Additional components are added to the WHEEL-AND-AXLE assembly in the chapter exercises.

PNEUMATIC-TEST-MODULE Assembly

Create the PNEUMATIC-TEST-MODULE assembly. The first component is the FLAT-PLATE. The FLAT-PLATE part is located in the SW-TUTORIAL 2023 Chapter 6 Models folder. Fix the FLAT-PLATE to the ⌐.

Modify the LINKAGE assembly. Insert the HEX-STANDOFF part. Insert the LINKAGE assembly into the PNEUMATIC-TEST-MODULE assembly.

Insert the AIR-RESERVOIR-SUPPORT assembly. Utilize the Linear Component Pattern and Feature Driven Component Pattern tools. Insert the FRONT-SUPPORT assembly. Utilize the Mirror Components tool to create a mirrored version of the FRONT-SUPPORT assembly.

Insert the WHEEL-AND-AXLE assembly. Utilize Component Properties and create a Flexible State. Utilize the ConfigurationManager to select the Flexible configuration for the AirCylinder assembly.

💡 Work between multiple part and sub-assembly documents to create the final assembly.

Activity: PNEUMATIC-TEST - MODULE Assembly

Create the PNEUMATIC-TEST-MODULE assembly.

214)   Click **New** from the Main menu.

215)   Double-click **Assembly** from the Templates tab.

Insert the FLAT-PLATE part.
216)   Double-click **FLAT-PLATE** from the SW-TUTORIAL 2023\Chapter 6 Models folder.

217)   Click **OK** ✔ from the Begin Assembly PropertyManager. FLAT-PLATE is fixed to the Origin.

Save the assembly. Enter name. Enter description.
218)   Click **Save As** from the drop-down Main menu.

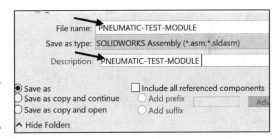

219)   Browse to the **SW-TUTORIAL 2023** folder

220)   Enter **PNEUMATIC-TEST-MODULE** for File name.

221)   Enter **PNEUMATIC-TEST-MODULE** for Description.

222)   Click **Save**.

Activity: Modify the LINKAGE Assembly

The LINKAGE assembly is located in the SW-TUTORIAL 2023\Chapter 6 Models\LINKAGE Assembly folder.

223)   Click **Open** from the Main menu.

224)   Double-click **LINKAGE** from the SW-TUTORIAL 2023\Chapter 6 Models\LINKAGE Assembly folder. The LINKAGE assembly is displayed.

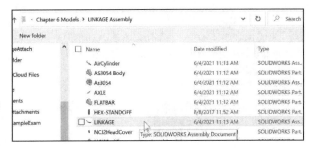

Insert the HEX-STANDOFF part into the LINKAGE assembly.

225)   Click the **Insert Components** tool from the Assembly toolbar.

226)   Double-click **HEX-STANDOFF** from the SW-TUTORIAL 2023\Chapter 6 Models\LINKAGE Assembly folder.

227)   Click a **position** to the front of the half Slot Cut as illustrated.

View the Temporary Axes.
228)   Click **View**, **Hide/Show**, check **Temporary Axes** from the Main menu.

229)   Click **View**, **Hide/Show**, check **Axes** from the Main menu.

Insert a Coincident mate.
230)   **Zoom-in** to view the Temporary Axis of the HEX-STANDOFF part.

231)   Click the **HEX-STANDOFF** tapped hole Temporary Axis as illustrated.

232)   **Zoom-in** to view the half Slot Cut Axis1 as illustrated.

233)   Hold the **Ctrl** key down.

234)   Click the **half Slot Cut Axis1**.

235)   Release the **Ctrl** key. The Mate pop-up menu is displayed.

236)   Click **Coincident** from the Mate pop-up menu.

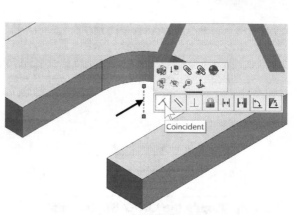

🔆 Reference geometry defines the shape or form of a surface or a solid. Reference geometry includes planes, axes, coordinate systems, and points.

Insert a Coincident mate.

237) Click the **HEX-STANDOFF** top face.

238) **Rotate** to view the BRACKET bottom face.

239) Hold the **Ctrl** key down.

240) Click the **BRACKET bottom face**.

241) Release the **Ctrl** key. The Mate pop-up menu is displayed.

242) Click **Coincident** from the Mate pop-up menu.

Insert a Parallel mate.

243) **Rotate** to view the model as illustrated.

244) Click the **HEX-STANDOFF** front face.

245) Hold the **Ctrl** key down.

246) Click the **BRACKET front face**.

247) Release the **Ctrl** key. The Mate pop-up menu is displayed.

248) Click **Parallel** ⟍ from the Mate pop-up menu.

Fit the model to the Graphics window.

249) Press the **f** key.

Display an Isometric view. Save the model.

250) Click **Isometric view** ▣.

251) Click **Save** 💾.

View the created Mates.

252) **Expand** the Mates folder. View the created mates.

Insert the second HEX-STANDOFF.

253) Click the **Insert Components** tool from the Assembly toolbar.

254) Double-click **HEX-STANDOFF** from the SW-TUTORIAL 2023\Chapter 6 Models\LINKAGE Assembly folder.

255) Click a **position** to the back right of the back half Slot Cut as illustrated.

Insert a Concentric mate.

256) **Rotate** the model to view the second HEX-STANDOFF.

257) **Zoom-in** on the Temporary Axis.

258) Click the **second HEX-STANDOFF** Tapped Hole Temporary Axis.

259) **Rotate** to view the back half Slot Cut Axis1.

260) Hold the **Ctrl** key down.

261) Click the **back half Slot Cut Axis1**.

262) Release the **Ctrl** key. The Mate pop-up menu is displayed.

263) Click **Coincident** from the Mate pop-up menu.

Insert a Coincident mate.

264) Click the **second HEX-STANDOFF top face**.

265) **Rotate** to view the BRACKET bottom face.

266) Hold the **Ctrl** key down.

267) Click the **BRACKET bottom face**.

268) Release the **Ctrl** key. The Mate pop-up menu is displayed.

269) Click **Coincident** from the Mate pop-up menu.

Insert a Parallel mate.

270) Click the **second HEX-STANDOFF** face as illustrated.

271) Hold the **Ctrl** key down.

272) Click the **BRACKET back right** face.

273) Release the **Ctrl** key. The Mate pop-up menu is displayed.

274) Click **Parallel** ⟍ from the Mate pop-up menu.

☆ Use Selection Filter toolbar to select the correct face, edge, Axis, etc.

Fit the model to the Graphics window. Display an Isometric view. Hide Axis and Temporary Axis.

275) Press the **f** key.

276) Click **View**, **Hide/Show**, check **Temporary Axes** from the Main menu.

277) Click **View**, **Hide/Show**, check **Axes** from the Main menu.

278) Click **Isometric view** 🔲 .

Save the assembly.

279) Click **Save** 💾 .

Insert the second AXLE.

280) Click the **Insert Components** 🗗 tool from the Assembly toolbar.

281) Double-click **AXLE** from the SW-TUTORIAL 2023\Chapter 6 Models\LINKAGE Assembly folder.

282) Click a **position** in the assembly as illustrated.

Insert a Concentric mate.

283) Click the **second AXLE cylindrical face**.

284) Hold the **Ctrl** key down.

285) Click the **FLATBAR bottom hole** face as illustrated.

286) Release the **Ctrl** key. The Mate pop-up menu is displayed.

287) Click **Concentric** from the Mate pop-up menu.

Insert a Coincident mate.

288) Click the second **AXLE Front Plane** from the FeatureManager.

289) Hold the **Ctrl** key down.

290) Click the **LINKAGE assembly Front Plane** from the FeatureManager.

291) Release the **Ctrl** key. The Mate pop-up menu is displayed.

292) Click **Coincident** from the Mate pop-up menu.

Fit the model to the Graphics window.

293) Press the **f** key.

Display an Isometric view. Save the assembly.

294) Click **Isometric view** from the Heads-up View toolbar.

Save the LINKAGE assembly.

295) Click **Save** .

Insert the first SHAFT-COLLAR on the second AXIS.

296)   Click the **Insert Components** Assembly tool.

297)   Double-click **SHAFT-COLLAR** from the SW-TUTORIAL 2023\Chapter 6 Models\LINKAGE Assembly folder.

298)   Click a **position** to the back of the second AXLE as illustrated.

Enlarge the view.
299)   Zoom-in on the **SHAFT-COLLAR** and the **AXLE** to enlarge the view.

Insert a Concentric mate.
300)   Click the inside **hole face** of the SHAFT-COLLAR.

301)   Hold the **Ctrl** key down.

302)   Click the **long cylindrical face** of the AXLE.

303)   Release the **Ctrl** key. The Mate pop-up menu is displayed.

304)   Click **Concentric** from the Mate pop-up menu.

The book is designed to expose the new SOLIDWORKS user to many different tools, techniques and procedures. It may not always use the most direct tool or process.

Insert a Coincident mate.

305) Click the **front face** of the SHAFT-COLLAR as illustrated.

306) **Rotate** the assembly to view the back face of the FLATBAR.

307) Hold the **Ctrl** key down.

308) Click the **back face** of the FLATBAR.

309) Release the **Ctrl** key. The Mate pop-up menu is displayed.

310) Click **Coincident** from the Mate pop-up menu.

Display the Isometric view.

311) Click **Isometric view** .

Insert the second SHAFT-COLLAR.

312) Click the **Insert Components** Assembly tool.

313) Double-click **SHAFT-COLLAR** from the SW-TUTORIAL 2023\Chapter 6 Models\LINKAGE Assembly folder.

314) Click a **position** to the front of the AXLE.

Enlarge the view.

315) **Zoom in** on the second SHAFT-COLLAR and the AXLE.

Insert a Concentric mate.

316) Click the **inside hole face** of the second SHAFT-COLLAR. Note the icon face feedback symbol.

317) Hold the **Ctrl** key down.

318) Click the **long cylindrical face** of the AXLE.

319) Release the **Ctrl** key. The Mate pop-up menu is displayed.

320) Click **Concentric** from the Mate pop-up menu.

Insert a Coincident mate.

321) Press the **f** key to fit the model to the Graphics window.

322) **Zoom in** on the front face of the FLATBAR as illustrated.

323) Click the **front face** of the FLATBAR.

324) **Rotate** to view the back face of the second SHAFT-COLLAR.

325) Hold the **Ctrl** key down.

326) Click the **back face** of the second SHAFT-COLLAR.

327) Release the **Ctrl** key. The Mate pop-up menu is displayed.

328) Click **Coincident** from the Mate pop-up menu.

Display an Isometric view.

329) Click **Isometric view** .

Save the LINKAGE assembly.

330) **Rebuild** the model.

331) Click **Save** . As an exercise, insert SCREWs between the AirCylinder assembly and the two HEX-STANDOFFs as illustrated.

Activity: PNEUMATIC-TEST-MODULE - Insert LINKAGE Assembly

Insert the LINKAGE assembly into the PNEUMATIC-TEST-MODULE assembly.

332) Click **Window**, **Tile Horizontally** from the Main menu. The PNEUMATIC-TEST-MODULE assembly should be open. If not, open the PNEUMATIC-TEST MODULE assembly document.

333) Rotate the two **FLATBARs** approximately 45º.

334) Click and drag the **LINKAGE**
LINKAGE<1> assembly icon into the PNEUMATIC-TEST-MODULE assembly.

335) Click a **position** above the FLAT-PLATE. The LINKAGE assembly should be in a rigid state.

336) **Maximize** the PNEUMATIC-TEST-MODULE Graphics window.

337) Click **View**, **Hide/Show**, un-check **Origins** from the Main menu. If required, click View, Hide/Show, un-check Planes from the Main menu.

Rotate model.

338) **Rotate** the model to view the Front HEX-STANDOFF part as illustrated.

Insert a Concentric mate.

339) Click the **Front HEX-STANDOFF Tapped** inside hole face as illustrated.

340) **Rotate** the model to view the FLAT-PLATE hole as illustrated. Hold the **Ctrl** key down.

341) Click the **FLAT-PLATE Hole** face in the 5th row, 4th column as illustrated.

342) Release the **Ctrl** key. The Mate pop-up menu is displayed.

343) Click **Concentric** from the Mate pop-up menu.

Insert a Parallel mate between two Planes.

344) Click the **PNEUMATIC-TEST-MODULE Front Plane** from the FeatureManager.

345) Hold the **Ctrl** key down.

346) Click the **LINKAGE assembly Front Plane** from the FeatureManager.

347) Release the **Ctrl** key. The Mate pop-up menu is displayed.

348) Click **Parallel** from the Mate pop-up menu.

Insert a Coincident mate.

349) Click the **Front HEX-STANDOFF** bottom face.

350) Hold the **Ctrl** key down.

351) Click the **FLAT-PLATE top face**.

352) Release the **Ctrl** key. The Mate pop-up menu is displayed.

353) Click **Coincident** from the Mate pop-up menu.

Display an Isometric view.

354) Click **Isometric view** .

Save the PNEUMATIC-TEST-MODULE assembly.

355) Click **Save** .

The LINKAGE assembly is fully defined and located on the FLAT-PLATE part. Insert the AIR-RESERVOIR-SUPPORT assembly. The AIR-RESERVOIR-SUPPORT assembly is located in the SW-TUTORIAL 2023\Chapter 6 Models folder.

Activity: PNEUMATIC-TEST-MODULE - Insert AIR-RESERVOIR-SUPPORT

Insert the AIR-RESERVOIR-SUPPORT assembly.

356) Click the **Insert Components** Assembly tool. The Insert Component PropertyManager is displayed.

357) Double-click the **AIR-RESERVOIR-SUPPORT** assembly from the SW-TUTORIAL 2023\Chapter 6 Models folder. The AIR-RESERVOIR SUPPORT assembly is in the Chapter 6 Models folder. Work from your local hard drive.

358) Click a **position** above the FLAT-PLATE as illustrated.

359) Click the **Rotate Component** Assembly tool. The Rotate Component PropertyManager is displayed. The Rotate ↺ icon is displayed in the Graphics window.

360) Click and drag the **AIR-RESERVOIR-SUPPORT** until the tank is parallel with the AirCylinder assembly as illustrated.

361) Click **OK** ✔ from the Rotate Component PropertyManager.

Insert a Concentric mate.

362) Click the **FLAT-PLATE back left inside hole face** as illustrated.

363) Hold the **Ctrl** key down.

364) Click the **fourth ANGLE-BRACKET inside hole** face.

365) Release the **Ctrl** key. The Mate pop-up menu is displayed.

366) Click **Concentric** from the Mate pop-up menu.

Insert a Coincident mate.

367) Click the **ANGLE-BRACKET bottom** face.

368) **Rotate** the assembly to view the FLAT-PLATE top face. Hold the **Ctrl** key down.

369) Click the **FLAT-PLATE top** face.

370) Release the **Ctrl** key. The Mate pop-up menu is displayed.

371) Click **Coincident** from the Mate pop-up menu.

Insert a Parallel mate.

372) Click **Left view** from the Heads-up View toolbar.

373) Click the **ANGLE-BRACKET** narrow face.

374) Hold the **Ctrl** key down.

375) Click the **FLAT-PLATE narrow** face.

376) Release the **Ctrl** key. The Mate pop-up menu is displayed.

377) Click **Parallel** from the Mate pop-up menu. The second Parallel mate is created.

Display an Isometric view.

378) Click **Isometric view** from the Heads-up View toolbar.

Save the PNEUMATIC-TEST-MODULE assembly.

379) Click **Save** .

Component Patterns in the Assembly

There are various methods to define a pattern in an assembly:

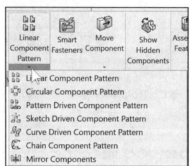

- Linear Component Pattern.

- Circular Component Pattern.

- Pattern Driven Component (Feature).

- Sketch Driven Component Pattern.

- Curve Driven Component Pattern.

- Chain Component Pattern.

- Mirror Components.

A Linear/Circular Component Pattern tool utilizes geometry in the assembly to arrange instances in a Linear or Circular pattern.

A Pattern Driven Component Pattern utilizes an existing feature pattern.

The SCREW part fastens the ANGLE-BRACKET part to the FLAT-PLATE part. Mate one SCREW to the first instance on the ANGLE-BRACKET Linear Pattern.

Utilize the Pattern Driven Component Pattern tool to create instances of the SCREW. Suppress the instances.

Utilize the Linear Component Pattern tool to copy the Pattern Driven Component Pattern of SCREWS to the second ANGLE-BRACKET part.

Drag the part by specific geometry to create a mate.

Activity: PNEUMATIC-TEST-MODULE - Component Pattern

Open the SCREW part.

380) Click **Open** from the Main menu.

381) Double-click **SCREW** from the SW-TUTORIAL 2023\Chapter 6 Models folder.

382) Un-suppress the **Fillet1** and **Chamfer1** feature.

383) Click **Window**, **Tile Horizontally** to display the SCREW and the PNEUMATIC-TEST-MODULE assembly.

Insert and mate the SCREW.
384) Click the **bottom circular edge** of the SCREW.

385) Drag the **SCREW** into the PNEUMATIC-TEST-MODULE assembly window. Note: Zoom in on the top circular edge of the ANGLE-BRACKET left hole.

386) Release the mouse pointer on the **top circular edge** of the ANGLE-BRACKET left hole. The mouse pointer displays the Coincident/Concentric feedback symbol.

387) **Return** to the PNEUMATIC-TEST-MODULE assembly window.

388) Click **Left view** from the Heads-up View toolbar.

The SCREW part is positioned in the left hole with a Coincident/Concentric mate.

Create a Pattern Driven Component Pattern.

389) Click the **Pattern Driven Component Pattern** tool from the Consolidate Assembly toolbar. The PropertyManager is displayed.

390) Click the **SCREW** component in the Graphics window. SCREW<1> is displayed in the Components to Pattern box.

391) Click inside the **Driving Feature or Component** box. **Expand** the AIR-RESERVOIR-SUPPORT assembly in the PNEUMATIC-TEST-MODULE fly-out FeatureManager.

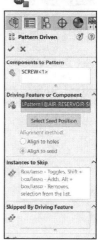

392) **Expand** ANGLE-BRACKET <1>. Click **LPattern1** in the fly-out FeatureManager. The SCREW is the seed feature.

393) Click **OK** ✔ from the PropertyManager. Six instances are displayed in the Graphics window. DerivedLPattern1 is displayed in the FeatureManager.

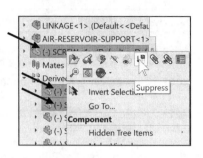

394) Click **SCREW<1>** from the FeatureManager.

395) Hold the **Ctrl** key down.

396) **Expand** DerivedLPattern1 from the FeatureManager.

397) Click the first two entries: **SCREW<2>** and **SCREW<3>**. Release the **Ctrl** key.

398) Right-click **Suppress**. The first three SCREWs are not displayed.

💡 The **Pattern Driven Component Pattern** PropertyManager contains an option to skip instances of a feature component in a pattern.

Activity: PNEUMATIC-TEST-MODULE - Linear Component Pattern

Create a Linear Component Pattern.

399) Click **Front view** 🔲 from the Heads-up View toolbar.

400) Click the **Linear Component Pattern** ⬚⬚ tool from the Assembly toolbar. The Linear Pattern PropertyManager is displayed.

401) Click inside the **Pattern Direction** box for Direction 1.

402) Click the long front edge of the **FRONT-PLATE** as illustrated. Edge<1> is displayed in the Pattern Direction box. The direction arrow points to the right.

403) Enter **177.80**mm, [7in] for Spacing.

404) Enter **2** for Instances.

405) Click inside the **Components to Pattern** box.

406) Click **DerivedLPattern1** from the fly-out FeatureManager.

407) Click **OK** ✔ from the Linear Pattern PropertyManager.

408) Click **Top view** 🔲 from the Heads-up View toolbar. View the results.

The Linear Component Pattern feature is displayed in the second ANGLE-BRACKET part. LocalLPattern1 is displayed in the FeatureManager.

Display an Isometric view. Save the PNEUMATIC TEST MODULE assembly.

409) Click **Isometric view** 🔲.

410) **View** the FeatureManager and the created features in their states.

411) Click **Save** 💾.

Hide the AIR-RESERVOIR-SUPPORT assembly.
412)   Right-click **AIR-RESERVOIR-SUPPORT<1>** in the FeatureManager.

413)   Click **Hide Components** 🖎. Utilize the Show components tool to display a component that has been hidden.

Hide the LINKAGE assembly.
414)   Right-click **LINKAGE<1>** in the FeatureManager.

415)   Click **Hide Components** 🖎.

Insert the FRONT-SUPPORT assembly into the PNEUMATIC-TEST-MODULE assembly. Mate the FRONT-SUPPORT assembly to the FLAT-PLATE part.

Utilize the Mirror Components tool to create a mirrored copy of the FRONT-SUPPORT assembly.

🔅 You can create new components by mirroring existing part or sub-assembly components. The new components can either be a copy or a mirror of the original components. A mirrored component is sometimes called a "right-hand" version of the original "left-hand" version.

🔅 Access the Mirror Components tool from the Consolidated Linear Component Pattern drop-down menu.

Activity: PNEUMATIC-TEST-MODULE - Insert FRONT-SUPPORT Assembly

---

Insert and mate the FRONT-SUPPORT assembly.

416)   Click the **Insert Components** 🗊 tool from the Assembly toolbar.

417)   Double-click **FRONT-SUPPORT** from the SW-TUTORIAL 2023\Chapter 6 Models folder.

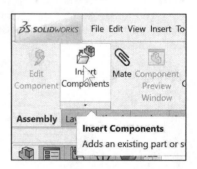

Hide the SCREWs in the FRONT-SUPPORT.

418) Click a **position** above the FLAT-PLATE as illustrated.

419) Right-click **FRONT-SUPPORT** in the FeatureManager.

420) Click **Open Assembly**.

421) Hide the SCREW parts and HEX-NUTS if required in the assembly.

Save the model.

422) Click **Save** 💾.

Return to the PNEUMATIC-TEST-MODULE assembly.

423) Press **Ctrl Tab**.

424) Select the **PNEUMATIC-TEST-MODULE** document.

Save the model.

425) Click **Save** 💾.

426) Click **Yes** to the Message, "Save the document and referenced models now?"

The PNEUMATIC-TEST-MODULE is displayed. The SCREW and HEX-NUT components are hidden in the FRONT-SUPPORT assembly.

Insert a Concentric mate.

427) Click the **FLAT-PLATE hole face** in the 3rd row, right most column (17th column).

428) Hold the **Ctrl** key down.

429) Click the **HEX-STANDOFF Tapped Hole** as illustrated.

430) Release the **Ctrl** key. The Mate pop-up menu is displayed.

431) Click **Concentric** from the Mate pop-up menu. The mate is created.

Insert a Coincident mate.

432)   Click the bottom face of the **HEX-STANDOFF** part.

433)   **Rotate** the model to view the top face of the FLAT-PLATE.

434)   Hold the **Ctrl** key down.

435)   Click the top face of the **FLAT-PLATE**.

436)   Release the **Ctrl** key. The Mate pop-up menu is displayed.

437)   Click **Coincident** from the Mate pop-up menu.

Insert a Parallel mate.

438)   Click the **FRONT-SUPPORT Front Plane** from the FeatureManager.

439)   **Show** the LINKAGE assembly.

440)   Hold the **Ctrl** key down.

441)   Click the **LINKAGE Front Plane** from the FeatureManager.

442)   Release the **Ctrl** key. The Mate pop-up menu is displayed.

443)   Click **Parallel** ⟍ from the Mate pop-up menu.

444)   **Hide** the LINKAGE assembly.

Save the PNEUMATIC-TEST-MODULE assembly.

445)   Click **Save** .

Mirrored Components

Create new components by mirroring existing parts or sub-assemblies. New components are created as copied geometry or mirrored geometry.

A mirrored component is sometimes called a "right-hand" version of the original "left hand" version.

The copied or mirrored component changes when the original component is modified.

A mirrored component creates a new document. The default document prefix is mirror. A copied component does not create a new document.

Suppressed components in the original sub-assembly are not mirrored or copied. The SCREW parts in the FRONT-SUPPORT assembly are not copied.

Left Hand Version (Original)   Right Hand Version (Mirrored)

Instance Component

Mirrored Component

Front

Instance Component

PNEUMATIC-TEST-MODULE

Activity: PNEUMATIC-TEST-MODULE Assembly - Mirrored Component

Insert a Mirrored Component.

446) Click the **Mirror Components** 📑 tool from the Consolidated Assembly toolbar. The Mirror Components PropertyManager is displayed.

Step 1: The Selections box is displayed. Select face/plane to mirror about and the components to be mirrored.

447) **Expand** the PNEUMATIC-TEST-MODULE fly-out FeatureManager. Click the **FLAT-PLATE Front Plane**. The Front Plane is displayed in the Mirror plane box.

448) Click the **FRONT-SUPPORT** assembly from the fly-out FeatureManager.

449) Click **Next** . The FRONT-SUPPORT assembly is displayed in the Components to Mirror box.

Step 2: Verify the orientation of the components to be mirrored and adjust accordingly using the buttons below.

450) Click the **Create opposite hand version** button. View the results in the Graphics window. View your options.

451) Click **Next** . Accept the default settings.

Step 3: The mirrored components selected in step 2 need new geometries. This may be new files or new configurations in existing files. Specify the new configuration or file name using the options below.

452) Click **Next** . This is for any Import Features. Accept the default settings.

453) Click **OK** from the Mirror Components PropertyManager. Click Yes if needed. View the results in the Graphics window.

Select suppressed components and their mates to be mirrored in the Components to Mirror box.

The MirrorFRONT-SUPPORT, TRIANGLE, ANGLE-13HOLE, and SCREW components are mated.

Utilize the Fix option to mate the MirrorFRONT-SUPPORT to its current location in the PNEUMATIC-TEST-MODULE. No other mates are required.

Activity: PNEUMATIC-TEST-MODULE - Fix MIRRORFRONT-SUPPORT

Fix the MirrorFRONT-SUPPORT.

454) Right-click **MirrorFRONT-SUPPORT** from the FeatureManager.

455) Click **Fix**. The MirrorFRONT assembly is fixed in the PNEUMATIC-TEST-MODULE assembly. The MirrorFRONT-SUPPORT does not move or rotate.

Display the LINKAGE assembly.

456) Right-click **LINKAGE** in the FeatureManager.

457) Click **Show Components**.

Save the PNEUMATIC-TEST-MODULE assembly.

458) Click **Save** 💾.

Reuse geometry in the assembly. Utilize the Mirror Component tool to create a left and right version of parts and assemblies.

Component Properties

Component Properties control the flexibility of the sub-assembly when inserted into an assembly. Components do not translate or rotate after insertion into the assembly.

The FLATBAR parts in the LINKAGE assembly do not rotate after insertion into the PNEUMATIC TEST MODULE assembly. The LINKAGE assembly is in the Rigid State.

Insert the WHEEL-AND-AXLE assembly into the PNEUMATIC-TEST-MODULE assembly. Modify the Mate state of the LINKAGE assembly from the Rigid State to the Flexible State. The LINKAGE assembly is free to rotate.

Modify the Mate State of the AirCylinder assembly from the Rigid State to the Flexible State. The AirCylinder assembly is free to translate.

By default, when you create a sub-assembly, it is rigid. Within the parent assembly, the sub-assembly acts as a single unit and its components do not move relative to each other.

Activity: PNEUMATIC-TEST-MODULE Assembly - Insert WHEEL-AND-AXLE Assembly

Insert the WHEEL-AND-AXLE assembly.

459) Click the **Insert Components** tool from the Assembly toolbar.

460) Double-click the **WHEEL-AND-AXLE** assembly from the SW-TUTORIAL 2023\Chapter 6 Models folder.

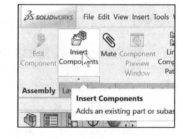

461) Click a **position** above the FLAT-PLATE part as illustrated.

Insert a Concentric mate.

462) Click the **AXLE-3000 cylindrical** face.

463) **Rotate** the model to view the inside TRIANGLE top hole face as illustrated.

464) Hold the **Ctrl** key down.

465) Click the inside **TRIANGLE top hole** face.

466) Release the **Ctrl** key. The Mate pop-up menu is displayed.

467) Click **Concentric** from the Mate pop-up menu.

Insert a Coincident mate.

468) Click the **WHEEL-AND-AXLE Front Plane** from the FeatureManager.

469) Hold the **Ctrl** key down.

470) Click the **LINKAGE Front Plane** from the FeatureManager.

471) Release the **Ctrl** key. The Mate pop-up menu is displayed.

472) Click **Coincident** from the Mate pop-up menu.

A Concentric mate is required between the left hole of the FLATBAR-3HOLE and right AXLE of the LINKAGE assembly.

DO NOT INSERT A CONCENTRIC MATE AT THIS TIME.

A Concentric mate will result in Mate errors.

The WHEEL-AND-AXLE is free to rotate in the PNEUMATIC-TEST-MODULE assembly. The LINKAGE assembly interferes with the WHEEL-AND-AXLE. The LINKAGE assembly is not free to rotate or translate. Mate errors will occur.

Sub-assemblies within the LINKAGE assembly are in a Rigid Mate state when inserted into the PNEUMATIC-TEST-MODULE assembly. Remove the Rigid Mate state and insert a Concentric mate between the LINKAGE assembly and the WHEEL-AND-AXLE.

Activity: PNEUMATIC-TEST-MODULE Assembly - Remove Rigid state

---

Remove the Rigid State.

473) Right-click the **LINKAGE** assembly from the FeatureManager.

474) Click **Component Properties** from the Context toolbar. The Component Properties dialog box is displayed.

475) Check **Flexible** in the Solve as box.

476) Click **OK** from the Component Properties dialog box. Note the icon next to the LINKAGE assembly in the FeatureManager.

477) Right-click the **AirCylinder** assembly from the FeatureManager.

478) Click **Component Properties**. **Flexible** is selected in the Solve as box.

479) **Flexible** is selected in the Referenced configuration box.

480) Click **OK** from the Component Properties dialog box.

Hide the FRONT-SUPPORT assembly.

481) Right-click **FRONT-SUPPORT** from the FeatureManager.

482) Click **Hide Components**.

Move the LINKAGE assembly.

483) Click and drag the **front FLATBAR-9HOLE** downward. The FLATBAR-9HOLE rotates about the left AXLE.

484) Click a **position** below the WHEEL-AND-AXLE.

485) Click **inside** the Graphics window to deselect.

Insert a Concentric mate.

486) Click the second bottom **AXLE** face of the LINKAGE assembly.

487) Hold the **Ctrl** key down.

488) Click the inside face of the **bottom hole** of the FLATBAR-3HOLE as illustrated.

489) Release the **Ctrl** key. The Mate pop-up menu is displayed.

490) Click **Concentric** from the Mate pop-up menu.

Display an Isometric view. Save the PNEUMATIC-TEST-MODULE assembly.

491) Click **Isometric view** .

492) Click **Save** .

The AirCylinder assembly is inserted in a Rigid state by default. The AirCylinder contains three configurations:

- Default

- Extended

- Flexible

Open the AirCylinder assembly. Modify the configuration to Flexible.

Open the LINKAGE assembly. Modify the AirCylinder Component Properties from Default to Flexible.

Activity: PNEUMATIC-TEST-MODULE Assembly - Review AirCylinder Configurations

Review the AirCylinder configurations.

493) Right-click **LINKAGE** from the PNEUMATIC-TEST-MODULE FeatureManager.

494) Click **Open Subassembly** from the Context toolbar.

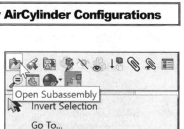

495) Right-click **AirCylinder** in the LINKAGE
FeatureManager.

496) Click **Open Sub-assembly**.

497) Click the **Configuration
Manager** tab at the top of
the AirCylinder
FeatureManager. Three
configurations are displayed:
Default, *Extended* and *Flexible*.
The current Default
configuration sets the Piston
Rod at 0mm.

Display the Extended Configuration.
498) Double-click **Extended**. The
Piston Rod of the AirCylinder
extends 25mm.

Display the Flexible Configuration.
499) Double-click **Flexible**.

500) Click the **Rod-Clevis** in the
Graphics window.

501) Click and drag the **Piston Rod** from left to right.

502) Click a **position** near its original location. The
AirCylinder remains in the Flexible configuration
for the rest of this project.

Update the LINKAGE assembly.
503) Click **Window**, **LINKAGE** from the Main menu.
The current configuration of the AirCylinder part
in the LINKAGE assembly is Default. Modify the
configuration.

504) Right-click **AirCylinder** from the LINKAGE assembly
FeatureManager.

505) Click **Component Properties**.

506) **Flexible** is selected in the Solve as section.

507) **Flexible** is selected in the Referenced configuration
section.

508) Click **OK**. The AirCylinder displays the Flexible
configuration next to its name in the FeatureManager.

Update the PNEUMATIC-TEST-MODULE.
509) Click **Window**, **PNEUMATIC-TEST-MODULE** assembly from the Main menu.

Move the ROD-CLEVIS.
510) Click and drag the **ROD-CLEVIS** to the right. The WHEEL rotates in a counterclockwise direction.

511) Click and drag the **ROD-CLEVIS** to the left. The WHEEL rotates in a clockwise direction.

Display the AIR-RESERVOIR-SUPPORT assembly.
512) Right-click **AIR-RESERVOIR-SUPPORT** in the FeatureManager.

513) Click **Show Components**.

Display the second FRONT-SUPPORT assembly.
514) Right-click **FRONT-SUPPORT** in the FeatureManager.

515) Click **Show Components**.

516) **Display** the SCREWs in the FRONT-SUPPORT.

Display an Isometric view. Save the PNEUMATIC-TEST-MODULE.

517) Click **Isometric view** .

518) Click **Save** .

519) Click **Yes** to update referenced documents.

Close all documents.
520) Click **Windows**, **Close All** from the Main menu.

Explore additional parts and assemblies at the end of this project. Add the WEIGHT, HOOK and FLATBAR parts.

 Additional details on Show components, Hide components, Linear Component Pattern, Circular Component Pattern, Pattern Driven Component Pattern, Sketch Driven Component Patter, Curve Driven Component Pattern, Mirror Components, Configurations, and Configuration Manager are available in SOLIDWORKS Help. Select Help, SOLIDWORKS Help topics.

Review the PNEUMATIC-TEST-MODULE Assembly.

The PNEUMATIC TEST MODULE assembly was created by combining five major mechanical sub-assemblies.

The PNEUMATIC TEST MODULE assembly utilized the FLAT-PLATE as the first component. The LINKAGE assembly, AIR-RESERVOIR-SUPPORT assembly, FRONT-SUPPORT assembly, MirrorFRONT-SUPPORT assembly and the WHEEL-AND-AXLE assembly were mated to the FLAT-PLATE part.

Work at the lowest assembly level. The LINKAGE assembly required two HEX-STANDOFF parts. The HEX-STANDOFF components were inserted into the LINKAGE assembly. The LINKAGE assembly was inserted into the PNEUMATIC-TEST-MODULE assembly.

The AIR-RESERVOIR-SUPPORT assembly was inserted into the PNEUMATIC-TEST-MODULE assembly. The SCREW component utilized a Feature Driven Component Pattern tool and the Linear Component Pattern tool to create multiple copies.

The FRONT-SUPPORT assembly was inserted into the PNEUMATIC-TEST-MODULE assembly. The Mirror Components tool created a mirrored version of the FRONT-SUPPORT assembly.

The WHEEL-AND-AXLE assembly was inserted into the PNEUMATIC-TEST-MODULE assembly. Component Properties created a Flexible state mate.

The LINKAGE assembly "AirCylinder" is in a Flexible state.

In the next section, create the final ROBOT assembly. Either use your created assemblies and components, or use the created assemblies and components provided in the SW-TUTORIAL 2023\Chapter 6 Models folder.

Final ROBOT Assembly

The final ROBOT assembly is comprised of three major sub-assemblies and the HEX-STANDOFF component:

- Robot Platform assembly.

- PNEUMATIC-TEST-MODULE assembly.

- Basic_integration assembly.

- 4 - HEX-STANDOFF components.

💡 The LINKAGE assembly is in a Flexible state.

💡 A Rebuild icon may be displayed in the final ROBOT Assembly FeatureManager and a flexible rebuild 🔃 icon at the sub-assembly level when the AirCylinder assembly is in the Flexible state.

All assemblies and components for the final ROBOT assembly are provided in the SW-TUTORIAL 2023\Chapter 6 Models folder.

💡 As an exercise apply PhotoView 360 to render the final ROBOT assembly.

Combine the created new assemblies and components to develop the final ROBOT assembly. Note: Step-by-step instructions are not provided in this section for the mates.

Activity: Create the ROBOT Assembly

Create the final ROBOT assembly.

521)   Click **New** 📄 from the Main menu.

522)   Double-click **Assembly** from the Templates tab. The Begin Assembly PropertyManager is displayed.

Insert the Robot-platform assembly.
523)   Double-click **Robot-platform** from the SW-TUTORIAL 2023\Chapter 6 Models folder.

524)   Click **OK** ✔ from the Begin Assembly PropertyManager. The Robot-platform is fixed to the assembly Origin.

Save the Assembly.
525)   Click **Save As** from the drop-down Main menu.

526)   Enter **ROBOT** for File name.

527)   Enter **Final ROBOT Assembly** for Description.

528)   Click **Save**.

Insert the PNEUMATIC-TEST-MODULE assembly.

529)   Click **Insert Components** 📇 from the Assembly toolbar. The Insert Components PropertyManager is displayed.

530) Double-click your **PNEUMATIC-TEST-MODULE** assembly. If you did not create the assembly, double-click the **PNEUMATIC-TEST-MODULE** assembly from the SW-TUTORIAL 2023\Chapter 6 Models\PNEUMATIC-TEST MODULE folder.

531) Click a **position** above the Robot-platform.

If an assembly or component is loaded in a Lightweight state, right-click the **assembly name** or **component name** from the FeatureManager. Click **Set to Resolved**.

Insert 4 HEX-STANDOFF components.

532) Click **Insert Components** from the Assembly toolbar.

533) Double-click the **HEX-STANDOFF** part from SW-TUTORIAL 2023\Chapter 6 Models folder.

534) **Pin** the Insert Components PropertyManager.

535) Click **four positions** as illustrated for the four HEX-STANDOFF components.

536) **Un-Pin** the Insert Components PropertyManager.

537) Click **OK** from the Insert Components PropertyManager.

Insert Coincident and Concentric mates between the four HEX-STANDOFF components and the PNEUMATIC-TEST-MODULE and the Robot-platform as illustrated. View the location of the HEX-STANDOFF components on the PNEUMATIC-TEST-MODULE assembly.

View the location of the HEX-STANDOFF
components on the Robot-platform as
illustrated.

538) **Expand** the Mates folder. View the
created mates.

Insert the basic_integration assembly.

539) Browse to the **SW-TUTORIAL
2023\Chapter 6 Models** folder.

540) Double-click the
basic_integration assembly.

541) Click a position to the **left** of the
assembly as illustrated.

Insert Coincident and Concentric
mates between the two screws in the
basic_integration assembly and the
PNEUMATIC-TEST-MODULE
assembly.

542) **Expand** the Mates folder. View the
created mates.

Save the ROBOT assembly.

543) Click **Save** 💾. You are finished
with the final ROBOT assembly.

Chapter Summary

In this chapter, you created and worked with multiple documents in sub-assemblies and assemblies. You developed sound assembly modeling techniques that utilized Standard mates, fixed components, symmetry, component patterns and mirrored components.

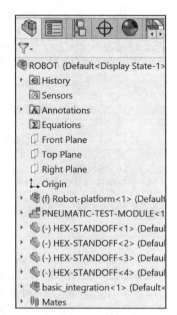

The WHEEL-AND-AXLE assembly combined the WHEEL-FLATBAR assembly with the AXLE-3000 part, SHAFTCOLLAR-500 part and HEX-ADAPTER part.

The WHEEL-FLATBAR assembly combined the 3HOLE-SHAFTCOLLAR assembly and 5HOLE-SHAFTCOLLAR assembly.

The PNEUMATIC-TEST-MODULE assembly combined four major mechanical sub-assemblies. The final ROBOT assembly was comprised of three major sub-assemblies and the HEX-STANDOFF component.

Organize your assemblies. For an exercise, create an assembly Layout Diagram of the final ROBOT assembly to determine the grouping of components and assemblies.

Utilize the Quick mate procedure on Standard mates, Cam mate, Profile Center mate, Slot mate, Symmetric mate and Width mate. To activate the Quick Mates functionality, click Tools, Customize. On the toolbars tab, under Context toolbar settings, select Show quick mates. Quick Mates is selected by default.

As an exercise apply PhotoView 360 to render the final ROBOT assembly.

Questions

1. What function does the Save as copy and continue check box perform when saving a part under a new name?

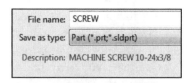

2. True or False. Never reuse geometry from one part to create another part.

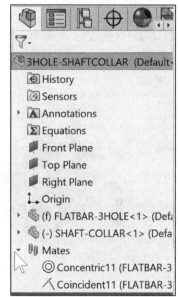

3. Describe four Assembly features utilized in this chapter.

4. True or False. A fixed (f) component cannot move and is locked to the Origin.

5. Describe the purpose of an Assembly Layout diagram.

6. Describe the difference between a Pattern Driven Component Pattern and a Linear Component Pattern.

7. Describe the difference between copied geometry and mirrored geometry utilizing the Mirror Components tool.

8. True or False. An assembly contains one or more configurations.

9. Identify how you can incorporate design intent into an assembly.

Additional projects are included in the exercise section of Chapter 6. Use the web to locate a similar assembly and to understand the parts needed in the assembly. View the provided sample parts. Create your own parts and final assembly. View the provided motion file to create the proper movement in SOLIDWORKS.

- Arbor Press Project
- Bench Vice Project
- Butterfly Valve Project
- Drill Guide Project
- Kant Twist Clamp Project
- Pipe Vice Project
- Pulley Project
- Quick Acting Clamp Project
- Radial Engine Project
- Shaper Tool Head Project
- Welder Arm Project

Exercises

Exercise 6.1: Butterfly Valve Project

Use the web to locate a similar assembly and to understand the parts needed in the assembly. View the provided sample parts. Create your own parts and final assembly. View the provided motion file to create the proper movement in SOLIDWORKS.

Create an ANSI Butterfly Valve assembly document. Insert all needed components and mates to assemble the assembly. Create and insert any additional components if needed.

Create a A-ANSI Landscape - Third Angle Isometric Exploded Drawing document with Explode lines of the assembly using your knowledge of SOLIDWORKS. Insert a BOM with Balloons. Insert all needed General notes and Custom Properties in the Title Block.

Arbor Press Project
Bench Vice Project
Butterfly Valve Project
Drill Guide Project
Kant Twist Clamp Project
Pipe Vice Project
Pulley Project
Quick Acting Clamp Project
Radial Engine Project
Shaper Tool Head Project
Welder Arm Project

ITEM NO.	PartNo	MATERIAL	DESCRIPTION	QTY.
1	1001	Gray Cast Iron	BODY	1
2	1002	Stainless Steel (ferritic)	SHAFT	1
3	1005	1060 Alloy	PLATE 2	1
4	1008	1060 Alloy	BOLT [small]	2
5	1003	1060 Alloy	PLATE	1
6	1004	1060 Alloy	BOLT	3
7	1007	Stainless Steel (ferritic)	ARM	1
8	1009	Stainless Steel (ferritic)	HANDLE	1
9	1010	1060 Alloy	HEX NUT	1

FINAL ASSEMBLY

SIZE A   DWG. NO. 2001   REV A

SCALE: 1:10   WEIGHT: SHEET 1 OF 3

Exercise 6.2: Quick Acting Clamp Project

Use the web to locate a similar assembly and to understand the parts needed in the assembly. View the provided sample parts. Create your own parts and final assembly. View the provided motion file to create the proper movement in SOLIDWORKS.

Create an ANSI Quick Acting clamp assembly document. Insert all needed components and mates to assemble the assembly. Create and insert any additional components if needed.

Create a C-ANSI Landscape - Third Angle Isometric Exploded Drawing document with Explode lines of the assembly using your knowledge of SOLIDWORKS. Insert a BOM with Balloons. Insert all needed General notes and Custom Properties in the Title Block.

- Arbor Press Project
- Bench Vice Project
- Butterfly Valve Project
- Drill Guide Project
- Kant Twist Clamp Project
- Pipe Vice Project
- Pulley Project
- Quick Acting Clamp Project
- Radial Engine Project
- Shaper Tool Head Project
- Welder Arm Project

ITEM NO.	PART NUMBER	DESCRIPTION	Material	QTY.
1	100	LEFT SUPPORT	Alloy Steel	1
2	WPJ-07200	PIN A	Alloy Steel	3
3	2133	ARM	Alloy Steel	2
4	34564	PIN B	Alloy Steel	1
6	34465	GRIP	Alloy Steel	1
6	WPJ-062477	HINGE-B	Alloy Steel	1
7	1eard	SHAFT	Alloy Steel	1
8	WPJ-05556yk	HINGE A	Alloy Steel	1
9	WPJ-1133	RIGHT SUPPORT	Alloy Steel	1
10	WPJ-03	HANDLE	Alloy Steel	1
11	WPJ-04	HANDLE RING	Alloy Steel	2

QUICK ACTING CLAMP

C 334-447 A

Exercise 6.3: Drill Guide Project

Use the web to locate a similar assembly and to understand the parts needed in the assembly. View the provided sample parts. Create your own parts and final assembly. View the provided motion file to create the proper movement in SOLIDWORKS.

Create an ANSI Drill Guide assembly document. Insert all needed components and mates to assemble the assembly. Create and insert any additional components if needed.

Create a C-ANSI Landscape - Third Angle Isometric Exploded Drawing document with Explode lines of the assembly using your knowledge of SOLIDWORKS. Insert a BOM with Balloons. Insert all needed General notes and Custom Properties in the Title Block.

- Arbor Press Project
- Bench Vice Project
- Butterfly Valve Project
- Drill Guide Project
- Kant Twist Clamp Project
- Pipe Vice Project
- Pulley Project
- Quick Acting Clamp Project
- Radial Engine Project
- Shaper Tool Head Project
- Welder Arm Project

ITEM NO.	PART NUMBER	DESCRIPTION	MATERIAL	QTY.
1	WPI-1000-01	BASE	2014 ALLOY	1
2	WPI-1000-02	ROTATOR	PLAIN CARBON STEEL	2
3	WPI-1000-03	ROD GUIDE	CAST ALLOY STEEL	2
4	WPI-1000-04	SLIDE	2014 ALLOY	1
5	WPI-1000-05	COLLAR	2014 ALLOY	2
6	WPI-1000-06	BUSHING	ALUMINUM BRONZE	2
7	WPI-1000-07	RETAINING RING B27.1 - NA2-45	ALLOY STEEL	2
8	WPI-1000-08	DRILL ADAPTOR	PLAIN CARBON STEEL	1
9	WPI-1000-09	THUMB SCREW .25-20x0.51 TYPE B, FLAT POINT-C	2014 ALLOY	4

Exercise 6.4: Pulley Project

Use the web to locate a similar assembly and to understand the parts needed in the assembly. View the provided sample parts. Create your own parts and final assembly. View the provided motion file to create the proper movement in SOLIDWORKS.

Create an ANSI Pully assembly document. Insert all needed components and mates to assemble the assembly. Create and insert any additional components if needed.

Create a A-ANSI Landscape - Third Angle Isometric Exploded Drawing document with Explode lines of the assembly using your knowledge of SOLIDWORKS. Insert a BOM with Balloons. Insert all needed General notes and Custom Properties in the Title Block.

Note: Additional components were added below.

- Arbor Press Project
- Bench Vice Project
- Butterfly Valve Project
- Drill Guide Project
- Kant Twist Clamp Project
- Pipe Vice Project
- Pulley Project
- Quick Acting Clamp Project
- Radial Engine Project
- 'Shaper Tool Head Project
- Welder Arm Project

ITEM NO.	PART NUMBER	DESCRIPTION	MATERIAL	QTY.
1	6	Pulley Base	Balsa	1
2	5	Bracket	Alloy Steel	2
3	2	Brace	Alloy Steel	4
4	8	Hex Nut	Alloy Steel	6
5	4	Support'	Alloy Steel	2
6	9	Screw Cap	Alloy Steel	4
7	10	Turn Screw	Alloy Steel	2
8	3	Bushing	Alloy Steel	2
9	1	Pulley	Alloy Steel	2
10	7	Washer	Alloy Steel	2
11	11	Belt1-1	Silicon Rubber	1

Exercise 6.5: Welder Arm Project

Use the web to locate a similar assembly and to understand the parts needed in the assembly. View the provided sample parts. Create your own parts and final assembly. View the provided motion file to create the proper movement in SOLIDWORKS.

Create an ANSI Welder Arm assembly document. Insert all needed components and mates to assemble the assembly. Create and insert any additional components if needed.

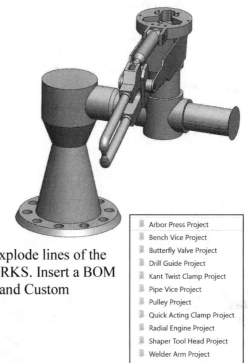

Create a C-ANSI Landscape - Third Angle Isometric Exploded Drawing document with Explode lines of the assembly using your knowledge of SOLIDWORKS. Insert a BOM with Balloons. Insert all needed General notes and Custom Properties in the Title Block.

Arbor Press Project
Bench Vice Project
Butterfly Valve Project
Drill Guide Project
Kant Twist Clamp Project
Pipe Vice Project
Pulley Project
Quick Acting Clamp Project
Radial Engine Project
Shaper Tool Head Project
Welder Arm Project

Exercise 6.6: Radial Engine Project

Use the web to locate a similar assembly and to understand the parts needed in the assembly. View the provided sample parts. Create your own parts and final assembly. View the provided motion file to create the proper movement in SOLIDWORKS.

Create an ANSI Radial Engine assembly document. Insert all needed components and mates to assemble the assembly. Create and insert any additional components if needed.

Create a C-ANSI Landscape - Third Angle Isometric Exploded Drawing document with Explode lines of the assembly using your knowledge of SOLIDWORKS. Insert a BOM with Balloons. Insert all needed General notes and Custom Properties in the Title Block.

Note: Additional components were added to the drawing.

Exercise 6.7: Bench Vice Project

Use the web to locate a similar assembly and to understand the parts needed in the assembly. View the provided sample parts. Create your own parts and final assembly. View the provided motion file to create the proper movement in SOLIDWORKS.

Create an ANSI Bench Vice assembly document. Insert all needed components and mates to assemble the assembly. Create and insert any additional components if needed.

- Arbor Press Project
- Bench Vice Project
- Butterfly Valve Project
- Drill Guide Project
- Kant Twist Clamp Project
- Pipe Vice Project
- Pulley Project
- Quick Acting Clamp Project
- Radial Engine Project
- Shaper Tool Head Project
- Welder Arm Project

Create a A-ANSI Landscape - Third Angle Isometric Exploded Drawing document with Explode lines of the assembly using your knowledge of SOLIDWORKS.
Insert a BOM with Balloons. Insert all needed General notes and Custom Properties in the Title Block.

ITEM NO.	PART NUMBER	DESCRIPTION	MATERIAL	QTY.
1	WPI - 0001	Base	Alloy Steel	1
2	WPI - 0002	Base Plate	Alloy Steel	2
3	WPI - 0003	Vice Jaw	Alloy Steel	1
4	WPI - 0004	Clamping Plate	Alloy Steel	1
5	WPI - 0005	Jaw Screw	Alloy Steel	1
6	WPI - 0006	Screw Bar	Alloy Steel	1
7	WPI - 0007	Bar Globes	Alloy Steel	2
8	WPI-008	Hex Screw M6x1.0x14-14WN	Alloy Steel	4
9	WPI-0009	Bolt	Alloy Steell	1
10	WPI- 00019	Blot1	Alloy Steel	2

Exercise 6.8: Kant Twist Clamp Project

Use the web to locate a similar assembly and to understand the parts needed in the assembly. View the provided sample parts. Create your own parts and final assembly. View the provided motion file to create the proper movement in SOLIDWORKS.

Create an ANSI Kant Twist clamp assembly document. Insert all needed components and mates to assemble the assembly. Create and insert any additional components if needed.

Create a C-ANSI Landscape - Third Angle Isometric Exploded Drawing document with Explode lines of the assembly using your knowledge of SOLIDWORKS. Insert a BOM with Balloons. Insert all needed General notes and Custom Properties in the Title Block.

Note: Additional components were added to the drawing.

ITEM NO.	PART NUMBER	DESCRIPTION	MATERIAL	QTY.
1	101	BASE	ALLOY STEEL	1
2	108	HINGE A	ALLOY STEEL	1
3	102	BRACE A	ALLOY STEEL	2
4	110	PIN A	ALLOY STEEL	4
5	104	GRIP A	ALLOY STEEL	1
6	114	CR-FHMI 0.19-32x0.8x0.5-N	ALLOY STEEL	2
7	103	BRACE B	ALLOY STEEL	2
8	111	PIN B	ALLOY STEEL	2
9	109	HINGE B	ALLOY STEEL	1
10	105	GRIP B	ALLOY STEEL	1
11	112	SHAFT	ALLOY STEEL	1
12	107	HANDLE SHAFT	ALLOY STEEL	1
13	106	HANDLE GRIP	RUBBER	1
14	115	CR-BHMS 0.19-32x0.25x0.25-N	ALLOY STEEL	1
15	113	CR-FHM1 0.19-32x0.5x0.5-N	ALLOY STEEL	1

Kant Twist Clamp

C WPI-1-101-112 A

Exercise 6.9: Arbor Press Assembly Project

Use the web to locate a similar assembly and to understand the parts needed in the assembly. View the provided sample parts. Create your own parts and final assembly. View the provided motion file to create the proper movement in SOLIDWORKS.

Create an ANSI Arbor Press assembly document. Insert all needed components and mates to assemble the assembly. Create and insert any additional components if needed.

Create a A-ANSI Landscape - Third Angle Isometric Exploded Drawing document with Explode lines of the assembly using your knowledge of SOLIDWORKS. Insert a BOM with Balloons. Insert all needed General notes and Custom Properties in the Title Block.

- Arbor Press Project
- Bench Vice Project
- Butterfly Valve Project
- Drill Guide Project
- Kant Twist Clamp Project
- Pipe Vice Project
- Pulley Project
- Quick Acting Clamp Project
- Radial Engine Project
- Shaper Tool Head Project
- Welder Arm Project

ITEM NO.	PART NUMBER	DESCRIPTION	Material	QTY.
1	1000-14	ARBOR PRESS SUBASSEMBLY		1
1.1	1000-6	COLUMN	AISI 1020	1
1.2	1000-2	FRONT COVER	AISI 1020	1
1.3	1000-7	HEX_SCREW_AI	AISI 1035 Steel (SS)	4
2	1000-1	TABLE PLATE	AISI 1035 Steel (SS)	1
3	1000-5	STUBBY PEG	AISI 1035 Steel (SS)	1
4	1000-3	SQUARE PEG	AISI 1035 Steel (SS)	1
5	1000-4	HANDLE	AISI 1035 Steel (SS)	1
6	1000-8	Part1^ARBOR PRESS	AISI 1035 Steel (SS)	1

Exercise 6.10: Shaper tool head Assembly Project

Use the web to locate a similar assembly and to understand the parts needed in the assembly. View the provided sample parts. Create your own parts and final assembly. View the provided motion file to create the proper movement in SOLIDWORKS.

Create an ANSI Shaper tool head assembly document. Insert all needed components and mates to assemble the assembly. Create and insert any additional components if needed.

Create a C-ANSI Landscape - Third Angle Isometric Exploded Drawing document with Explode lines of the assembly using your knowledge of SOLIDWORKS. Insert a BOM with Balloons. Insert all needed General notes and Custom Properties in the Title Block.

📄	Arbor Press Project
📄	Bench Vice Project
📄	Butterfly Valve Project
📄	Drill Guide Project
📄	Kant Twist Clamp Project
📄	Pipe Vice Project
📄	Pulley Project
📄	Quick Acting Clamp Project
📄	Radial Engine Project
📄	Shaper Tool Head Project
📄	Welder Arm Project

Exercise 6.11: Quick Acting Clamp Project

Use the web to locate a similar assembly and to understand the parts needed in the assembly. View the provided sample parts. Create your own parts and final assembly. View the provided motion file to create the proper movement in SOLIDWORKS.

Create an ANSI Quick acting clamp assembly document. Insert all needed components and mates to assemble the assembly. Create and insert any additional components if needed.

Create a A-ANSI Landscape - Third Angle Isometric Exploded Drawing document with Explode lines of the assembly using your knowledge of SOLIDWORKS. Insert a BOM with Balloons. Insert all needed General notes and Custom Properties in the Title Block.

- Arbor Press Project
- Bench Vice Project
- Butterfly Valve Project
- Drill Guide Project
- Kant Twist Clamp Project
- Pipe Vice Project
- Pulley Project
- Quick Acting Clamp Project
- Radial Engine Project
- Shaper Tool Head Project
- Welder Arm Project

ITEM NO.	PART NUMBER	DESCRIPTION	Material	QTY.
1	10	LEFT SUPPORT	Alloy Steel	1
2	07	PIN A	Alloy Steel	3
3	01	ARM	Alloy Steel	2
4	08	PIN B	Alloy Steel	1
5	-02	GRIP	Alloy Steel	1
6	-06	HINGE B	Alloy Steel	1
7	-09	SHAFT	Alloy Steel	1
8	-05	HINGE A	Alloy Steel	1
9	-11	RIGHT SUPPORT	Alloy Steel	1
10	-03	HANDLE	Alloy Steel	1
11	-04	HANDLE PINS	Alloy Steel	2

QUICK ACTING CLAMP

Notes:

CHAPTER 7 - CSWA INTRODUCTION AND DRAFTING COMPETENCIES

Introduction

The CSWA Academic exam is provided either in a single 3-hour segment, or 2 - 90 minute segments.

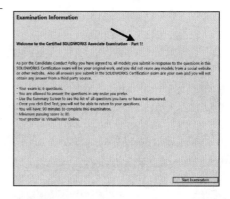

Part 1 of the CSWA Academic exam is 90 minutes, minimum passing score is 80, with 6 questions. There are two questions in the **Basic Part Creation and Modification** category, two questions in the **Intermediate Part Creation and Modification** category and two questions in the **Assembly Creation and Modification** category.

Part 2 of the CSWA Academic exam is 90 minutes, minimum passing score is 80 with 8 questions. There are three questions on the CSWA Academic exam in the **Drafting Competencies** category, three questions in the **Advanced Part Creation and Modification** category and two questions in the **Assembly Creation and Modification** category.

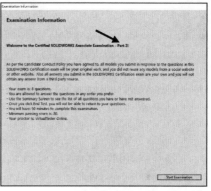

You need to pass both segments to obtain your CSWA Certification. Once you pass a segment, you will not have to take it again. You can take the segments in any order. If you fail a segment, there is a 14-day waiting period before retaking the failed segment.

The CSWA 3-hour Academic exam consists of 14 questions worth a total of 240 points. All exams cover the same material. You need to wait 14 days before you can retake the exam.

The CSWA exam consists of 14 questions in the following categories and subject areas:

- **Drafting Competencies**: (Three questions - multiple choice - 5 points each).

  - Questions on general drawing views: Projected, Section, Break, Crop, Detail, Alternate Position, etc.

- **Basic Part Creation and Modification**: (Two questions - one multiple choice/one single answer - 15 points each).

  - Sketch Planes:

    - Front, Top, Right.

  - 2D Sketching:

    - Geometric relations and dimensioning.

  - Extruded Boss/Base Feature.

  - Extruded Cut feature.

  - Modification of basic part.

    - Sketch, Feature, Pattern, etc.

- **Intermediate Part Creation and Modification**: (Two questions - one multiple-choice/one single answer - 15 points each).

  - Sketch Planes:

    - Front, Top, Right, face, plane, etc.

  - 2D Sketching:

    - Geometric relations and dimensioning.

  - Extruded Boss/Base Feature.

  - Extruded Cut Feature.

  - Revolved Boss/Base Feature.

  - Mirror, Chamfer, and Fillet Feature.

  - Circular and Linear Pattern Feature.

  - Plane Feature.

  - Modification of Intermediate part:

    - Sketch, Feature, Pattern, etc.

If your school is an academic certification provider, your instructor can allocate free exam credits for the CSWA (Segment 1 & 2), CSWA-SD, CSWA-S, and CSWA-AM certifications. The instructor will require your .edu email address.

Advanced Part Creation and Modification: (Three questions - one multiple-choice/two single answers - 15 points each).

- Sketch Planes:

    - Front, Top, Right, Face, Created Plane etc.

    - 2D Sketching or 3D Sketching.

- Sketch Tools:

    - Circle, Line, Rectangle, Offset Entities, Convert Entities, etc.

- Extruded Boss/Base Feature.

- Extruded Cut Feature.

- Revolved Boss/Base Feature.

- Mirror and Fillet Feature.

- Circular and Linear Pattern Feature.

- Shell Feature.

- Plane Feature.

- Rib Feature.

- Modifications of geometry.

Visit https://3dexperience.virtualtester.com/#home to take a free sample online CSWA exam.

There is a 30-day waiting period before retaking the sample online CSWA exam.

Note: The sample online CSWA exam is provided in the book (.pdf) format along with the SOLIDWORKS models, solutions, and additional practice exam problems.

- **Assembly Creation and Modification**: (Two different assemblies - four questions - two multiple choice/two single answers - 30 points each).

  - Download the Assembly zip file.

  - Save the zip file to a local hard drive.

  - Un-zip the components.

  - Create a new Assembly document.

  - Insert the first component. Fix the first component to the origin of the assembly.

  - Insert all needed components. Follow the provided stated assembly conditions.

  - Insert the needed mates: Coincident, Concentric, Perpendicular, Parallel, Tangent, Distance, Angle, Advanced Distance, Advanced Angle, and Aligned, Anti-Aligned option.

  - Use the Measure tool.

  - Modifications of geometry.

In the Assembly Creation and Modification category, expect to see five to eight components. There are two dimension modification questions based on the first (multiple-choice) question. You should be within 1% of the multiple-choice answer before you go on to the modification single answer section.

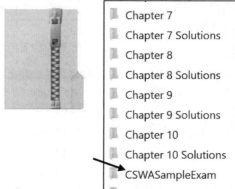

Taking the CSWA Exam (Segment 1 or 2)

Go to the 3DEXPERIENCE® Certification Center at https://3dexperience.virtualtester.com/#home. The 3DEXPERIENCE® Certification Center is where you are able to login to manage your certificates, take certifications, and make changes to your account settings.

If your school is an academic certification provider, your instructor can allocate a free CSWA Segment 1 and Segment 2 exam credit. The instructor will require your .edu email address.

Download the TesterPRO Client.

Execute the Tester PRO Client.

Read the License Agreement. Agree to the Candidate Conduct Policy. Click **I Agree**.

Install the Tester PRO Client. Click **install**.

Click **Finish**.

Select test **Language**.

Click **Continue**.

Create an account. If you already have an account, select "**I already have a VirtualTester UserID and password**".

Login. **Click Continue**.

Click **SOLIDWORKS**.

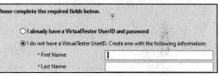

Select CSWA Academic - Part 1 or Part 2. This will depend on the exam credit that your instructor provided. Click **Start Exam**.

Select **language**. Click **Start Examination**.

Click **Yes**.

Read the instructions. Agree to the Confidentiality Agreement and Candidate Conduct Policy. **Click I Agree**.

Click **Start Examination**.

Objectives

Drafting Competencies is one of the categories on the CSWA exam.

There are three questions (total) on the CSWA exam in the Drafting Competencies category. Each question is worth five (5) points. Drafting Competency questions are addressed in part 2 of the CSWA exam.

The three questions are in a multiple-choice single answer format. You are allowed to answer the questions in any order you prefer. Use the Summary Screen during the exam to view the list of all questions you have or have not answered.

In the Drafting Competencies category of the exam, you are **not required** to create or perform an analysis on a part, assembly, or drawing but you are required to have general drafting/drawing knowledge and understanding of various drawing view methods.

On the completion of the chapter, you will be able to:

- Identify the procedure to create a named drawing view:

  - Projected view, Section view, Break view, Crop view, Detail, Alternate Position view, etc.

In the Basic Part Creation and Modification, Intermediate Part Creation and Modification, Advanced Part Creation and Modification and Assembly Creation and Modification categories, you are required to read and interpret various types of drawing views and understand various types of drawing annotations.

Download all needed model files from the SDC Publications website (www.SDCpublications.com/downloads/978-1-63057-556-4).

Note: The sample online CSWA exam is provided in the book (.pdf) format along with the SOLIDWORKS models, solutions, and additional practice exam problems.

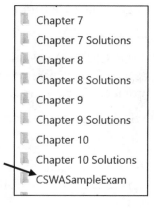

Procedure to Create a Named Drawing view

You need the ability to identify the procedure to create a named drawing view: Standard 3 View, Model View, Projected View, Auxiliary View, Section View, Removed Section, Detail View, Broken-out Section, Break, Crop View and Alternate Position View.

Create a Section view in a drawing by cutting the parent view with a section line. The Section view can be a straight cut section or an offset section defined by a stepped section line. The section line can also include concentric arcs.

Create an Aligned Section view in a drawing through a model, or portion of a model, that is aligned with a selected section line segment. The Aligned Section view is similar to a Section view, but the section line for an aligned section comprises two or more lines connected at an angle.

Create a Crop view, sketch a closed profile such as a circle or spline. The view outside the closed profile disappears as illustrated.

🔅 Crop any drawing view except a Detail view, a view from which a Detail view has been created, or an Exploded view.

Create a Detail view in a drawing to display a portion of a view, usually at an enlarged scale. This detail may be of an orthographic view, a non-planar (isometric) view, a Section view, a Crop view, an Exploded assembly view, or another Detail view.

Section View

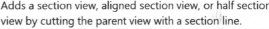

Adds a section view, aligned section view, or half section view by cutting the parent view with a section line.

Detail View

Adds a detail view to show a portion of a view, usually at an enlarged scale.

Tutorial: Drawing Named Procedure 7-1

Identify the drawing name view and understand the procedure to create the name view.

1. **View** the illustrated drawing views. The top drawing view is a Break view. The Break view is created by adding a break line to a selected view.

💡 Broken views make it possible to display the drawing view in a larger scale on a smaller size drawing sheet. Reference dimensions and model dimensions associated with the broken area reflect the actual model values.

💡 In views with multiple breaks, the Break line style must be the same.

Tutorial: Drawing Named Procedure 7-2

Identify the drawing name view and understand the procedure to create the name view.

1. **View** the illustrated drawing views. The top drawing view is a Section view of the bottom view. The Section view is created by cutting the parent view with a cutting section line.

💡 Create a Section view in a drawing by cutting the parent view with a section line. The section view can be a straight cut section or an offset section defined by a stepped section line. The section line can also include Concentric arcs.

Tutorial: Drawing Named Procedure 7-3

Identify the drawing name view and understand the procedure to create the name view.

1. **View** the illustrated drawing views. The view to the right is an Auxiliary view of the Front view. Select a reference edge to create an auxiliary view as illustrated.

💡 An Auxiliary view is similar to a Projected view, but it is unfolded normal to a reference edge in an existing view.

Tutorial: Drawing Named Procedure 7-4

Identify the drawing name view and understand the procedure to create the name view.

1. **View** the illustrated drawing views. The right drawing view is an Aligned half Section view of the view to the left. The Section view is created by using two lines connected at an angle. Create an Aligned half Section view in a drawing through a model, or portion of a model, that is aligned with a selected section line segment.

🔅 The Aligned Section view is very similar to a Section View, with the exception that the section line for an aligned half section is comprised of two or more lines connected at an angle.

Tutorial: Drawing Named Procedure 7-5

Identify the drawing name view and understand the procedure to create the name view.

1. **View** the illustrated drawing views. The left drawing view is a Detail view of the Section view. The Detail view is created by sketching a circle with the Circle Sketch tool. Click and drag for the location.

SECTION A-A

The Detail view ⒶA tool provides the ability to add a Detail view to display a portion of a view, usually at an enlarged scale.

🔅 To create a profile other than a circle, sketch the profile before clicking the Detail view tool. Using a sketch entity tool, create a closed profile around the area to be detailed.

Tutorial: Drawing Named Procedure 7-6

Identify the drawing name view and understand the procedure to create the name view.

1. **View** the illustrated drawing views. The top drawing view is a Broken-out Section view of the bottom drawing view. The Broken-out Section View is part of an existing drawing view, not a separate view. Create the Broken-out Section view with a closed profile, usually by using the Spline Sketch tool. Material is removed to a specified depth to expose inner details.

Tutorial: Drawing Named Procedure 7-7

Identify the drawing name view and understand the procedure to create the name view.

1. **View** the illustrated drawing view. The top drawing view is a Crop view. The Crop view is created by a closed sketch profile such as a circle, or spline as illustrated.

The Crop View provides the ability to crop an existing drawing view. You cannot use the Crop tool on a Detail view, a view from which a Detail view has been created, or an Exploded view.

Use the Crop tool to save steps. Example: instead of creating a Section View and then a Detail view, then hiding the unnecessary Section view, use the Crop tool to crop the Section view directly.

In the exam, you are allowed to answer the questions in any order. Use the Summary Screen during the exam to view the list of all questions you have or have not answered.

Tutorial: Drawing Named Procedure 7-8

Identify the drawing name view and understand the procedure to create the name view.

1. **View** the illustrated drawing view. The drawing view is an Alternate Position View. The

    Alternate Position view tool ▣ provides the ability to superimpose an existing drawing view precisely on another. The alternate position is displayed with phantom lines.

Intended Audience

The intended audience is anyone with a minimum of 6 - 9 months of SOLIDWORKS experience and basic knowledge of engineering fundamentals and practices.

SOLIDWORKS recommends that you review their SOLIDWORKS Tutorials on Parts, Assemblies and Drawings as a prerequisite and have at least 45 hours of classroom time learning SOLIDWORKS or using SOLIDWORKS with basic engineering design principles and practices.

To prepare for the CSWA exam, it is recommended that you first perform the following:

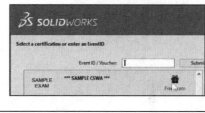

- Take the free sample online CSWA exam. Visit https://3dexperience.virtualtester.com/#home.

- Practice creating models from the isometric working drawings sections of any Technical Drawing or Engineering Drawing Documentation textbooks.

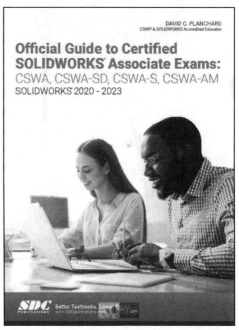

Additional references to help you prepare are as follows:

- **Official Guide to Certified SOLIDWORKS Associate Exams: CSWA, CSWA-SD, CSWA-S, CSWA-AM; 2020 - 2023,** 2017 - 2019, 2015 - 2012. **Note:** There are over 70 practice questions and examples (initial and final models) in the Official GUIDE to Certified SOLIDWORKS Exam book for the CSWA certification.

- **Engineering Drawing and Design**, Jensen & Helsel, Glencoe, 1990.

Summary

Drafting Competencies is one of the categories in the CSWA exam.

There are three questions on the CSWA exam in this category. Each question is worth five (5) points. The three questions are in a multiple-choice single answer format.

If your school is an academic certification provider, your instructor can allocate free exam credits for the CSWA (Segment 1 & 2), CSWA-SD, CSWA-S, and CSWA-AM certifications. The instructor will require your .edu email address.

Visit https://3dexperience.virtualtester.com/#home to take the free sample online CSWA exam.

There is a 30-day waiting period before retaking the free sample online exam.

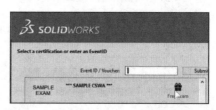

Basic Part Creation and Modification and Intermediate Part Creation and Modification is the next chapter in the book.

The chapter covers the knowledge to create and modify models for these categories from detailed dimensioned illustrations.

The complexity of the models along with the features progressively increases throughout the chapter to simulate the final types of models that could be provided on the exam.

Homework Questions

1. Identify the illustrated Drawing view.

- A: Projected
- B: Alternative Position
- C: Extended
- D: Aligned Section

2. Identify the illustrated Drawing view.

- A: Crop
- B: Break
- C: Broken-out Section
- D: Aligned Section

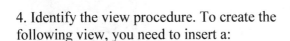

3. Identify the illustrated Drawing view.

- A: Section
- B: Crop
- C: Broken-out Section
- D: Aligned

4. Identify the view procedure. To create the following view, you need to insert a:

- A: Rectangle Sketch tool
- B: Closed Profile: Spline
- C: Open Profile: Circle
- D: None of the above

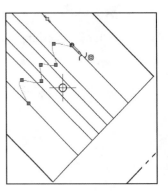

5. Identify the view procedure. To create the following view, you need to insert a:

- A: Open Spline

- B: Closed Spline

- C: 3 Point Arc

- D: None of the above

6. Identify the illustrated view type.

- A: Crop

- B: Section

- C: Projected

- D: Detail

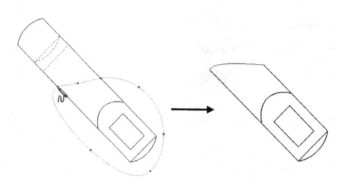

7. To create View B from Drawing View A insert which View type?

- A: Crop

- B: Section

- C: Aligned Section

- D: Projected

A   B

8. To create View B it is necessary to sketch a closed spline on View A and insert which View type?

- A: Broken out Section

- B: Detail

- C: Section

- D: Projected

9. To create View B it is necessary to sketch a closed spline on View A and insert which View type?

- A: Horizontal Break

- B: Detail

- C: Section

- D: Broken out Section

A   B

☼ Screen shots from an older CSWA exam for the Drafting Competencies category. Read each question carefully. Use SOLIDWORKS help if needed.

Zoom in on the part or view if needed.

Alternative Position View:

Screen shots from an older CSWA exam for the *Drafting Competencies* category. Read each question carefully. Use SOLIDWORKS help if needed.

Broken out Section View:

Section View:

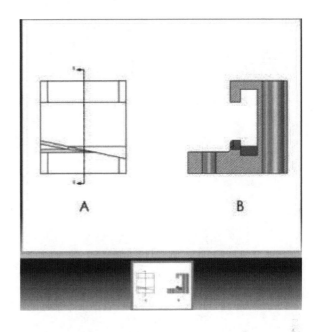

Notes:

Chapter 8 - Basic Part and Intermediate Part Creation and Modification

Chapter Objective

Provide in-depth coverage of the Basic Part Creation and Modification and Intermediate Part Creation and Modification categories on the CSWA exam. Knowledge to create and modify models in these categories from detailed dimensioned illustrations.

Introduction

The main difference between the Basic Part Creation and Modification category and the Intermediate Part Creation and Modification or the Advance Part Creation and Modification category is the complexity of the sketches and the number of dimensions and geometric relations along with an increase in the number of features.

Segment 1 of the CSWA Academic exam has two questions in the Basic Part Creation and Modification category and two questions in the Intermediate Part Creation and Modification category.

Segment 2 of the CSWA Academic exam has no questions in the Basic Part Creation and Modification category and no questions in the Intermediate Part Creation and Modification category.

Note: All answers are in the MMGS unit system. Decimal place 2.

On the completion of the chapter, you will be able to:

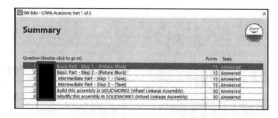

- Read and understand an Engineering document used in the CSWA exam.

- Identify the Sketch plane, part Origin location, part dimensions, geometric relations and design intent of the sketch and feature.

- Build a part from a detailed dimensioned illustration using the following SOLIDWORKS tools and features:

  - 2D & 3D sketch tools, Extruded Boss/Base, Extruded Cut, Fillet, Mirror, Revolved Base, Chamfer, Reference geometry, Plane, Axis, Calculate the overall mass of the created part and locate the Center of mass for the created part relative to the Origin.

The complexity of the models along with the features progressively increases throughout this chapter to simulate the final types of models that would be provided on the exam.

Engineering Documentation Practices

A 2D drawing view is displayed in the Basic Part Creation and Modification, Intermediate Part Creation and Modification, Advance Part Creation and Modification, and Assembly Creation and Modification categories of the CSWA exam to clarify dimensions and details.

The ability to interpret a 2D drawing view is required.

- Example 1: *8X Ø.19 EQ. SP.* Eight holes with a .19in. diameter are required that are equally (.55in.) spaced.

- Example 2: *R2.50 TYP.* Typical radius of 2.50. The dimension has a two decimal place precision.

- Example 3: ⊽. The Depth/Deep ⊽ symbol with a 1.50 dimension associated with the hole. The hole Ø.562 has a three decimal place precision.

- Example 4: *A+40*. A is provided to you on the CSWA exam. A + 40mm.

  💡 N is a Detail view of the M-M Section view.

- Example 5: *ØB*. Diameter of B. B is provided to you on the exam.

- Example 6: ⑥. Parallel.

- Example 7: ⑦ The faces are Coincident.

- Example 8: ⑧ Tangent.

  💡 Set the correct units in document properties. Set the correct material in the FeatureManager.

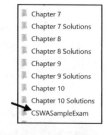

💡 Download all needed model files from the SDC Publication website www.SDCpublications.com/downloads/978-1-63057-556-4.

💡 Visit https://3dexperience.virtualtester.com/#home to take the free sample online CSWA exam. There is a 30-day waiting period before retaking the sample online CSWA exam.

The free sample online CSWA exam is included in the book along with the SOLIDWORKS models, solutions and additional questions.

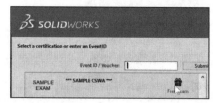

Document Properties

Create a new part.

Set Unit system and Decimal place.

💡 All answers in the CSWA exam are in the MMGS unit system. Decimal place: 2.

💡 If you don't find your answer (within 1%) in the multiple-choice single answer format section - recheck your model for precision and accuracy.

Build a Basic Part from a Detailed illustration

Tutorial: Volume/Center of Mass 8-1

Build this model in SOLIDWORKS. Calculate the volume of the part and locate the Center of mass with the provided information.

1. **Create** a new part in SOLIDWORKS.

2. **Build** the illustrated dimensioned model. The model displays all edges on perpendicular planes. Think about the steps to build the model. Insert two features: Extruded Base (Boss-Extrude1) and Extruded Cut (Cut-Extrude1). The part Origin is located in the front left corner of the model. Think about your Base Sketch plane. Keep your Base Sketch simple.

3. **Set** document properties for the model.

4. Create **Sketch1**. Select the Front Plane as the Sketch plane. Sketch1 is the Base sketch. Sketch1 is the profile for the Extruded Base (Boss-Extrude1) feature. Insert the required geometric relations and dimensions.

5. Create the **Extruded Base** feature. Boss-Extrude1 is the Base feature. Blind is the default End Condition in Direction 1. Depth = 2.25in. Identify the extrude direction to maintain the location of the Origin.

6. Create **Sketch2**. Select the Top right face as the Sketch plane for the second feature. Sketch a square. Sketch2 is the profile for the Extruded Cut feature. Insert the required geometric relations and dimensions.

7. Create the **Extruded Cut** feature. Select Through All for End Condition in Direction 1.

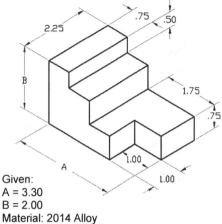

Given:
A = 3.30
B = 2.00
Material: 2014 Alloy
Density = .101 lb/in^3
Units: IPS
Decimal place: 2

8. **Assign** 2014 Alloy material to the part. Material is required to locate the Center of mass.

9. **Calculate** the volume. The volume = 8.28 cubic inches.

💡 There are numerous ways to build the models in this chapter. A goal is to display different design intents and techniques.

10. **Locate** the Center of mass. The location of the Center of mass is derived from the part Origin.

- X: 1.14 inch

- Y: 0.75 inch

- Z: -1.18 inch

11. **Save** the part and name it Volume-Center of mass 8-1.

12. **Close** the model.

💡 The principal axes and Center of mass are displayed graphically on the model in the Graphics window.

Tutorial: Volume/Center of Mass 8-2

Build this model. Calculate the volume of the part and locate the Center of mass with the provided information.

1. **Create** a new part in SOLIDWORKS.

2. **Build** the illustrated dimensioned model. The model displays all edges on perpendicular planes. Think about the steps that are required to build this model. Remember, there are numerous ways to create the models in this chapter.

Given:
A = 100
B = 40
Material: Brass
Density = .0085 g/mm^3
Units: MMGS
Decimal place: 2

Origin

The CSWA exam is timed. Work efficiently.

View the provided Part FeatureManagers. Both FeatureManagers create the same illustrated model. In Option1, there are four sketches and four features (Extruded Base and three Extruded Cuts) that are used to build the model.

In Option2, there are three sketches and three features (Extruded Boss/Base) that are used to build the model. Which FeatureManager is better? In a timed exam, optimize your time and use the least number of features through mirror, pattern, symmetry, etc.

Create the model using the Option2 Part FeatureManager.

3. **Set** document properties for the model.

4. Create **Sketch1**. Select the Top Plane as the Sketch plane. Sketch a Corner rectangle. Insert the required dimensions.

5. Create the **Extruded** feature. Boss-Extrude1 is the Base feature. Blind is the default End Condition in Direction 1. Depth = 10mm.

6. Create **Sketch2**. Select the back face of Boss-Extrude1.

7. Select **Normal To** view. Sketch2 is the profile for the second Extruded feature. Insert the required geometric relations and dimensions as illustrated.

8. Create the second Extruded feature (**Boss-Extrude2**). Blind is the default End Condition in Direction 1. Depth = 20mm. Note the direction of the extrude towards the front of the model.

9. Create **Sketch3**. Select the left face of Boss-Extrude1 as the Sketch plane. Sketch3 is the profile for the third Extrude feature. Insert the required geometric relations and dimensions.

10. Create the third Extruded feature (**Boss-Extrude3**). Blind is the default End Condition in Direction 1. Depth = 20mm.

11. **Assign** brass material to the part.

12. **Calculate** the volume of the model. The volume = 130,000.00 cubic millimeters.

13. **Locate** the Center of mass. The location of the Center of mass is derived from the part Origin.

- X: 43.46 millimeters

- Y: 15.00 millimeters

- Z: -37.69 millimeters

Density = 0.01 grams per cubic millimeter

Mass = 1105.00 grams

Volume = 130000.00 cubic millimeters

Surface area = 23400.00 square millimeters

Center of mass: (millimeters)
   X = 43.46
   Y = 15.00
   Z = -37.69

14. **Save** the part. Name it Volume-Center of mass 8-2.

15. **Calculate** the volume of the model using the IPS unit system. Decimal place: 2. The volume = 7.93 cubic inches.

16. **Locate** the Center of mass using the IPS unit system. Decimal place 2: The location of the Center of mass is derived from the part Origin.

- X: 1.71 inches

- Y: 0.59 inches

- Z: -1.48 inches

17. **Save** the part. Name it Volume-Center of mass 8-2-IPS.

18. **Close** the model.

Density = 0.31 pounds per cubic inch

Mass = 2.44 pounds

Volume = 7.93 cubic inches

Surface area = 36.27 square inches

Center of mass: (inches)
   X = 1.71
   Y = 0.59
   Z = -1.48

Tutorial: Mass-Volume 8-3

Build this model. Calculate the overall mass of the illustrated model with the provided information.

1. **Create** a new part in SOLIDWORKS.

2. **Build** the illustrated model. The model displays all edges on perpendicular planes. Think about the steps required to build the model. Apply the Mirror Sketch tool to the Base sketch. Insert an Extruded Base (Boss-Extrude1) and Extruded-Cut (Cut-Extrude1) feature.

3. **Set** document properties for the model.

Given:
A = 50, B = 50, C = 120
Material: 6061 Alloy
Density = .0027 g/mm^3
Units: MMGS
Decimal place: 2

To activate the Mirror Sketch tool, click **Tools**, **Sketch Tools**, **Mirror** from the Main menu. The Mirror PropertyManager is displayed.

4. Create **Sketch1**. Select the Front Plane as the Sketch plane. Apply the Mirror Sketch tool. Select the construction geometry to mirror about as illustrated. Select the Entities to mirror. Insert the required geometric relations and dimensions.

Construction geometry is ignored when the sketch is used to create a feature. Construction geometry uses the same line style as centerlines.

When you create a new part or assembly, the three default Planes (Front, Right and Top) are aligned with specific views.

5. Create the **Boss-Extrude1** feature.
   Boss-Extrude1 is the Base feature.
   Apply the Mid Plane End Condition in
   Direction 1 for symmetry.
   Depth = 50mm.

6. Create **Sketch2**. Select the right face
   for the Sketch plane. Sketch2 is the
   profile for the Extruded Cut feature.
   Insert the required geometric relations
   and dimensions. Apply construction
   geometry.

7. Create the **Extruded Cut** feature.
   Through All is the selected End
   Condition in Direction 1.

8. **Assign** 6061 Alloy material to the
   part.

9. **Calculate** the overall mass. The
   overall mass = 302.40 grams.

10. **Save** the part. Name it Mass-Volume
    8-3.

11. **Close** the model.

Tutorial: Mass-Volume 8-4

Build this model. Calculate the overall
mass of the part and locate the Center of
mass with the provided information.

1. **Create** a new part in SOLIDWORKS.

2. **Build** the illustrated model. All edges
   of the model are not located on
   Perpendicular planes. Think about the
   steps required to build the model.
   Insert two features: Extruded Base
   (Boss-Extrude1) and Extruded Cut
   (Cut-Extrude1).

3. **Set** document properties for the
   model.

Mass = 302.40 grams

Volume = 112000.00 cubic millimeters

Surface area = 26200.00 square millimeters

Center of mass: (millimeters)
   X = 0.00
   Y = 19.20
   Z = 0.00

Given:
A = 3.00, B = 1.00
Material: 6061 Alloy
Density = .097 lb/in^3
Units: IPS
Decimal place: 2

4. Create **Sketch1**. Select the Right Plane as the Sketch plane. Apply construction geometry. Insert the required geometric relations and dimensions.

5. Create the **Extruded Base** feature. Boss-Extrude1 is the Base feature. Apply symmetry. Select Mid Plane as the End Condition in Direction 1. Depth = 3.00in.

6. Create **Sketch2**. Select the Right Plane as the Sketch plane. Select the Line Sketch tool. Insert the required geometric relations. Sketch2 is the profile for the Extruded Cut feature.

7. Create the **Extruded Cut (Cut-Extrude1)** feature. Apply symmetry. Select Mid Plane as the End Condition in Direction 1. Depth = 1.00in.

8. **Assign** 6061 Alloy material to the part.

9. **Calculate** overall mass. The overall mass = 0.87 pounds.

10. **Locate** the Center of mass. The location of the Center of mass is derived from the part Origin.

*   X: 0.00 inches

*   Y: 0.86 inches

*   Z: 0.82 inches

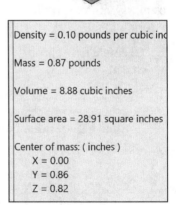

Density = 0.10 pounds per cubic inc

Mass = 0.87 pounds

Volume = 8.88 cubic inches

Surface area = 28.91 square inches

Center of mass: (inches)
    X = 0.00
    Y = 0.86
    Z = 0.82

In this category an exam question could read: Build this model. Locate the Center of mass with respect to the part Origin.

- A: X = 0.10 inches, Y = -0.86 inches, Z = -0.82 inches

- B: X = 0.00 inches, Y = 0.86 inches, Z = 0.82 inches

- C: X = 0.15 inches, Y = -0.96 inches, Z = -0.02 inches

- D: X = 1.00 inches, Y = -0.89 inches, Z = -1.82 inches

The correct answer is B.

11. **Save** the part. Name it Mass-Volume 8-4.

12. **Close** the model.

As an exercise, modify the Mass-Volume 8-4 part using the MMGS unit system. Assign Nickel as the material. Calculate the overall mass. The overall mass of the part = 1236.20 grams. Save the part. Name it Mass-Volume 8-4-MMGS.

Tutorial: Mass-Volume 8-5

Build this model. Calculate the overall mass of the part and locate the Center of mass with the provided information.

Given:
A = 100, B = 50, C = 60
Material: Alloy Steel
Density = .007 g/mm^3
Units: MMGS
Decimal Place: 2

1. **Create** a new part in SOLIDWORKS.

2. **Build** the illustrated model. Insert five sketches and five features to build the model: Extruded Base, three Extruded Cut features and a Mirror feature.

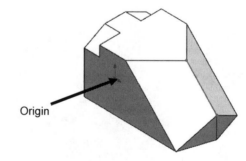

Origin

There are numerous ways to build the models in this chapter. A goal is to display different design intents and techniques.

3. **Set** document properties for the model.

4. Create **Sketch1**. Select the Front Plane as the Sketch plane. Sketch a rectangle. Insert the required relations and dimensions. The part Origin is located in the lower left corner of the sketch.

5. Create the **Extruded Base (Boss-Extrude1)** feature. Apply symmetry. Select the Mid Plane End Condition for Direction 1. Depth = 60mm.

6. Create **Sketch2**. Select the left face of Boss-Extrude1 as the Sketch plane. Insert the required geometric relations and dimensions.

7. Create the first **Extruded Cut** feature. Blind is the default End Condition in Direction 1. Depth = 15mm. Note the direction of the extrude feature.

8. Create **Sketch3**. Select the bottom face of Boss-Extrude1 for the Sketch plane. Insert the required geometric relations and dimension.

9. Create the second **Extruded Cut** feature. Blind is the default End Condition in Direction 1. Depth = 20mm.

10. Create **Sketch4**. Select Front Plane as the Sketch plane. Sketch a diagonal line. Sketch4 is the direction of extrusion for the third Extruded Cut feature. Insert the required dimension.

11. Create **Sketch5**. Select the top face of Boss-Extrude1 as the Sketch plane. Sketch5 is the sketch profile for the third Extruded Cut feature. Apply construction geometry. Insert the required geometric relations and dimensions.

12. Create the third **Extruded Cut** feature. Select Through All for End Condition in Direction 1.

13. Select **Sketch4** in the Graphics window for Direction of Extrusion. Line1@Sketch4 is displayed in the Cut-Extrude PropertyManager.

14. Create the **Mirror** feature. Mirror the three Extruded Cut features about the Front Plane. Use the fly-out FeatureManager.

15. **Assign** Alloy Steel material to the part.

16. **Calculate** the overall mass. The overall mass = 1794.10 grams.

17. **Locate** the Center of mass. The location of the Center of mass is derived from the part Origin.

- X = 41.17 millimeters

- Y = 22.38 millimeters

- Z = 0.00 millimeters

View the triad location of the Center of mass for the part.

18. **Save** the part. Name it Mass-Volume 8-5.

19. **Close** the model.

Set document precision from the Document Properties dialog box or from the Dimension PropertyManager. You can also address Callout value, Tolerance type, and Dimension Text symbols in the Dimension PropertyManager.

💡 You are allowed to answer the questions in any order you prefer. Use the Summary Screen during the CSWA exam to view the list of all questions you have or have not answered.

There are no Surfacing or Boundary feature questions on the CSWA exam at this time.

System Options	Document Properties

Drafting Standard
⊞ Annotations
⊞ Dimensions
  Virtual Sharps
⊞ Tables
⊞ DimXpert
Detailing
Grid/Snap
Units
Model Display
Material Properties
Image Quality
Sheet Metal MBD
Sheet Metal
Weldments
Plane Display
Configurations

Unit system
- ○ MKS (meter, kilogram, second)
- ○ CGS (centimeter, gram, second)
- ○ MMGS (millimeter, gram, second)
- ● IPS (inch, pound, second)
- ○ Custom

Type	Unit	Decimals
Basic Units		
Length	inches	.12
Dual Dimension Length	inches	.12
Angle	degrees	.12

Build Additional Basic Parts

Tutorial: Mass-Volume 8-6

Build this model. Calculate the overall mass of the part and locate the Center of mass with the provided information.

1. **Create** a new part in SOLIDWORKS.

2. **Build** the illustrated model. Think about the required steps to build this part. Insert four features: Extruded Base, two Extruded Cuts, and a Fillet. All Thru Holes.

Given:
A = 4.00
B = R.50
Material: 6061 Alloy
Density = .0975 lb/in^3
Units: IPS
Decimal places: 2

💡 There are numerous ways to build the models in this chapter. A goal is to display different design intents and techniques.

3. **Set** document properties for the model.

4. Create **Sketch1**. Select the Right Plane as the Sketch plane. The part Origin is located in the lower left corner of the sketch. Insert the required geometric relations and dimensions.

5. Create the **Extruded Base (Boss-Extrude1)** feature. Apply symmetry. Select the Mid Plane End Condition for Direction 1. Depth = 4.00in.

6. Create **Sketch2**. Select the top flat face of Boss-Extrude1 as the Sketch plane. Sketch a circle. The center of the circle is located at the part Origin. Insert the required dimension.

7. Create the first **Extruded Cut** feature. Select Through All for End Condition in Direction 1.

8. Create **Sketch3**. Select the front vertical face of Extrude1 as the Sketch plane. Sketch a circle. Insert the required geometric relations and dimensions.

9. Create the second **Extruded Cut** feature. Select Through All for End Condition in Direction 1.

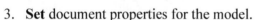

10. Create the **Fillet** feature. Constant radius is selected by default. Fillet the top two edges as illustrated. Radius = .50in.

Radius: 0.5in

A Fillet feature removes material. Selecting the correct radius value is important to obtain the correct mass and volume answer in the exam.

11. **Assign** the defined material.

12. **Calculate** the overall mass. The overall mass = 0.66 pounds.

13. **Locate** the Center of mass. The location of the Center of mass is derived from the part Origin.

- X: 0.00 inches

- Y: 0.90 inches

- Z: -1.46 inches

Mass = 0.66 pounds

Volume = 6.76 cubic inches

Surface area = 36.99 square inches

Center of mass: (inches)
   X = 0.00
   Y = 0.90
   Z = -1.46

In this category an exam question could read: Build this model. Locate the Center of mass relative to the part Origin.

- A: X = -2.63 inches, Y = 4.01 inches, Z = -0.04 inches

- B: X = 4.00 inches, Y = 1.90 inches, Z = -1.64 inches

- C: X = 0.00 inches, Y = 0.90 inches, Z = -1.46 inches

- D: X = -1.69 inches, Y = 1.00 inches, Z = 0.10 inches

The correct answer is C. Note: Tangent edges and Origin is displayed for educational purposes.

14. **Save** the part. Name it Mass-Volume 8-6.

15. **Close** the model.

As an exercise, calculate the overall mass of the. Use MMGS unit system.

Assign 2014 Alloy material.

Decimal place: 2.

The overall mass of the part = 310.17 grams.

Save the part. Name it Mass-Volume 8-6-MMGS.

Mass = 310.17 grams

Volume = 110774.26 cubic millimeters

Surface area = 23865.83 square millimeters

Center of mass: (millimeters)
   X = 0.00
   Y = 22.83
   Z = -37.11

Tutorial: Mass-Volume 8-7

Build this model. Calculate the overall mass of the part and locate the Center of mass with the provided information.

1. **Create** a new part in SOLIDWORKS.

2. **Build** the illustrated model. Insert two features: Extruded Base (Boss-Extrude1) and Revolved Boss.

3. **Set** document properties for the model.

🔅 Tangent edges and the Origin are displayed for educational purposes.

4. Create **Sketch1**. Select the Top Plane as the Sketch plane. Apply construction geometry. Apply the Tangent Arc and Line Sketch tool. Insert the required geometric relations and dimensions.

5. Create the **Extruded Base** feature. Blind is the default End Condition. Depth = 8mm.

6. Create **Sketch2**. Select the Front Plane as the Sketch plane. Apply construction geometry for the Revolved Boss feature. Insert the required geometric relations and dimension.

Given:
A = 60, B = 40, C = 8
Material: Cast Alloy Steel
Density = .0073 g/mm^3
Units: MMGS
Decimal place: 2

7. Create the **Revolved Boss** feature. The default angle is 360deg. Select the centerline for Axis of Revolution.

8. **Assign** the defined material to the part.

9. **Calculate** the overall mass. The overall mass = 229.46 grams.

10. **Locate** the Center of mass. The location of the Center of mass is derived from the part Origin.

- X = -46.68 millimeters

- Y = 7.23 millimeters

- Z = 0.00 millimeters

In this category an exam question could read: Build this model. What is the overall mass of the part? Decimal place:2. Unit system: MMGS.

- A: 229.46 grams

- B: 249.50 grams

- C: 240.33 grams

- D: 120.34 grams

The correct answer is A.

11. **Save** the part. Name it Mass-Volume 8-7.

12. **Close** the model.

Tangent edges and Origin are displayed for educational purposes.

🔅 When you create a new part or assembly, the three default Planes (Front, Right and Top) are aligned with specific views. The Plane you select for the Base sketch determines the orientation of the part.

Mass = 229.46 grams

Volume = 31433.02 cubic millimeters

Surface area = 9459.63 square millimeters

Center of mass: (millimeters)
  X = -46.68
  Y = 7.23
  Z = 0.00

Tutorial: Basic/Intermediate Part 8-1

Build this model. Calculate the overall mass of the part and locate the Center of mass with the provided information.

1. **Create** a new part in SOLIDWORKS.

2. **Build** the illustrated model. Think about the various features that create the part. Insert seven features and a plane to build this part: Extruded-Thin1, Boss-Extrude1, Cut-Extrude1, Cut-Extrude2 and three Fillets. Apply reference construction planes to build the circular features. All Thru Holes.

3. **Set** document properties for the model.

4. Create **Sketch1**. Select the Front Plane as the Sketch plane. Apply construction geometry as the reference line for the 30deg angle. Insert the required geometric relations and dimensions. Note the location of the Origin.

5. Create the **Extrude-Thin1** feature. This is the Base feature. Apply symmetry. Select Mid Plane for End Condition in Direction 1 to maintain the location of the Origin. Depth = 52mm. Thickness = 12mm.

Use the Thin Feature option to control the extrude thickness, not the Depth.

Given:
A = 52, B = 58
ALL-ROUNDS R 4MM
Material: 6061 Alloy
Density: .0027 g/mm^3
Units: MMGS
Decimal place: 2

6. Create **Plane1**. Plane1 is the Sketch plane for the Extruded Boss (Boss-Extrude1) feature. Select the midpoint and the top face as illustrated. Plane1 is located in the middle of the top and bottom faces. Select Parallel Plane at Point for option.

 Plane1 uses the Depth dimension of 32mm.

7. Create **Sketch2**. Select Plane1 as the Sketch plane. Use the Normal To view tool. Sketch a circle to create the Extruded Boss feature. Insert the required geometric relations.

 The Normal To view tool rotates and zooms the model to the view orientation normal to the selected plane, planar face, or feature.

8. Create the **Extruded Boss** feature. Apply Symmetry. Select Mid Plane for End Condition in Direction 1. Depth = 32mm.

9. Create **Sketch3**. Select the top circular face of Boss-Extrude1 as the Sketch plane. Sketch a circle. Insert the required geometric relation and dimension.

There are numerous ways to create the models in this chapter. A goal is to display different design intents and techniques.

10. Create the first **Extruded Cut** feature. Select Through All for End Condition in Direction 1.

11. Create **Sketch4**. Select the top face of Extrude-Thin1 as the Sketch plane. Apply construction geometry. Insert the required geometric relations and dimensions.

12. Create the second **Extruded Cut** feature. Select Through All for End Condition in Direction 1.

13. Create the **Fillet1** feature. Fillet the left and right edges of Extrude-Thin1 as illustrated. Radius = 12mm.

14. Create the **Fillet2** feature. Fillet the top and bottom edges of Extrude-Thin1 as illustrated. Radius = 4mm.

15. Create the **Fillet3** feature. Fillet the rest of the model, six edges as illustrated. Radius = 4mm.

16. **Assign** the defined material.

17. **Calculate** the overall mass of the part. The overall mass = 300.65 grams.

18. **Locate** the Center of mass. The location of the Center of mass is derived from the part Origin.

- X: 34.26 millimeters

- Y: -29.38 millimeters

- Z: 0.00 millimeters

19. **Save** the part. Name it Part-Modeling 8-1.

20. **Close** the model.

Mass = 300.65 grams

Volume = 111352.30 cubic millimeters

Surface area = 23765.33 square millimeters

Center of mass: (millimeters)
  X = 34.26
  Y = -29.38
  Z = 0.00

As an exercise, modify the Fillet2 and Fillet3 radius from 4mm to 2mm. Modify the Fillet1 radius from 12m to 10mm. Modify the material from 6061 Alloy to ABS.

Modify the Sketch1 angle from 30deg to 45deg. Modify the Extrude depth from 32mm to 38mm. Recalculate the location of the Center of mass with respect to the part Origin.

- X = 27.62 millimeters

- Y = -40.44 millimeters

- Z = 0.00 millimeters

21. **Save** the part. Name it Part-Modeling 8-1-Modify.

In the exam, you can answer the questions in any order. Use the Summary Screen during the exam to view the list of all questions you have or have not answered.

Mass = 123.60 grams

Volume = 121173.81 cubic millimeters

Surface area = 25622.46 square millimeters

Center of mass: (millimeters)
  X = 27.62
  Y = -40.44
  Z = 0.00

Tutorial: Basic/Intermediate-Part 8-2

Build this model. Calculate the volume of the part and locate the Center of mass with the provided information.

1. **Create** a new part in SOLIDWORKS.

2. **Build** the illustrated model. Think about the various features that create this model. Insert five features and a plane to build this part: Extruded Base, two Extruded Bosses, Extruded Cut and a Rib. Insert a reference plane to create the Boss-Extrude2 feature. All Thru Holes.

Given:
A = Ø3.00, B = 1.00
Material: 6061 Alloy
Density: .097 lb/in^3
Units: IPS
Decimal place: 2

3. **Set** document properties for the model.

4. Create **Sketch1**. Select the Top Plane as the Sketch plane. Sketch a center rectangle. Use the horizontal construction line as the Plane1 reference. Insert the required relations and dimensions.

5. Create the **Extruded Base** feature. Blind is the default End Condition in Direction 1. Depth = 1.00in. Note the extrude direction is downward.

💡 Create planes to aid in the modeling for the exam. Use planes to sketch, to create a section view, for a neutral plane in a draft feature, and so on.

💡 The created plane is displayed 5% larger than the geometry on which the plane is created, or 5% larger than the bounding box. This helps reduce selection problems when planes are created directly on faces or from orthogonal geometry.

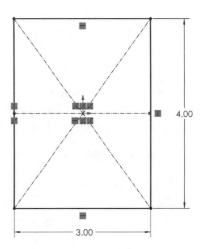

6. Create **Plane1**. Plane1 is the Sketch plane for the Extruded Boss feature. Show Sketch1. Select the horizontal construction line in Sketch1 and the top face of Boss-Extrude1. Angle = 48deg.

🔆 Click **View**, **Hide/Show**, **Sketches** from the Menu bar to displayed sketches in the Graphics window.

🔆 The Normal To view tool rotates and zooms the model to the view orientation normal to the selected plane, planar face, or feature.

7. Create **Sketch2**. Select Plane1 as the Sketch plane. Create the Extruded Boss profile. Insert the required geometric relations and dimension. Note: Dimension to the front **top edge** of Boss-Extrude1.

8. Create the first **Extruded Boss** feature. Select the Up To Vertex End Condition in Direction 1. Select the back top right vertex point as illustrated.

9. Create **Sketch3**. Select the back angled face of Boss-Extrude2 as the Sketch plane. Sketch a circle. Insert the required geometric relations.

10. Create the third **Extruded Boss** feature. Blind is the default End Condition in Direction 1. Depth = 3.00in.

11. Create **Sketch4**. Select the front face of Boss-Extrude3 as the Sketch plane. Sketch a circle. Sketch4 is the profile for the Extruded Cut feature. Insert the required geometric relation and dimension.

The part Origin is displayed in blue.

12. Create the **Extruded Cut** feature. Select Through All for End Condition in Direction 1.

13. Create **Sketch5**. Select the Right Plane as the Sketch plane. Insert a Parallel relation to partially define Sketch5. Sketch5 is the profile for the Rib feature. Sketch5 does not need to be fully defined. Sketch5 locates the end conditions based on existing geometry.

14. Create the **Rib** feature. Thickness = 1.00in.

 The Rib feature is a special type of extruded feature created from open or closed sketched contours. The Rib feature adds material of a specified thickness in a specified direction between the contour and an existing part. You can create a rib feature using single or multiple sketches.

15. **Assign** 6061 Alloy material.

16. **Calculate** the volume. The volume = 30.65 cubic inches.

17. **Locate** the Center of mass. The location of the Center of mass is derived from the part Origin.

- X: 0.00 inches
- Y: 0.73 inches
- Z: -0.86 inches

18. **Save** the part. Name it Part-Modeling 8-2.

19. **Close** the model.

As an exercise, modify the Rib1 feature from 1.00in to 1.25in. Modify the Extrude depth from 3.00in to 3.25in. Modify the material from 6061 Alloy to Copper.

Modify the Plane1 angle from 48deg to 30deg. Recalculate the volume of the part. The new volume = 26.94 cubic inches.

20. **Save** the part. Name it Part-Modeling 8-2-Modify.

Summary

Basic Part Creation and Modification and Intermediate Part Creation and Modification are two of the categories on the CSWA exam.

There are two questions on the CSWA Academic exam (Part 1) in the Basic Part Creation and Modification category. One question is in a multiple-choice single answer format and the other question (Modification of the model) is in the fill in the blank format. Each question is worth fifteen (15) points for a total of thirty (30) points.

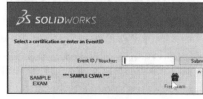

The main difference between the Basic Part Creation and Modification category and the Intermediate Part Creation and Modification or the Advance Part Creation and Modification category is the complexity of the sketches and the number of dimensions and geometric relations along with an increase in the number of features.

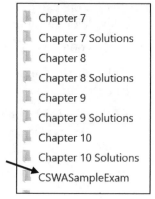

Chapter 7

Chapter 7 Solutions

Chapter 8

Chapter 8 Solutions

Chapter 9

Chapter 9 Solutions

Chapter 10

Chapter 10 Solutions

CSWASampleExam

If your school is an academic certification provider, your instructor can allocate free exam credits for the CSWA (Segment 1 & 2), CSWA-SD, CSWA-S, and CSWA-AM certifications. The instructor will require your .edu email address.

Homework Questions

Question 1:

Build the model from the provided information. Set document properties. Identify the correct Sketch planes, apply the correct Sketch and Feature tools and apply material.

Calculate the overall mass of the part, volume, and locate the Center of mass with the provided illustrated information.

- Material: 6061 Alloy.

- Units: MMGS.

- All Thru Holes.

- Decimal place: 2.

When you create a new part or assembly, the three default planes (Front, Right and Top) are aligned with specific views. The plane you select for the Base sketch determines the orientation of the part.

Question 2:

Build the model from the provided information. Set document properties. Apply material.

Calculate the overall mass of the part, volume, and locate the Center of mass with the provided information.

- Material: 6061 Alloy.

- Units: MMGS.

- Decimal place: 2.

- All Thru Holes.

Question 3:

Build the model from the provided information. Set document properties. Apply material.

Calculate the overall mass of the part, volume, and locate the Center of mass with the provided information.

- Material: 6061.

- Units: IPS.

- Decimal place: 2.

- View the provided drawing for details.

Question 4:

Build the model from the provided information.

Set document properties. Apply material.

Calculate the overall mass of the part, volume, and locate the Center of mass with the provided information.

- Material: 6061.

- Units: IPS.

- Decimal place: 2.

- View the provided drawing views for details.

Question 5:

Build the model from the provided information.

Set document properties.

Apply material.

Calculate the overall mass in grams. Calculate the volume of the part in cubic millimeters with the provided information.

- Decimal place: 2.

- Material: AISI 304.

- Units: MMGS.

- All Holes ⊤ 25mm.

- All Rounds 5mm.

- All Holes Ø4mm.

⌖ As a general rule, add fillets at the end of the modeling procedure.

Front views

Top and Auxiliary view

Calculate the mass:

A = **888.48grams**

B = 990.50grams

C = 788.48grams

D = 809.57grams

Back view

If you don't find your answer (within 1%) in the multiple-choice single answer format section, recheck your solid model for precision and accuracy. It could be as simple as missing a few fillets.

Calculate the volume:

A = 122259.43 cubic millimeters
B = 133359.47 cubic millimeters
C = 111059.43 cubic millimeters
D = 145059.49 cubic millimeters

Question 5A:

Open Homework 8-5 from the SW-TUTORIAL 2023/SOLIDWORKS CSWA Models Folder/Chapter 8 folder. Center a new coordinate system with the provided illustration. The new coordinate system location is at the front right bottom point (vertex) of the model.

Enter the Center of Mass:

X = -80.39 millimeters

Y = 15.93 millimeters

Z = -22.65 millimeters

Question 5B:

Modify the model. Calculate the overall mass and volume of the part.

- Modify all fillets (rounds) to 7mm.

- Modify the overall length to 140mm.

- Modify material to 1060 alloy.

Enter the mass:

309.75 grams

Enter the volume:

114721.22 cubic millimeters

If you don't find your answer (within 1%) in the multiple-choice single answer format section, recheck your solid model for precision and accuracy. It could be as simple as missing a few fillets.

Question 6:

Build the illustrated model.

Set document properties.

Apply material.

Calculate the overall mass in grams. Calculate the volume of the part in cubic millimeters.

- Decimal place: 2.

- Material: 1060 Alloy.

- Units: MMGS.

- TYP $\emptyset 12$.

Front view

Top view

Calculate the mass:

A = 610.92 grams

B = 509.92 grams

C = 701.93 grams

D = 619.34 grams

Calculate the volume:

A = 188860.93 cubic millimeters
B = 206660.93 cubic millimeters
C = 198880.65 cubic millimeters
D = 230021.67 cubic millimeters

Question 6A:

Open Homework 8-6 from the SW-TUTORIAL 2023/SOLIDWORKS CSWA Models Folder/Chapter 8 folder. Modify the model.

Calculate the overall mass in grams. Calculate the volume of the part in cubic millimeters. Decimal place: 2.

- Modify material to Plain Carbon Steel.

- Modify TYP Hole diameter from TYP 12 to TPY 10.

Enter the mass:

1465.70 grams

Enter the volume:

187910.60 cubic millimeters

Right view

Mass properties of Homework 8-6
    Configuration: Default
    Coordinate system: -- default --

Density = 0.00 grams per cubic millimeter

Mass = 509.92 grams

Volume = 188860.93 cubic millimeters

Surface area = 32545.06 square millimeters

Center of mass: (millimeters)
    X = 31.39
    Y = 16.55
    Z = 1.37

Configuration: modified
    Coordinate system: -- default --

Density = 0.01 grams per cubic millimeter

Mass = 1465.70 grams

Volume = 187910.60 cubic millimeters

Surface area = 32373.16 square millimeters

Center of mass: (millimeters)
    X = 31.29
    Y = 16.45
    Z = 1.33

Question 6B:

Create a new coordinate system.

The new coordinate system location is at the front left bottom point (vertex) of the model.

Enter the Center of Mass:

X = 31.39 millimeters

Y = 16.45 millimeters

Z = -48.67 millimeters

Question 7:

Build the illustrated model.

Calculate the overall mass in grams. Calculate the volume of the part in cubic millimeters.

- Decimal place: 2.

- Material: Plain Carbon Steel.

- Units: MMGS.

- The part is **not symmetrical** about the Front Plane.

Top view   Front view

Calculate the mass:

A = 4311.50 grams

B = 4079.32 grams

C = 4234.30 grams

D = 5322.00 grams

Calculate the volume:

A = 522989.22 cubic millimeters
B = 555655.11 cubic millimeters
C = 591233.34 cubic millimeters
D = 655444.00 cubic millimeters

Right view

Question 7A:

Open Homework 8-7 from the SW-TUTORIAL 2023\SOLIDWORKS CSWA Models Folder\Chapter 8 folder.

Center a new coordinate system with the provided illustration.

The new coordinate system location is at the back right bottom point (vertex) of the model.

Mass properties of Homework 8-7
Configuration: Default
Coordinate system: -- default --

Density = 0.01 grams per cubic millimeter

Mass = 4079.32 grams

Volume = 522989.22 cubic millimeters

Surface area = 92824.57 square millimeters

Center of mass: (millimeters)
X = 35.91
Y = 0.00
Z = 0.70

Enter the Center of Mass:

X = -64.09 millimeters

Y = 75.00 millimeters

Z = 40.70 millimeters

Mass = 4079.32 grams

Volume = 522989.22 cubic millimeters

Surface area = 92824.57 square millimeters

Center of mass: (millimeters)
X = -64.09
Y = 75.00
Z = 40.70

Question 8:

Build the provided model.

Set document properties.

Calculate the overall mass and volume of the part. There are no illustrated dimensions.

Open Homework 8-8 from the SW-TUTORIAL 2023\SOLIDWORKS CSWA Models Folder\Chapter 8 folder.

Use the rollback bar to obtain features and dimensions. Think about the various ways that this model can be built.

- Decimal place: 2.

- Material: 1060 Alloy.

- Units: MMGS.

Mass = 8892.98 grams

Volume = 3293697.11 cubic millimeters

Surface area = 1691889.56 square millimeters

Center of mass: (millimeters)
X = -2.94
Y = 9.80
Z = 0.00

Sample screen shots from an older CSWA exam for a simple part. Click on the additional views to understand the part and to provide information. Read each question carefully. Understand the dimensions, center of mass and units. Apply needed materials.

Screen shots from an exam

Zoom in on the part or view if needed.

Chapter 9 - CSWA Advanced Part Creation and Modification

Objectives

Provide in-depth coverage of the Advanced Part Creation and Modification category on the CSWA exam. Knowledge to create and modify models in this category from detailed dimensioned illustrations.

Introduction

The main difference between the Advanced Part Creation and Modification and the Basic Part Creation and Modification or the Intermediate Part Creation and Modification category is the complexity of the sketches and the number of dimensions and geometric relations along with an increased number of features.

There are three questions on the CSWA exam (part 2) in this category. The first question is in a multiple-choice single answer format and the other two questions (Modification of the model) are in the fill in the blank format.

Each question is worth fifteen (15) points for a total of forty-five (45) points.

You are required to build a model with six or more features and to answer a question either on the overall mass, volume, or the location of the Center of mass for the created model relative to the default part Origin location. You are then requested to modify the model and answer fill in the blank format questions.

On the completion of the chapter, you will be able to:

- Read and understand an Engineering document used in the CSWA exam.

Screen shots from an exam

- Identify the Sketch plane, part Origin location, part dimensions, geometric relations and design intent of the sketch and feature.

- Build an advanced part from a detailed dimensioned illustration using the following tools and features:

    - 2D & 3D Sketch tools, Extruded Boss/Base, Extruded Cut, Fillet, Chamfer, Mirror, Revolved Boss/Base, Linear & Circular Pattern, Plane, Axis, and Revolved Cut.

- Locate the Center of mass relative to the part Origin.

- Create a new coordinate system.

- Locate the Center of mass relative to a created Coordinate system.

In the *Advanced Part Creation and Modification* category, there are two dimension modification questions based on the first (multiple choice) question. You should be within 1% of the multiple-choice answer before you go on to the modification single answer section.

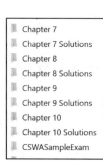

Chapter 7
Chapter 7 Solutions
Chapter 8
Chapter 8 Solutions
Chapter 9
Chapter 9 Solutions
Chapter 10
Chapter 10 Solutions
CSWASampleExam

Download all needed model files from the SDC Publication website www.SDCpublications.com/downloads/978-1-63057-556-4.

Build an Advanced Part from a Detailed Dimensioned Illustration

Tutorial: Advanced Part 9-1

An exam question in this category could read: Build this part. Calculate the overall mass and locate the Center of mass of the illustrated model.

1. **Create** a new part in SOLIDWORKS.

2. **Build** the illustrated model. Insert seven features: Extruded Base, two Extruded Bosses, two Extruded Cuts, a Chamfer and a Fillet.

Given:
A = 2.00, B = Ø.35
Material: 1060 Alloy
Density: 0.097 lb/in^3
Units: IPS
Decimal place:2

Think about the steps that you would take to build the illustrated part. Identify the location of the part Origin. Start with the back base flange. Review the provided dimensions and annotations in the illustration.

The main difference between the *Advanced Part Creation and Modification* and the *Basic Part Creation and Modification* or the *Intermediate Part Creation and Modification* category is the complexity of the sketches and the number of dimensions and geometric relations along with an increased number of features.

Origin

💡 All SOLIDWORKS models for the next few chapters (initial and final) are provided in the SOLIDWORKS CSWA model folder.

3. **Set** document properties for the model.

4. Create **Sketch1**. Sketch1 is the Base sketch. Select the Front Plane as the Sketch plane. Apply construction geometry. Sketch a horizontal and vertical centerline. Sketch four circles. Insert an Equal relation. Insert a Symmetric relation about the vertical and horizontal centerlines. Sketch two top angled lines and a tangent arc. Apply the Mirror Sketch tool. Complete the sketch. Insert the required geometric relations and dimensions.

💡 In a Symmetric relation, the selected items remain equidistant from the centerline, on a line perpendicular to the centerline. Sketch entities to select: a centerline and two points, lines, arcs or ellipses.

Chapter 7
Chapter 7 Final Solutions
Chapter 8
Chapter 8 Final Solutions
Chapter 9
Chapter 9 Final Solutions
Chapter 10
Chapter 10 Final Solutions
CSWA Sample Exam

▾ 🔩 Boss-Extrude1
　　� Sketch1
▾ 🔩 Boss-Extrude2
　　� Sketch2
▾ 🔩 Boss-Extrude3
　　� Sketch3
▾ 🔟 Cut-Extrude1
　　� Sketch4
▾ 🔟 Cut-Extrude2
　　� Sketch5
　 🔵 Chamfer1
　 🔷 Fillet1

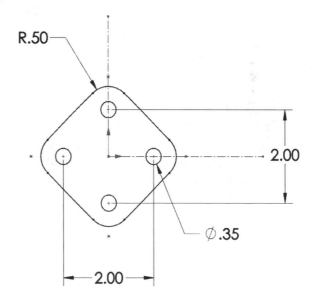

R.50

2.00

Ø.35

2.00

In the exam, you are allowed to answer the questions in any order. Use the Summary Screen during the exam to view the list of all questions you have or have not answered.

The Sketch Fillet tool rounds the selected corner at the intersection of two sketch entities, creating a tangent arc.

5. Create the **Extruded Base** feature. Boss-Extrude1 is the Base feature. Blind is the default End Condition in Direction 1. Depth = .40in.

6. Create **Sketch2**. Select the front face of Boss-Extrude1 as the Sketch plane. Sketch a circle. Insert the required geometric relation and dimension.

7. Create the first **Extruded Boss** feature. Blind is the default End Condition in Direction 1. The Extrude feature is the tube between the two flanges. Depth = 1.70in. Note: 1.70in = 2.60in - (.50in + .40in).

The complexity of the models along with the features progressively increases throughout this chapter to simulate the final types of parts that could be provided on the CSWA exam.

When you create a new part or assembly, the three default Planes (Front, Right and Top) are aligned with specific views. The Plane you select for the Base sketch determines the orientation of the part.

There are no Surfacing or Boundary feature questions on the CSWA exam at this time.

The book is designed to expose the new user to many tools, techniques and procedures. It may not always use the most direct tool or process.

Origin

8. Create **Sketch3**. Select the front circular face of Boss-Extrude2 as the Sketch plane. Sketch a horizontal and vertical centerline. Sketch the top two circles. Insert an Equal and Symmetric relation between the two circles. Mirror the top two circles about the horizontal centerline. Insert dimensions to locate the circles from the Origin. Apply either the 3 Point Arc or the Centerpoint Arc Sketch tool. The center point of the Tangent Arc is aligned with a Vertical relation to the Origin. Complete the sketch.

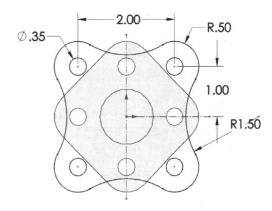

🔆 Use the Centerpoint Arc Sketch tool to create an arc from a centerpoint, a start point, and an end point.

🔆 Apply the Tangent Arc Sketch tool to create an arc, tangent to a sketch entity.

The Arc PropertyManager controls the properties of a sketched Centerpoint Arc, Tangent Arc, and 3 Point Arc.

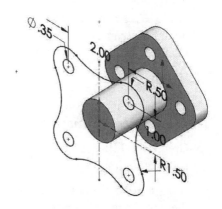

9. Create the second **Extruded Boss** feature. Blind is the default End Condition in Direction 1. Depth = .50in.

10. Create **Sketch4**. Select the front face of the Extrude feature as the Sketch plane. Sketch a circle. Insert the required geometric relation and dimension.

11. Create the first **Extruded Cut** feature. Select the Through All End Condition for Direction 1.

12. Create **Sketch5**. Select the front face of the Extrude feature as the Sketch plane. Sketch a circle. Insert the required geometric relation and dimension.

13. Create the second **Extruded Cut** feature. Blind is the default End Condition for Direction1. Depth = .10in.

14. Create the **Chamfer** feature. In order to have the outside circle 1.50in, select the inside edge of the sketched circle. Create an Angle distance chamfer.
Distance = .10in.
Angle = 45deg.

The Chamfer feature creates a beveled feature on selected edges, faces or a vertex.

15. Create the **Fillet** feature. Fillet the two edges as illustrated. Radius = .10in.

16. **Assign** 1060 Alloy material to the part. Material is required to calculate the overall mass of the part.

17. **Calculate** the overall mass. The overall mass = 0.59 pounds.

18. **Locate** the Center of mass. The location of the Center of mass is relative to the part Origin.

Density = 0.10 pounds per cubic inch

Mass = 0.59 pounds

Volume = 6.01 cubic inches

Surface area = 46.61 square inches

Center of mass: (inches)
  X = 0.00
  Y = 0.00
  Z = 1.51

- X: 0.00

- Y: 0.00

- Z: 1.51

19. **Save** the part. Name it Advanced Part 9-1.

20. **Close** the model.

Tutorial: Advanced Part 9-2

An exam question in this category could read: Build this part in SOLIDWORKS. Calculate the overall mass and locate the Center of mass of the illustrated model.

Given:
A = 70, B = 76
Material: 6061 Alloy
Density: .0027 g/mm^3
Units: MMGS
Decimal place: 2

1. **Create** a new part in SOLIDWORKS.

2. **Build** the illustrated dimensioned model. Insert eight features: Extruded Base, Extruded Cut, Circular Pattern, two Extruded Bosses, Extruded Cut, Chamfer, and Fillet.

Think about the steps that you would take to build the illustrated part. Review the provided information. Start with the six hole flange.

 Tangent Edges are displayed for educational purposes.

3. **Set** document properties for the model.

4. Create **Sketch1**. Sketch1 is the Base sketch. Select the Front Plane as the Sketch plane. Sketch two circles. Insert the required geometric relations and dimensions.

5. Create the **Extruded Base** feature. Blind is the default End Condition in Direction 1. Depth = 10mm. Note the direction of the extrude feature to maintain the Origin location.

6. Create **Sketch2**. Select the front face of Boss-Extrude1 as the Sketch plane. Sketch2 is the profile for first Extruded Cut feature. The Extruded Cut feature is the seed feature for the Circular Pattern. Apply construction reference geometry. Insert the required geometric relations and dimensions.

7. Create the **Extruded Cut** feature. Cut-Extrude1 is the first bolt hole. Select Through All for End Condition in Direction 1.

8. Create the **Circular Pattern** feature. Default Angle = 360deg. Number of instances = 6. Select the center axis for the Pattern Axis box.

💡 The Circular Pattern PropertyManager is displayed when you pattern one or more features about an axis.

9. Create **Sketch3**. Select the front face of the Extrude feature as the Sketch plane. Sketch two circles. Insert a Coradial relation on the inside circle. The two circles share the same center point and radius. Insert the required dimension.

10. Create the first **Extruded Boss (Boss-Extrude2)** feature. The Boss-Extrude2 feature is the connecting tube between the two flanges. Blind is the default End Condition in Direction 1. Depth = 48mm.

11. Create **Sketch4**. Select the front circular face of Extrude3 as the Sketch plane. Sketch a horizontal and vertical centerline from the Origin. Sketch the top and bottom circles symmetric about the horizontal centerline. Dimension the distance between the two circles and their diameter. Create the top center point arc with the center point Coincident to the top circle. The start point and the end point of the arc are horizontal. Sketch the two top angled lines symmetric about the vertical centerline. Apply symmetry. Mirror the two lines and the center point arc about the horizontal centerline. Insert the left and right tangent arcs with a center point Coincident with the Origin. Complete the sketch.

12. Create the second **Extruded Boss** (Boss-Extrude3) feature. Blind is the default End Condition in Direction 1. Depth = 12mm.

13. Create **Sketch5**. Select the front face of the Extrude feature as the Sketch plane. Sketch a circle. The part Origin is located in the center of the model. Insert the required dimension.

14. Create the second **Extruded Cut** feature. Blind is the default End Condition in Direction 1. Depth = 25mm.

15. Create the **Chamfer** feature. Chamfer1 is an Angle distance chamfer. Chamfer the inside edge of the Extrude feature as illustrated. Distance = 3mm. Angle = 45deg.

16. Create the **Fillet** feature. Fillet the two edges of Extrude1. Radius = 2mm.

17. **Assign** 6061 Alloy material to the part.

18. **Calculate** the overall mass of the part. The overall mass = 276.97 grams.

19. **Locate** the Center of mass. The location of the Center of mass is relative to the part Origin.

- X: 0.00 millimeters

- Y: 0.00 millimeters

- Z: 21.95 millimeters

20. **Save** the part. Name it Advanced Part 9-2.

21. **Close** the model.

In the Advanced Part Modeling category, an exam question could read "Build this model. Locate the Center of mass with respect to the part Origin."

- A: X = 0.00 millimeters, Y = 0.00 millimeters, Z = 21.95 millimeters

- B: X = 21.95 millimeters, Y = 10.00 millimeters, Z = 0.00 millimeters

- C: X = 0.00 millimeters, Y = 0.00 millimeters, Z = -27.02 millimeters

- D: X= 1.00 millimeters, Y = -1.01 millimeters, Z = -0.04 millimeters

The correct answer is A.

Mass = 276.97 grams

Volume = 102579.75 cubic millimeters

Surface area = 29703.14 square millimeters

Center of mass: (millimeters)
   X = 0.00
   Y = 0.00
   Z = 21.95

Calculate the Center of Mass Relative to a Created Coordinate System Location

In the Simple Part Modeling chapter, you located the Center of mass relative to the default part Origin. In the Advanced Part Modeling category, you may need to locate the Center of mass relative to a created coordinate system location. The exam model may display a created coordinate system location. Example:

☀ The SOLIDWORKS software displays positive values for (X, Y, Z) coordinates for a reference coordinate system. The CSWA exam displays either a positive or negative sign in front of the (X, Y, Z) coordinates to indicate direction (-X, +Y, -Z).

The following section reviews creating a Coordinate System location for a part.

Tutorial: Coordinate location 9-1

Use the Mass Properties tool to calculate the Center of mass for a part located at a new coordinate location through a point. Decimal place: 2. Units: MMGS.

1. **Open** the Plate-3-Point part from the SW-TUTORIAL 2023\SOLIDWORKS CSWA Folder\Chapter 9 location. View the location of the part Origin.

2. **Locate** the Center of mass. The location of the Center of mass is relative to the part Origin.

- X = 28.01 millimeters

- Y = 11.01 millimeters

- Z = -3.49 millimeters

Create a new coordinate system location as illustrated. Locate the new coordinate system location at the center of the center hole. Decimal place: 2. The center of the CBORE hole is under defined. Note: It appears center on the Base-Extrude.

3. **Exit** the Mass Properties dialog box.

4. Right-click the **front face** of Base-Extrude.

5. Click **Sketch** from the Context toolbar.

6. Click the **edge** of the center hole as illustrated.

7. Click **Convert Entities** from the Sketch toolbar. The center point for the new coordinate location is displayed.

8. **Exit** the sketch. Sketch4 is displayed.

9. Click the **Coordinate System** tool from the Consolidated Reference Geometry toolbar. The Coordinate System PropertyManager is displayed.

10. Click the **center point** of the center hole in the Graphics window. Point2@Sketch4 is displayed in the Selections box as the Origin.

11. Click **OK** from the Coordinate System PropertyManager. Coordinate System1 is displayed.

12. **View** the new coordinate location at the center of the center hole.

View the Mass Properties of the part with the new coordinate location.

13. Click the **Mass Properties** tool from the Evaluate tab.

14. Select **Coordinate System1** from the Output box. The Center of mass relative to the new location is located at the following coordinates: X = .21 millimeters, Y = .31 millimeters, Z = -3.49 millimeters

☀️ To reverse the direction of an axis, click its **Reverse Axis Direction** button in the Coordinate System PropertyManager.

Tutorial: Coordinate location 9-2

Create a new coordinate system location. Locate the new coordinate system at the top back point as illustrated.

Coordinate System1

1. **Open** the Plate-1 part from SW-TUTORIAL 2023\SOLIDWORKS CSWA Folder\Chapter 9 location.

2. **View** the location of the part Origin.

3. Drag the **Rollback bar** under the Base-Extrude feature in the FeatureManager.

4. Click the **Coordinate System** tool from the Consolidated Reference Geometry toolbar. The Coordinate System PropertyManager is displayed.

5. Click the **back left vertex** as illustrated. Address direction if needed.

6. Click the **top back horizontal** edge as illustrated. Do not select the midpoint. Address direction if needed.

7. Click the **back left vertical** edge as illustrated. Address direction if needed.

8. Click **OK** from the Coordinate System PropertyManager. Coordinate System1 is displayed in the FeatureManager and in the Graphics window.

9. Drag the **Rollback bar** to the bottom of the FeatureManager.

10. **Calculate** the Center of mass relative to the new coordinate system.

11. Select **Coordinate System1**. The Center of mass relative to the new location is located at the following coordinates:

X = -28.01 millimeters

Y = 10.99 millimeters

Z = 3.51 millimeters

12. **Close** the model.

🔆 Define a
Coordinate system for a
part or assembly. Apply
a Coordinate system
with the Measure and
Mass Properties tool.

Tutorial: Advanced Part 9-3

An exam question in this
category could read "Build this
part in SOLIDWORKS.
Calculate the overall mass and
locate the Center of mass of the
illustrated model."

1. **Create** a new part in
   SOLIDWORKS.

2. **Build** the illustrated
   dimensioned model. Insert
   thirteen features: Extrude-
   Thin1, Fillet, two Extruded
   Cuts, Circular Pattern, two
   Extruded Cuts, Mirror,
   Chamfer, Extruded Cut,
   Mirror, Extruded Cut and
   Mirror.

Think about the steps that you
would take to build the illustrated
part. Review the provided
information. The depth of the left
side is 50mm. The depth of the
right side is 60mm.

🔆 There are numerous ways to build the models in
this chapter. A goal is to display different design intents
and techniques.

Given:
A = 110, B = 55,
C = 5 X 45Ø CHAMFER
Material: 5MM, 6061 Alloy
Density: .0027 g/mm^3
Units: MMGS
ALL HOLES 6MM
Decimal place: 2

💡 If the inside radius = 5mm and the material thickness = 5mm, then the outside radius = 10mm.

3. **Set** document properties for the model.

4. Create **Sketch1**. Sketch1 is the Base sketch. Select the Top Plane as the Sketch plane. Apply the Line and Sketch Fillet Sketch tools. Apply construction geometry. Insert the required geometric relations and dimensions.

5. Create the **Extrude-Thin1** feature. Extrude-Thin1 is the Base feature. Apply symmetry in Direction 1. Depth = 60mm. Thickness = 5mm. Check the Auto-fillet corners box. Radius = 5mm.

💡 The Auto-fillet corners option creates a round at each edge where lines meet at an angle.

6. Create the **Fillet** feature. Fillet1 is a full round fillet. Fillet the three illustrated faces: top, front and bottom.

7. Create **Sketch2**. Select the right face as the Sketch plane. Wake-up the center point. Sketch a circle. Insert the required relation and dimension.

8. Create the first **Extruded Cut** feature. Select Up To Next for the End Condition in Direction 1.

💡 The Up To Next End Condition extends the feature from the sketch plane to the next surface that intercepts the entire profile. The intercepting surface must be on the same part.

9. Create **Sketch3**. Select the right face as the Sketch plane. Create the profile for the second Extruded Cut feature. This is the seed feature for CirPattern1. Apply construction geometry to locate the center point of Sketch3. Insert the required relations and dimensions.

10. Create the second **Extruded Cut** feature. Select Up To Next for the End Condition in Direction 1.

11. Create the **Circular Pattern** feature. Number of Instances = 4. Default angle = 360deg.

12. Create **Sketch4**. Select the left outside face of Extrude-Thin1 as the Sketch plane. Apply the Line and Tangent Arc Sketch tool to create Sketch4. Insert the required geometric relations and dimensions.

13. Create the third **Extruded Cut** feature. Select Up To Next for End Condition in Direction 1. The Slot on the left side of Extrude-Thin1 is created.

14. Create **Sketch5**. Select the left outside face of Extrude-Thin1 as the Sketch plane. Sketch two circles. Insert the required geometric relations and dimensions.

15. Create the fourth **Extruded Cut** feature. Select Up To Next for End Condition in Direction 1.

There are numerous ways to create the models in this chapter. A goal is to display different design intents and techniques.

16. Create the first **Mirror** feature. Mirror the top two holes about the Top Plane.

17. Create the **Chamfer** feature. Create an Angle distance chamfer. Chamfer the selected edges as illustrated. Distance = 5mm. Angle = 45deg.

18. Create **Sketch6**. Select the front face of Extrude-Thin1 as the Sketch plane. Insert the required geometric relations and dimensions.

19. Create the fifth **Extruded Cut** feature. Select Through All for End Condition in Direction 1.

20. Create the second **Mirror** feature. Mirror Extrude5 about the Right Plane.

21. Create **Sketch7**. Select the front face of Extrude-Thin1 as the Sketch plane. Apply the 3 Point Arc Sketch tool. Apply the min First Arc Condition option. Insert the required geometric relations and dimensions.

22. Create the last **Extruded Cut** feature. Through All is the End Condition in Direction 1 and Direction 2.

23. Create the third **Mirror** feature. Mirror the Extrude feature about the Top Plane as illustrated.

24. **Assign** the material to the part.

25. **Calculate** the overall mass of the part. The overall mass = 132.45 grams.

26. **Locate** the Center of mass relative to the part Origin:

- X: 1.83 millimeters

- Y: -0.27 millimeters

- Z: -35.38 millimeters

Density = 0.00 grams per cubic millimeter

Mass = 132.45 grams

Volume = 49055.56 cubic millimeters

Surface area = 24219.80 square millimeters

Center of mass: (millimeters)
  X = 1.83
  Y = -0.27
  Z = -35.38

27. **Save** the part. Name it Advanced Part 9-3.

28. **Close** the model.

All questions on the exam are in a multiple-choice single answer or fill in the blank format. In the Advanced Part Modeling category, an exam question could read: "Build this model. Calculate the overall mass of the part with the provided information."

Coordinate system: — Default
Density = 0.00 grams per cubic millimeter
Mass = 132.45 grams
Volume = 49055.56 cubic millimeters
Surface area = 24219.80 square millimeters
Center of mass: (millimeters)
X = 1.83
Y = -0.27
Z = -35.38

- A: 159.34 grams, B: 155.19 grams, C: 132.45 grams, D: 163.91 grams

The correct answer is C.

Use the Options button in the Mass Properties dialog box to apply custom settings to units.

Tutorial: Advanced Part 9-3A

An exam question in this category could read "Build this part in SOLIDWORKS. Locate the Center of mass." Note the coordinate system location of the model as illustrated.

Where do you start? Build the model as you did in the Tutorial: Advanced Part 9-3. Create Coordinate System1 to locate the Center of mass.

A = 110, B = 55, C = 5 X 45Ø CHAMFER
Material: 5MM, 6061 Alloy
Density: .0027 g/mm^3
Units: MMGS
ALL HOLES 6MM
Decimal place: 2

1. **Open** Advanced Part 9-3 that you created in the previous exercise.

Create the illustrated coordinate system location.

2. Show **Sketch2** from the FeatureManager design tree.

3. Click the **center point** of Sketch2 in the Graphics window as illustrated.

4. Click the **Coordinate System** tool from the Consolidated Reference Geometry toolbar. The Coordinate System PropertyManager is displayed. Point2@Sketch2 is displayed in the Origin box.

5. Click **OK** from the Coordinate System PropertyManager. Coordinate System1 is displayed.

6. **Locate** the Center of mass based on the location of the illustrated coordinate system. Select Coordinate System1.

- X: -53.17 millimeters

- Y: -0.27 millimeters

- Z: -15.38 millimeters

7. **Save** the part. Name it Advanced Part 9-3A.

8. **Close** the model.

Mass = 132.45 grams

Volume = 49055.56 cubic millimeters

Surface area = 24219.80 square millimeters

Center of mass: (millimeters)
 X = -53.17
 Y = -0.27
 Z = -15.38

Tutorial: Advanced Part 9-3B

Build this part in SOLIDWORKS. Locate the Center of mass. View the location of the coordinate system. The coordinate system is located at the left front point of the model.

Build the illustrated model as you did in the Tutorial: Advanced Part 9-3. Create Coordinate System1 to locate the Center of mass for the model.

1. **Open** Advanced Part 9-3 that you created in the previous exercise.

Create the illustrated coordinate system.

2. Click the **vertex** as illustrated for the Origin location.

Given:
A = 110, B = 55,
C = 5 X 45Ø CHAMFER
Material: 5MM, 6061 Alloy
Density: .0027 g/mm^3
Units: MMGS
ALL HOLES 6MM
Decimal place: 2

3. Click the **Coordinate System** tool from the Consolidated Reference Geometry toolbar. The Coordinate System PropertyManager is displayed. Vertex<1> is displayed in the Origin box.

4. Click the **bottom horizontal edge** as illustrated. Edge<1> is displayed in the X Axis Direction box.

5. Click the **left back vertical edge** as illustrated. Edge<2> is displayed in the Y Axis Direction box.

6. Click **OK** from the Coordinate System PropertyManager. Coordinate System1 is displayed.

7. **Locate** the Center of mass based on the location of the illustrated coordinate system. Select Coordinate System1.

- X: 1.83 millimeters

- Y: -0.27 millimeters

- Z: -35.38 millimeters

8. **Save** the part. Name it Advanced Part 9-3B.

9. **Close** the model.

In the Advanced Part Modeling category, an exam question could read "Build this model. Locate the Center of mass."

- A: X = 56.83 millimeters, Y = -29.73 millimeters, Z = 35.38 millimeters

- B: X = 22.80 millimeters, Y = -21.27 millimeters, Z = -45.54 millimeters

- C: X = -59.20 millimeters, Y = -0.27 millimeters, Z = -15.54 millimeters

- D: X= -11.80 millimeters, Y = 3.05 millimeters, Z = -0.14 millimeters

The correct answer is A.

Mass = 132.45 grams

Volume = 49055.56 cubic millimeters

Surface area = 24219.80 square millimeters

Center of mass: (millimeters)
 X = 56.83
 Y = -29.73
 Z = 35.38

Tutorial: Advanced Part 9-4

An exam question in this category could read "Build this part. Calculate the overall mass and locate the Center of mass of the illustrated model."

1. **Create** a new part in SOLIDWORKS.

2. **Build** the illustrated dimensioned model. Insert twelve features and a Reference plane: Extrude-Thin1, two Extruded Bosses, Extruded Cut, Extruded Boss, Extruded Cut, Plane1, Mirror and five Extruded Cuts.

Think about the steps that you would take to build the illustrated part. Create an Extrude-Thin1 feature as the Base feature.

3. **Set** document properties for the model. Review the given information.

Given:
A = Ø19
Material: Gray Cast Iron
Density: .0072 g/mm^3
Units: MMGS
ALL HOLES THROUGH UNLESS
OTHERWISE NOTED
Decimal place:2

4. Create **Sketch1**. Sketch1 is the Base sketch. Select the Right Plane as the Sketch plane. Apply construction geometry. Insert the required geometric relations and dimensions. Sketch1 is the profile for Extrude-Thin1. Note the location of the Origin.

5. Create the **Extrude-Thin1** feature. Apply symmetry. Select Mid Plane as the End Condition in Direction 1. Depth = 64mm. Thickness = 19mm.

6. Create **Sketch2**. Select the top narrow face of Extrude-Thin1 as the Sketch plane. Sketch three lines: two vertical and one horizontal and a tangent arc. Insert the required geometric relations and dimensions.

7. Create the **Boss-Extrude1** feature. Blind is the default End Condition in Direction 1. Depth = 18mm.

8. Create **Sketch3**. Select the Right Plane as the Sketch plane. Sketch a rectangle. Insert the required geometric relations and dimensions. Note: 61mm = (19mm - 3mm) x 2 + 29mm.

9. Create the **Boss-Extrude2** feature. Select Mid Plane for End Condition in Direction 1. Depth = 38mm. Note: 2 x R19.

10. Create **Sketch4**. Select the Right Plane as the Sketch plane. Sketch a vertical centerline from the top midpoint of the sketch. The centerline is required for Plane1. Plane1 is a Reference plane. Sketch a rectangle symmetric about the centerline. Insert the required relations and dimensions. Sketch4 is the profile for Extrude3.

11. Create the first **Extruded Cut** feature. Extrude in both directions. Select Through All for End Condition in Direction 1 and Direction 2.

12. Create **Sketch5**. Select the inside face of the Extrude feature for the Sketch plane. Sketch a circle from the top midpoint. Sketch a construction circle. Construction geometry is required for future features. Complete the sketch.

13. Create the **Extruded Boss** (Boss-Extrude3) feature. Blind is the default End Condition. Depth = 19mm.

14. Create **Sketch6**. Select the inside face for the Sketch plane. Show Sketch5. Select the construction circle in Sketch5. Apply the Convert Entities Sketch tool.

15. Create the second **Extruded Cut** feature. Select the Up To Next End Condition in Direction 1.

💡 There are numerous ways to create the models in this chapter. A goal is to display different design intents and techniques.

💡 Tangent edges and Origin are displayed for educational purposes.

16. Create **Plane1**. Apply symmetry. Create Plane1 to mirror Cut-Extrude2 and Boss-Extrude3. Create a Parallel Plane at Point. Select the midpoint of Sketch4 and Face<1> as illustrated. Point1@Sketch4 and Face<1> is displayed in the Selections box.

17. Create the **Mirror** feature. Mirror Cut-Extrude2 and Boss-Extrude3 about Plane1.

The Mirror feature creates a copy of a feature (or multiple features) mirrored about a face or a plane. You can select the feature or you can select the faces that comprise the feature.

18. Create **Sketch7**. Select the top front angled face of Extrude-Thin1 as the Sketch plane. Apply the Centerline Sketch tool. Insert the required geometric relations and dimensions.

19. Create the third **Extruded Cut** feature. Select Through All for End Condition in Direction 1. Select the edge for the vector to extrude.

Click on the additional views during the CSWA exam to better understand the part and provided information. Read each question carefully. Identify the dimensions, center of mass and units. Apply needed material.

20. Create **Sketch8**. Select the top front angled face of Extrude-Thin1 as the Sketch plane. Sketch a centerline. Sketch two vertical lines and a horizontal line. Select the top arc edge. Apply the Convert Entities Sketch tool. Apply the Trim Sketch tool to remove the unwanted arc geometry. Insert the required geometric relations and dimension.

21. Create the fourth **Extruded Cut** feature. Blind is the default End Condition in Direction 1. Depth = 6mm.

22. Create **Sketch9**. Create a Cbore with Sketch9 and Sketch10. Select the top front angled face of Extrude-Thin1 as illustrated for the Sketch plane. Extrude8 is the center hole in the Extrude-Thin1 feature. Sketch a circle. Insert the required geometric relations and dimension.

23. Create the fifth **Extrude Cut** feature. Blind is the default End Condition. Depth = 9mm. Note: This is the first feature for the Cbore.

24. Create **Sketch10**. Select the top front angled face of Extrude-Thin1 as the Sketch plane. Sketch a circle. Insert the required geometric relation and dimension. Note: A = Ø19.

25. Create the sixth **Extruded Cut** feature. Select the Up To Next End Condition in Direction 1. The Cbore is complete.

In the exam, you can answer the questions in any order. Use the Summary Screen during the exam to view the list of all questions you have or have not answered.

26. Create **Sketch11**. Select
    the front angle face of
    the Extrude feature for
    the Sketch plane. Sketch
    two circles. Insert the
    required geometric
    relations and dimensions.

27. Create the last **Extruded
    Cut** feature. Select the
    Up To Next End
    Condition in Direction 1.

The FilletXpert
automatically calls the
FeatureXpert when it has
trouble placing a fillet on the
specified geometry.

28. **Assign** material to the part.

29. **Calculate** the overall mass of the part. The
    overall mass = 2536.59 grams.

Mass = 2536.59 grams

Volume = 352304.50 cubic millimeters

Surface area = 61252.90 square millimeters

Center of mass: (millimeters)
   X = 0.00
   Y = 34.97
   Z = -46.67

- **Locate** the Center of mass relative to the part
  Origin:

- X: 0.00 millimeters

- Y: 34.97 millimeters

- Z: -46.67 millimeters

30. **Save** the part. Name it Advanced Part 9-4.

Tutorial: Advanced Part 9-4A

An exam question in this category could read - Build this part. Locate the Center of mass for the illustrated coordinate system.

Where do you start? Build the illustrated model as you did in the Tutorial: Advanced Part 9-3. Create Coordinate System1 to locate the Center of mass for the model.

1. **Open** Advanced Part 9-4 that you created in the previous exercise.

Create the illustrated Coordinate system.

2. Click the **Coordinate System** tool from the Consolidated Reference Geometry toolbar. The Coordinate System PropertyManager is displayed.

3. Click the **bottom right vertex** of Extrude-Thin1 as illustrated. Vertex<1> is displayed in the Origin box.

4. Click **OK** from the Coordinate System PropertyManager. Coordinate System1 is displayed.

5. **Locate** the Center of mass based on the location of the illustrated coordinate system. Select Coordinate System1.

- X: -32.00 millimeters
- Y: 34.97 millimeters
- Z: 93.33 millimeters

6. **Save** the part.

7. Name it Advanced Part 9-4A.

8. **View** the Center of mass with the default coordinate system.

9. **Close** the model.

Mass properties of Advanced Part 4-6
   Configuration: Default
   Coordinate system: Coordinate System1

Density = 0.01 grams per cubic millimeter

Mass = 2747.98 grams

Volume = 352304.50 cubic millimeters

Surface area = 61252.90 square millimeters

Center of mass: (millimeters)
   X = -32.00
   Y = 34.97
   Z = 93.33

Summary

Advanced Part Creation and Modification is one of the categories on the CSWA exam. The main difference between the Advanced Part Creation and Modification and the Basic Part Creation and Modification or the Intermediate Part Creation and Modification category is the complexity of the sketches and the number of dimensions and geometric relations along with an increased number of features.

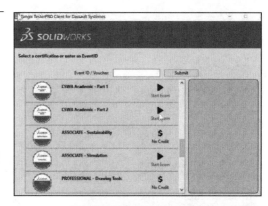

There are three questions on the CSWA exam in this category. One question is in a multiple-choice single answer format and the other two questions (Modification of the model) are in the fill in the blank format.

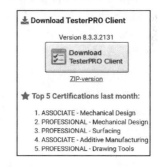

Visit https://3dexperience.virtualtester.com/#home to take the free sample online CSWA exam. There is a 30-day waiting period before retaking the free sample online exam.

If your school is an academic certification provider, your instructor can allocate free exam credits for the CSWA (Segment 1 & 2), CSWA-SD, CSWA-S, and CSWA-AM certifications. The instructor will require your .edu email address.

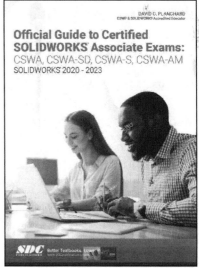

You need to pass both segments (Segment 1 & 2) to obtain your CSWA Certification. Once you pass a segment, you will not have to take it again. You can take the segments in any order. If you fail a segment, there is a 14-day waiting period before retaking the failed segment.

Homework Questions

Question 1:

In Tutorial: Advanced Part 9-1 you created the illustrated part. **Note**: Model located in the SW-TUTORIAL 2023\SOLIDWORKS CSWA Model Folder\Chapter 9 folder.

Modify the Base flange thickness from .40in to .50in. Modify the Chamfer feature angle from 45deg to 33deg. Modify the Fillet feature radius from .10in to .125in. Modify the material from 1060 Alloy to Nickel.

Calculate the overall mass of the part, volume, and locate the Center of mass with the provided information.

Given:
A = 2.00, B = Ø.35
Material: 1060 Alloy
Density: 0.097 lb/in^3
Units: IPS
Decimal places: 2

Question 2:

In the tutorial: Advanced Part 9-2 you created the illustrated part. **Note**: Model located in the SW-TUTORIAL 2023\SOLIDWORKS CSWA Model Folder\Chapter 9 folder.

Modify the CirPattern1 feature. Modify the number of instances from 6 to 8. Modify the seed feature from an 8mm diameter to a 6mm diameter.

Calculate the overall mass, volume, and the location of the Center of mass relative to the part Origin.

Given:
A = 70, B = 76
Material: 6061 Alloy
Density: .0027 g/mm^3
Units: MMGS
Decimal place: 2

Question 3:

Build the illustrated model in SOLIDWORKS. Set document properties, identify the correct Sketch planes, apply the correct Sketch and Feature tools, and apply material.

Calculate the overall mass of the part, volume and locate the Center of mass with the provided information.

- Material: 6061 Alloy.

- Units: MMGS.

- Decimal place: 2.

Origin

Origin

Question 4:

Create Coordinate System1 to locate the Center of mass for the model as illustrated. **Note:** Model located in the SW-TUTORIAL 2023\SOLIDWORKS CSWA Model Folder\Chapter 9 folder.

Material: 6061 Alloy.

- Units: MMGS.

- Decimal place: 2.

Question 5:

Build the illustrated model in SOLIDWORKS.

Calculate the overall mass of the part, volume, and locate the Center of mass with the provided information.

- Material: 6061 Alloy

- Units: MMGS

- Decimal place: 2

Question 6:

Create Coordinate System1 to locate the Center of mass for the model as illustrated. **Note:** Model located in the SW-TUTORIAL 2023\SOLIDWORKS CSWA Model Folder\Chapter 9 folder.

Question 7:

Build the provided model using SOLIDWORKS.

Note: Model located in the SW-TUTORIAL 2023\SOLIDWORKS CSWA Model Folder\Chapter 9 folder.

Open Homework Part 9-7.

Utilize the rollback bar to view the sketches and features. Note: There are many ways to create the model.

Calculate the overall mass and volume of part with the provided information.

- Decimal place: 2.

- Material: Case Stainless Steel.

- Units: MMGS.

Mass properties of Homework Part 9-7
    Configuration: Default
    Coordinate system: -- default --

Density = 0.01 grams per cubic millimeter

Mass = 3442.00 grams

Volume = 447012.75 cubic millimeters

Surface area = 67768.84 square millimeters

Center of mass: (millimeters)
    X = 0.00
    Y = 50.00
    Z = 0.00

💡 Sample screen shots from an older CSWA exam for an Advanced Modeling part. Click on the additional views to understand the part and provided information. Read each question carefully. Understand the dimensions, center of mass and units. Apply needed materials.

Screen shots from an exam

💡 Zoom in on the part if needed.

Notes:

Chapter 10 - CSWA - Assembly Creation and Modification

Objectives

Assembly Creation and Modification is one of the five categories on the CSWA exam. In the last two chapters, a Basic, Intermediate or Advanced model was the focus. The Assembly Creation and Modification (Bottom-up) category addresses an assembly with numerous components.

The chapter covers the general concepts and terminology used in Assembly modeling and then addresses the core elements that are aligned to the CSWA exam. Knowledge to insert mates and to create a new Coordinate system location is required in this category.

There are four questions on the CSWA Academic exam (2 questions in part 1, 2 questions in part 2) in the Assembly Creation and Modification category: (2) different assemblies - (4) questions - (2) multiple choice/(2) single answer - 30 points each.

The first question is in a multiple-choice single answer format. You should be within 1% of the multiple-choice answer before you move on to the modification single answer section, (fill in the blank format).

You are required to download the needed components (6 - 7) from the provided zip file to create the assembly.

💡 Download all needed model files from the SDC Publication website www.SDCpublications.com/downloads/978-1-63057-556-4.

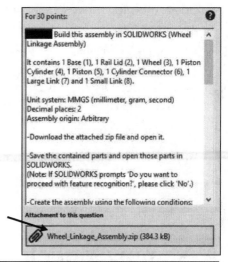

For 30 points:

▮▮▮▮ Build this assembly in SOLIDWORKS (Wheel Linkage Assembly)

It contains 1 Base (1), 1 Rail Lid (2), 1 Wheel (3), 1 Piston Cylinder (4), 1 Piston (5), 1 Cylinder Connector (6), 1 Large Link (7) and 1 Small Link (8).

Unit system: MMGS (millimeter, gram, second)
Decimal places: 2
Assembly origin: Arbitrary

-Download the attached zip file and open it.

-Save the contained parts and open those parts in SOLIDWORKS.
(Note: If SOLIDWORKS prompts 'Do you want to proceed with feature recognition?', please click 'No'.)

-Create the assembly using the following conditions:

Attachment to this question

📎 Wheel_Linkage_Assembly.zip (384.3 kB)

On the completion of the chapter, you will be able to:

- Specify Document Properties.

- Identify the first fixed component in an assembly.

- Create a Bottom-up assembly with the following mates:

  - Coincident, Concentric, Perpendicular, Parallel, Tangent, Distance, Angle, Advanced Distance, Advanced Angle, and Aligned, Anti-Aligned options.

- Apply the Measure tool.

- Apply the Mirror Component tool.

- Locate the Center of mass relative to the assembly Origin.

- Locate the Center of mass relative to a created Coordinate system.

Assembly Modeling

There are two key Assembly Modeling techniques:

- Top-down, "In-Context" assembly modeling.

- Bottom-up assembly modeling.

In Top-down assembly modeling, one or more features of a part are defined by something in an assembly, such as a sketch or the geometry of another component. The design intent comes from the top, and moves down into the individual components, hence the name Top-down assembly modeling.

 Mate the first component with respect to the assembly reference planes.

Bottom-up assembly modeling is a traditional method that combines individual components. Based on design criteria, the components are developed independently. The three major steps in a Bottom-up design approach are:

1. Create each part independent of any other component in the assembly.

2. Insert the parts into the assembly.

3. Mate the components in the assembly as they relate to the physical constraints of your design.

Build an Assembly from a Detailed Dimensioned illustration

An exam question in this category could read; Build this assembly in SOLIDWORKS. Locate the Center of mass of the model with respect to the illustrated coordinate system. Decimal place: 2. Unit system: IPS.

The assembly contains the following: (1) Clevis component, (3) Axle components, (2) 5 Hole Link components, (2) 3 Hole Link components, and (6) Collar components. All holes Ø.190 THRU unless otherwise noted. Angle A = 150deg. Angle B = 120deg.

Note: The location of the illustrated coordinate system.

In the exam, download the zip file of the components. Unzip the components. Do not use feature recognition when you open the downloaded components. This is a timed exam.

- Clevis, (Item 1): Material: 6061 Alloy. The two (5) Hole Link components are positioned with equal Angle mates, (150deg) to the Clevis component.

- Axle, (Item 2): Material: AISI 304. The first Axle component is mated Concentric and Coincident to the Clevis. The second and third Axle components are mated Concentric and Coincident to the 5 Hole Link and the 3 Hole Link components respectively.

- 5 Hole Link, (Item 3): Material: 6061 Alloy. Material thickness = .100in. Radius = .250in. Five holes located 1in. on center. The 5 Hole Link components are positioned with equal Angle mates, (120deg) to the 3 Hole Link components.

- 3 Hole Link, (Item 4): Material: 6061 Alloy. Material thickness = .100in. Radius = .250in. Three holes located 1in. on center. The 3 Hole Link components are positioned with equal Angle mates, (120deg) to the 5 Hole Link components.

- Collar, (Item 5): Material: 6061 Alloy. The Collar components are mated Concentric and Coincident to the Axle and the 5 Hole Link and 3 Hole Link components respectively.

Think about the steps that you would take to build the illustrated assembly. Identify the first fixed component. Insert the required mates.

Locate the Center of mass of the model with respect to the illustrated coordinate system. In this example, start with the Clevis component.

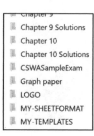 Do not use feature recognition when you open the downloaded components for the assembly in the CSWA exam. This is a timed exam. Manage your time. You do not need the additional feature information.

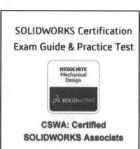 View the provided SOLIDWORKS Certification Exam Guide and Practice test located in the CSWA SampleExam folder. SOLIDWORKS models, solutions and additional exercises are provided.

Tutorial: Assembly Model 10-1

Build the assembly in SOLIDWORKS.

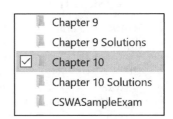

1. **Download** the needed components from the SW-TUTORIAL 2023\Chapter 10 folder.

2. **Create** a new assembly in SOLIDWORKS.

3. Click **Cancel** in the Open dialog box.

4. Click **Cancel** ✖ from the Begin Assembly PropertyManager.

5. **Set** document properties. Decimal place: 2. Unit system: IPS.

6. **Insert** the Clevis part.

7. **Fix** the component to the assembly Origin. Click OK from the Insert Component PropertyManager. The Clevis is displayed in the Assembly FeatureManager and in the Graphics window.

The first component or sub-assembly should be fixed **(f)** to the origin, fully defined to the assembly document or mated to an axis about the assembly origin.

💡 Only insert the required mates (timed exam) to obtain the needed Mass properties information.

8. **Insert** the Axle part above the Clevis component as illustrated.

9. **Insert** a Concentric mate between the inside cylindrical face of the Clevis and the outside cylindrical face of the Axle.

10. **Insert** a Coincident mate between the Right Plane of the Clevis and the Right Plane of the Axle.

11. **Insert** the 5 Hole Link part.

12. **Rotate** the component as illustrated.

13. **Insert** a Concentric mate between the outside cylindrical face of the Axle and the inside cylindrical face of the 5 Hole Link. Concentric2 is created.

14. **Insert** a Coincident mate between the right face of the Clevis and the left face of the 5 Hole Link. Coincident2 is created.

15. **Insert** an Angle mate between the bottom face of the 5 Hole Link and the back face of the Clevis. Angle = 30deg. If needed, click the Flip direction box and or the Aligned box.

💡 Depending on the component orientation, select the Flip direction option and/or enter the supplement of the angle.

16. **Insert** the second Axle part. Locate the second Axle component near the end of the 5 Hole Link as illustrated.

17. **Insert** a Concentric mate between the inside cylindrical face of the 5 Hole Link and the outside cylindrical face of the Axle. Concentric3 is created.

18. **Insert** a Coincident mate between the Right Plane of the assembly and the Right Plane of the Axle. Coincident3 is created.

19. **Insert** the 3 Hole Link part.

20. **Rotate** the component as illustrated.

21. **Insert** a Concentric mate between the outside cylindrical face of the Axle and the inside cylindrical face of the 3 Hole Link. Concentric4 is created.

22. **Insert** a Coincident mate between the right face of the 5 Hole Link and the left face of the 3 Hole Link.

23. **Insert** an Angle mate between the bottom face of the 5 Hole Link and the bottom face of the 3 Hole Link. Angle = 60deg. Angle2 is created.

💡 Depending on the component orientation, select the Flip direction option and/or enter the supplement of the angle when needed.

💡 Apply the Measure tool to check the angle.

24. **Insert** the third Axle part.

25. **Insert** a Concentric mate between the inside cylindrical face of the 3 Hole Link and the outside cylindrical face of the Axle.

26. **Insert** a Coincident mate between the Right Plane of the assembly and the Right Plane of the Axle.

27. **Insert** the Collar part. Locate the Collar near the first Axle component.

28. **Insert** a Concentric mate between the inside cylindrical face of the Collar and the outside cylindrical face of the first Axle.

29. **Insert** a Coincident mate between the right face of the 5 Hole Link and the left face of the Collar.

30. **Insert** the second Collar part. Locate the Collar near the second Axle component.

31. **Insert** a Concentric mate between the inside circular face of the second Collar and the outside circular face of the second Axle.

32. **Insert** a Coincident mate between the right face of the 3 Hole Link and the left face of the second Collar.

33. **Insert** the third Collar part. Locate the Collar near the third Axle component.

- ▸ 🔗 Mates
  - ◎ Concentric1 (Clevis<1>,Axle<1>)
  - ⋀ Coincident1 (Clevis<1>,Axle<1>)
  - ◎ Concentric2 (Axle<1>,5 Hole Link<1>)
  - ⋀ Coincident2 (Clevis<1>,5 Hole Link<1>)
  - ◖ Angle1 (5 Hole Link<1>,Clevis<1>)
  - ◎ Concentric3 (5 Hole Link<1>,Axle<2>)
  - ⋀ Coincident3 (Axle<2>,Right Plane)
  - ◎ Concentric4 (Axle<2>,3 Hole Link<1>)
  - ⋀ Coincident4 (5 Hole Link<1>,3 Hole Link
  - ◖ Angle2 (3 Hole Link<1>,5 Hole Link<1>)
  - ◎ Concentric5 (3 Hole Link<1>,Axle<3>)
  - ⋀ Coincident5 (Axle<3>,Right Plane)
  - ◎ Concentric6 (Axle<1>,Collar<1>)
  - ⋀ Coincident6 (5 Hole Link<1>,Collar<1>)
  - ◎ Concentric7 (Axle<2>,Collar<2>)
  - ⋀ Coincident7 (3 Hole Link<1>,Collar<2>)

34. **Insert** a Concentric mate between the inside cylindrical face of the Collar and the outside cylindrical face of the third Axle.

35. **Insert** a Coincident mate between the right face of the 3 Hole Link and the left face of the third Collar.

36. **Mirror** the components. Mirror the three Collars, 5 Hole Link and 3 Hole Link about the Right Plane. If using an older version of SOLIDWORKS, check the Recreate mates to new components box. Click Next in the Mirror Components PropertyManager. View your options.

Click **Insert, Mirror Components** from the Main menu or click the **Mirror Components** tool from the Linear Component Pattern Consolidated toolbar.

If using an older release of SOLIDWORKS, no check mark in the Components to Mirror box indicates that the components are copied. The geometry of a copied component is unchanged from the original, only the orientation of the component is different.

Create the coordinate system location for the assembly.

Boss-Extrude1 of Clevis<1>

37. Select the front right **vertex** of the Clevis component as illustrated.

38. Click the **Coordinate System** tool from the Reference Geometry Consolidated toolbar. The Coordinate System PropertyManager is displayed.

39. Click the **right bottom edge** of the Clevis component.

40. Click the **front bottom edge** of the Clevis component as illustrated.

41. Address the **direction** for X, Y, Z as illustrated.

42. Click **OK** from the Coordinate System PropertyManager. Coordinate System1 is displayed.

43. **Locate** the Center of mass based on the location of the illustrated coordinate system. Select Coordinate System1.

- X: 1.79 inches

- Y: 0.25 inches

- Z: 2.61 inches

44. **Save** the part. Name it Assembly Modeling 10-1.

45. **Close** the model.

Mass properties of Assem1
　　Configuration: Default
　　Coordinate system: Coordinate System1

Mass = 0.14 pounds

Volume = 1.20 cubic inches

Surface area = 27.04 square inches

Center of mass: (inches)
　X = 1.79
　Y = 0.25
　Z = 2.61

There are numerous ways to create the models in this chapter. A goal in this text is to display different design intents and techniques.

💡 Apply the Measure tool if needed between components.

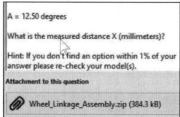

A = 12.50 degrees

What is the measured distance X (millimeters)?

Hint: If you don't find an option within 1% of your answer please re-check your model(s).

Attachment to this question

📎 Wheel_Linkage_Assembly.zip (384.3 kB)

Tutorial: Assembly Model 10-2

An exam question in this category could read:
Build this assembly in SOLIDWORKS. Locate
the Center of mass of the model with the
illustrated coordinate system. Decimal place: 2.
Unit system: MMGS.

The assembly contains the following: (2)
U-Bracket components, (4) Pin components
and (1) Square block component.

- U-Bracket, (Item 1): Material: AISI 304.
  Two U-Bracket components are combined
  together Concentric to opposite holes of the
  Square block component. The second
  U-Bracket component is positioned with an
  Angle mate to the right face of the first
  U-Bracket and a Parallel mate between the
  top face of the first U-Bracket and the top
  face of the Square block component. Angle
  A = 125deg.

- Square block, (Item 2): Material: AISI 304.
  The Pin components are mated Concentric
  and Coincident to the 4 holes in the Square
  block (no clearance). The depth of each
  hole = 10mm.

- Pin, (Item 3): Material: AISI 304. The Pin components are mated Concentric to the hole (no clearance). The end face of the Pin components are Coincident to the outer face of the U-Bracket components. The Pin component has a 5mm spacing between the Square block component and the two U-Bracket components.

Think about the steps that you would take to build the illustrated assembly. Identify the first fixed component. This is the Base component of the assembly. Insert the required Standard mates. Locate the Center of mass of the model with respect to the illustrated coordinate system. In this example, start with the U-Bracket part.

Create the assembly.

1. **Download** the needed components from the SW-TUTORIAL 2023\Chapter 10 folder.

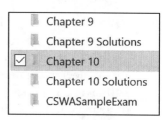

2. **Create** a new assembly in SOLIDWORKS. The created models are displayed in the Open documents box.

3. Click **Cancel** in the Open dialog box.

4. Click **Cancel** ✖ from the Begin Assembly PropertyManager.

5. **Set** document properties. Decimal place:2. Units: MMGS.

6. **Insert** the first U-Bracket component into the assembly document.

7. **Fix** the component to the assembly Origin. Click OK from the PropertyManager. The U-Bracket is displayed in the Assembly FeatureManager and in the Graphics window.

8. **Insert** the Square block above the U-Bracket component as illustrated.

9. **Insert** the first Pin part. Locate the first Pin to the front of the Square block.

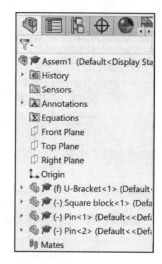

10. **Insert** the second Pin part. Locate the second Pin to the back of the Square block.

11. **Insert** the third Pin part. Locate the third Pin to the left side of the Square block.

12. **Rotate** the Pin as illustrated.

13. **Insert** the fourth Pin part. Locate the fourth Pin to the right side of the Square block.

14. **Rotate** the Pin as illustrated.

15. **Insert** a Concentric mate between the inside cylindrical face of the Square block and the outside cylindrical face of the first Pin. Concentric1 is created.

16. **Insert** a Coincident mate between the inside back circular face of the Square block and the flat back face of the first Pin. Coincident1 mate is created.

17. **Insert** a Concentric mate between the inside cylindrical face of the Square block and the outside cylindrical face of the second Pin. Concentric2 is created.

18. **Insert** a Coincident mate between the inside back circular face of the Square block and the front flat face of the second Pin. Coincident2 mate is created.

19. **Insert** a Concentric mate between the inside cylindrical face of the Square block and the outside cylindrical face of the third Pin. Concentric3 is created.

20. **Insert** a Coincident mate between the inside back circular face of the Square block and the right flat face of the third Pin. Coincident3 mate is created.

21. **Insert** a Concentric mate between the inside circular face of the Square block and the outside cylindrical face of the fourth Pin. Concentric4 is created.

22. **Insert** a Coincident mate between the inside back circular face of the Square block and the left flat face of the fourth Pin. Coincident4 mate is created.

> ▶ 🎓 🏴 (f) U-Bracket<1> (Default<<Default>_
> ▶ 🎓 🏴 (-) Square block<1> (Default<<Defau
> ▶ 🎓 🏴 (-) Pin<1> (Default<<Default>_Appea
> ▶ 🎓 🏴 (-) Pin<2> (Default<<Default>_Appea
> ▶ 🎓 🏴 (-) Pin<3> (Default<<Default>_Appea
> ▶ 🎓 🏴 (-) Pin<4> (Default<<Default>_Appea
> ▼ 🔗 Mates
>   ◎ Concentric1 (Square block<1>,Pin<1>
>   ⋏ Coincident1 (Square block<1>,Pin<1>
>   ◎ Concentric2 (Square block<1>,Pin<2>
>   ⋏ Coincident2 (Square block<1>,Pin<2>
>   ◎ Concentric3 (Square block<1>,Pin<3>
>   ⋏ Coincident3 (Square block<1>,Pin<3>
>   ◎ Concentric4 (Square block<1>,Pin<4>
>   ⋏ Coincident4 (Square block<1>,Pin<4>

23. **Insert** a Concentric mate between the inside right cylindrical face of the Cut-Extrude feature on the U-Bracket and the outside cylindrical face of the right Pin. Concentric5 is created.

24. **Insert** a Coincident mate between the Right Plane of the Square block and the Right Plane of the assembly. Coincident5 is created.

25. **Insert** the second U-Bracket part above the assembly. Position the U-Bracket as illustrated.

26. **Insert** a Concentric mate between the inside cylindrical face of the second U-Bracket component and the outside cylindrical face of the second Pin. The mate is created.

27. **Insert** a Coincident mate between the outside circular edge of the second U-Bracket and the back flat face of the second Pin. The mate is created.

There are numerous ways to mate the models in this chapter. A goal is to display different design intents and techniques.

28. **Insert** an Angle mate between the top flat face of the first U-Bracket component and the front narrow face of the second U-Bracket component as illustrated. Angle1 is created. An Angle mate is required to obtain the correct Center of mass.

29. **Insert** a Parallel mate between the top flat face of the first U-Bracket and the top flat face of the Square block component.

30. **Expand** the Mates folder and the components from the FeatureManager. View the created mates.

Create the coordinate location for the assembly.

31. Select the front **bottom left vertex** of the first U-Bracket component as illustrated.

32. Click the **Coordinate System** tool from the Reference Geometry Consolidated toolbar. The Coordinate System PropertyManager is displayed.

33. Click **OK** from the Coordinate System PropertyManager. Coordinate System1 is displayed.

34. **Locate** the Center of mass based on the location of the illustrated coordinate system. Select Coordinate System1.

- X: 31.54 millimeters

- Y: 85.76 millimeters

- Z: -45.00 millimeters

35. **Save** the part.

36. Name it Assembly Modeling 10-2.

37. **Close** the model.

If you don't find an option within 1% of your answer on the exam re-check your assembly.

Click on the additional views during the CSWA exam to better understand the assembly/component. Read each question carefully. Identify the dimensions, center of mass and units. Apply needed material.

Tutorial: Assembly Model 10-3

An exam question in this category could read: Build this assembly in SOLIDWORKS. Locate the Center of mass using the illustrated coordinate system. Decimal place: 2. Unit: MMGS.

The assembly contains the following: (1) WheelPlate component, (2) Bracket100 components, (1) Axle40 component, (1) Wheel1 component and (4) Pin-4 components.

- WheelPlate, (Item 1): Material: AISI 304. The WheelPlate contains 4-Ø10 holes. The holes are aligned to the left Bracket100 and the right Bracket100 components. All holes are THRU ALL. The thickness of the WheelPlate = 10 mm.

- Bracket100, (Item 2): Material: AISI 304. The Bracket100 component contains 2-Ø10 holes and 1- Ø16 hole. All holes are through-all.

- Wheel1, (Item 3): Material AISI 304: The center hole of the Wheel1 component is Concentric with the Axle40 component. There is a 3mm gap between the inside faces of the Bracket100 components and the end faces of the Wheel hub.

- Axle40, (Item 4): Material AISI 304: The end faces of the Axle40 are Coincident with the outside faces of the Bracket100 components.

- Pin-4, (Item 5): Material AISI 304: The Pin-4 components are mated Concentric to the holes of the Bracket100 components (no clearance). The end faces are Coincident to the WheelPlate bottom face and the Bracket100 top face.

Think about the steps that you would take to build the illustrated assembly. Identify the first fixed component. This is the Base component of the assembly. Insert the required Standard mates.

Locate the Center of mass of the illustrated model with respect to the referenced coordinate system.

The referenced coordinate system is located at the bottom, right, midpoint of the Wheelplate. In this example, start with the WheelPlate part.

1. **Download** the needed components from the SW-TUTORIAL 2023\Chapter 10 folder.

2. **Create** a new assembly in SOLIDWORKS.

3. Click **Cancel** in the Open dialog box.

4. Click **Cancel** ✖ from the Begin Assembly PropertyManager.

5. **Set** the document properties. Decimal place: 2. Units: MMGS.

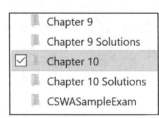

6. **Insert** the first component. Insert the WheelPlate. Fix the component to the assembly Origin. The WheelPlate is displayed in the Assembly FeatureManager and in the Graphics window. The WheelPlate component is fixed.

7. **Insert** the first Bracket100 part above the WheelPlate component as illustrated.

8. **Insert** a Concentric mate between the inside front left cylindrical face of the Bracket100 component and the inside front left cylindrical face of the WheelPlate. Concentric1 is created.

9. **Insert** a Concentric mate between the inside front right cylindrical face of the Bracket100 component and the inside front right cylindrical face of the WheelPlate. Concentric2 is created.

10. **Insert** a Coincident mate between the bottom flat face of the Bracket100 component and the top flat face of the WheelPlate component. Coincident1 is created.

11. **Insert** the Axle40 part above the first Bracket100 component as illustrated.

12. **Insert** a Concentric mate between the outside cylindrical face of the Axle40 component and the inside cylindrical face of the Bracket100 component. Concentric3 is created.

13. **Insert** a Coincident mate between the flat face of the Axle40 component and the front outside edge of the first Bracket100 component. Coincident2 is created.

To verify that the distance between holes of mating components is equal, utilize Concentric mates between pairs of cylindrical hole faces.

14. **Insert** the first Pin-4 part above the Bracket100 component.

15. **Insert** the second Pin-4 part above the Bracket100 component.

16. **Insert** a Concentric mate between the outside cylindrical face of the first Pin-4 component and the inside front left cylindrical face of the Bracket100 component. Concentric4 is created.

17. **Insert** a Coincident mate between the flat top face of the first Pin-4 component and the top face of the first Bracket100 component. Coincident3 is created.

18. **Insert** a Concentric mate between the outside cylindrical face of the second Pin-4 component and the inside front right cylindrical face of the Bracket100 component. Concentric5 is created.

19. **Insert** a Coincident mate between the flat top face of the second Pin-4 component and the top face of the first Bracket100 component. Coincident4 is created.

20. **Insert** the Wheel1 part as illustrated.

21. **Insert** a Concentric mate between the outside cylindrical face of Axle40 and the inside front cylindrical face of the Wheel1 component. Concentric6 is created.

22. **Insert** a Coincident mate between the Front Plane of Axle40 and the Front Plane of Wheel1. Coincident5 is created.

23. **Mirror** the components. Mirror the Bracket100 and the two Pin-4 components about the Front Plane.

Click **Insert, Mirror Components** from the Menu bar menu or click the **Mirror Components** tool from the Linear Component Pattern Consolidated toolbar.

Create the coordinate location for the assembly.

24. Click the **Coordinate System** tool from the Reference Geometry Consolidated toolbar. The Coordinate System PropertyManager is displayed.

25. **Select** the right bottom midpoint as the Origin location.

26. **Select** the bottom right edge as the X axis direction reference as illustrated.

27. Click **OK** from the Coordinate System PropertyManager. Coordinate System1 is displayed.

28. **Locate** the Center of mass based on the location of the illustrated coordinate system. Select Coordinate System1.

- X: = 0.00 millimeters

- Y: = 37.14 millimeters

- Z: = -50.00 millimeters

If you don't find an option within 1% of your answer on the exam re-check your assembly.

29. **Save** the part.

30. Name it Assembly Modeling 10-3.

31. **Close** the model.

Mass = 3797.32 grams

Volume = 474665.19 cubic millimeters

Surface area = 130119.83 square millimeters

Center of mass: (millimeters)
    X = 0.00
    Y = 37.14
    Z = -50.00

Summary

Assembly Creation and Modification is one of the five categories on the CSWA exam. In the last two chapters, a Basic, Intermediate or Advanced model was the focus. The Assembly Creation and Modification (Bottom-up) category addresses an assembly with numerous components.

The chapter covers the general concepts and terminology used in Assembly modeling and then addresses the core elements that are aligned to the CSWA exam. Knowledge to insert mates and to create a new Coordinate system location is required in this category.

There are four questions on the CSWA Academic exam (2 questions in part 1, 2 questions in part 2) in the Assembly Creation and Modification category: (2) different assemblies - (4) questions - (2) multiple choice/(2) single answer - 30 points each.

The first question is in a multiple-choice single answer format. You should be within 1% of the multiple-choice answer before you move on to the modification single answer section, (fill in the blank format).

You are required to download the needed components (6 - 7) from the provided zip file to create the assembly.

Homework Questions

1. Build this assembly from the provided information in SOLIDWORKS.

Calculate the overall mass and volume of the assembly. Decimal place: 2. Units: MMGS.

Locate the Center of mass using the illustrated coordinate system.

The assembly contains the following: one Base100 component, one Yoke component, and one AdjustingPin component.

- Base100, (Item 1): Material 1060 Alloy. The distance between the front face of the Base100 component and the front face of the Yoke = 60mm.
- Yoke, (Item 2): Material 1060 Alloy. The Yoke fits inside the left and right square channels of the Base100 component (no clearance). The top face of the Yoke contains a Ø12mm through all hole.
- AdjustingPin, (Item 3): Material 1060 alloy. The bottom face of the AdjustingPin head is located 40mm from the top face of the Yoke component. The AdjustingPin component contains an Ø5mm Through All hole.

The Coordinate system is located in the lower left corner of the Base100 component. The X axis points to the right.

2. Build the assembly from the provided information in SOLIDWORKS. Calculate the overall mass and volume of the assembly. Locate the Center of mass using the illustrated coordinate system. The assembly contains the following: three MachinedBracket components and two Pin-5 components. Unit system: MMGS. Decimal place: 2.

💡 Insert the Base component, float the component, then mate the first component with respect to the assembly reference planes.

- MachinedBracket, (Item 1): Material 6061 Alloy. The MachineBracket component contains two Ø10mm through all holes. Each MachinedBracket component is mated with two Angle mates. The Angle mate = 45deg. The top edge of the notch is located 20mm from the top edge of the MachinedBracket.
- Pin-5, (Item 2): Material Titanium. The Pin-5 component is 5mms in length and equal in diameter. The Pin-5 component is mated Concentric to the MachinedBracket (no clearance). The end faces of the Pin-5 component is Coincident with the outer faces of the MachinedBracket. There is a 1mm gap between the Machined Bracket components.

3. Build the assembly from the provided information in SOLIDWORKS. **Note**: Download the assembly from the SW-TUTORIAL 2023/SOLIDWORKS CSWA Model Folder/Chapter 10 folder. Use the assembly to better understand the angles.

Calculate the overall mass and volume of the assembly. Locate the Center of mass using the illustrated coordinate system. The illustrated assembly contains the following components: three Machined-Bracket components and two Pin-6 components. Unit system: MMGS. Decimal place: 2.

Insert the Base component, float the component, then mate the first component with respect to the assembly reference planes.

- Machined-Bracket, (Item 1): Material 6061 Alloy. The Machine-Bracket component contains two Ø10mm through all holes. Each Machined-Bracket component is mated with two Angle mates. The Angle mate = 45deg. The top edge of the notch is located 20mm from the top edge of the MachinedBracket.
- Pin-6, (Item 2): Material Titanium. The Pin-6 component is 5mms in length and equal in diameter. The Pin-5 component is mated Concentric to the Machined-Bracket (no clearance). The end faces of the Pin-6 component is Coincident with the outer faces of the Machined-Bracket. There is a 1mm gap between the Machined-Bracket components.

Sample screen shots from an older CSWA exam for an assembly. Click on the additional views to understand the assembly and provided information. Read each question carefully. Understand the dimensions, center of mass and units. Apply needed materials.

Zoom in on the part if needed.

Use the Mass Properties tool to locate the new coordinate system.

Screen shots from an exam

💡 Zoom in on the part if needed.

💡 If needed, use the Measure tool in the assembly on the components for distance or angle.

If needed, use the Measure tool in the assembly on the components for distance or angle.

Chapter 11

Additive Manufacturing - 3D Printing Fundamentals

Below are the desired outcomes and usage competencies based on the completion of Chapter 11.

Desired Outcomes:	Usage Competencies:
• Knowledge of Additive manufacturing.	• Examine the advantages and disadvantages of Additive manufacturing.
• Comprehend various 3D printer technologies and terminologies.	• Discuss 3D printer technology: Fused Filament Fabrication (FFF), STereoLithography (SLA), and Selective Laser Sintering (SLS).
• Familiarity with filament materials: PLA (Polylactic acid), FPLA (Flexible Polylactic acid), ABS (Acrylonitrile butadiene styrene), PVA (Polyvinyl alcohol), Nylon 618, and Nylon 645.	• Select the correct filament material. • Create an STL (*.stl) file, Additive Manufacturing file (*.amf), and a 3D Manufacturing Format file (*.3mf).
	• Select optimum build orientation and slicer parameters.
• Understand the following file types: STL (*.stl), Additive Manufacturing (*.amf), and 3D Manufacturing format (*.3mf).	• Discuss 3D printer terminology: Bed adhesion, support, percent infill, infill pattern, layer height, cooling, hot end, etc. • Address fit tolerance for interlocking parts.
• Awareness of slicer and 3D printer parameters.	• Information on the Certified SOLIDWORKS Associate Additive Manufacturing (CSWA-AM) exam.

Notes:

Chapter 11 - Additive Manufacturing - 3D Printing Fundamentals

Chapter Objective

Provide a basic understanding between Additive vs. Subtractive manufacturing. Discuss Fused Filament Fabrication (FFF), STereoLithography (SLA), and Selective Laser Sintering (SLS) printer technology. Select suitable filament material. Comprehend 3D printer terminology. Knowledge of preparing, saving, and printing a model on a Fused Filament Fabrication 3D printer. Information on the Certified SOLIDWORKS Associate Additive Manufacturing (CSWA-AM) exam and general topic areas.

On the completion of this chapter, you will be able to:

- Discuss Additive vs Subtractive manufacturing.

- Review 3D printer technology: Fused Filament Fabrication (FFF), STereoLithography (SLA), and Selective Laser Sintering (SLS).

- Select the correct filament material:

  o PLA (Polylactic acid)

  o FPLA (Flexible Polylactic acid)

  o ABS (Acrylonitrile butadiene styrene)

  o PVA (Polyvinyl alcohol)

  o Nylon 618

  o Nylon 645

- Create an STL (*.stl) file, Additive Manufacturing (*.amf) file and a 3D Manufacturing format (*.3mf) file.

- Prepare Geometric Code, "G-Code".

- Comprehend general 3D printer terminology.

- Understand optimum build orientation.

- Enter slicer parameters:

  o   Raft, brim, skirt, layer height, percent infill, infill pattern, wall thickness, fan speed, print speed, bed temperature, and extruder (hot end) temperature.

- Address tolerance for interlocking parts.

- Print directly from SOLIDWORKS.

- Knowledge of the Certified SOLIDWORKS Associate Additive Manufacturing exam.

Additive vs. Subtractive Manufacturing

In April 2012, *The Economist* published an article on 3D printing. In the article they stated that this was the "beginning of a third industrial revolution, offering the potential to revolutionize how the world makes just about everything."

Avi Reichental, who was President and CEO, 3D Systems stated, "With 3D printing, complexity is free. The printer doesn't care if it makes the most rudimentary shape or the most complex shape, and that is completely turning design and manufacturing on its head as we know it."

Over the past five years, companies are now using 3D printing to evaluate more concepts in less time to improve decisions early in product development. As the design process moves forward, technical decisions are iteratively tested at every step to guide decisions big and small, to achieve improved performance, lower manufacturing costs, delivering higher quality and more successful product introductions. In pre-production, 3D printing is enabling faster first article production to support marketing and sales functions, and early adopter customers. And in final production processes, 3D printing is enabling higher productivity, increased flexibility, reduced warehouse and other logistics costs, economical customization, improved quality, reduced product weight, and greater efficiency in a growing number of industries.

Technology for 3D printing continues to advance in three key areas: **printers** and **printing methods**, **design software**, and **materials** used in printing.

Already, 3D printing is being used in the medical industry to help save lives and in some space exploration efforts. But how will 3D printing affect the average, middle-class person in the future? Low-cost 3D printers are addressing this consumer market.

Additive manufacturing is the process of joining materials to create an object from a 3D model, usually adding layer over layer.

Subtractive manufacturing relies upon the removal of material to create something. The blacksmith hammered away at heated metal to create a product. Today, a Computer Numerical Control CNC machine cuts and drills and otherwise removes material from a larger initial block to create a product.

Additive manufacturing, sometimes known as ***rapid prototyping***, can be slower than Subtractive manufacturing. Both take skill in creating the Geometric Code "G-code" and understanding the machine limitations.

Fused Filament Fabrication (FFF) and Fused Deposition Modeling (FDM) are used interchangeably in the book.

The (CSWA-AM) exam covers the Ultimaker 3 (FFF) and the FormLabs Form 2 (SLA) machine as examples.

3D Printer Technology

Stages of 3D printing

There are three basic stages to preparing files for 3D printing.

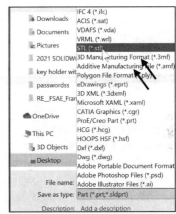

3D modeling

The process of developing a mathematical representation of any surface of an object (either inanimate or living) in three dimensions via software such as SOLIDWORKS. There are many other software applications for 3D modeling. These applications all have their own file format.

3D file export

Once the model is created, it needs to be saved and exported as a STereoLithography (*.stl), 3D Manufacturing Format (*.3mf), or an Additive Manufacturing File (*.amf). These file formats are recognized by many 3D printers (Slicers).

Slicing the file

The (*.3mf), (*amf), or (*.stl) file is imported into the 3D printer slicer. The slicer takes the file and translates the model into individual layers. It then generates the machine Geometric Code (G-code) that the 3D printer requires.

Slicer software allows the user to calibrate 3D printer settings: filament type, part orientation, extruder speed, extruder temperature, bed temperature, cooling fan rate, raft type, support type, percent infill, infill pattern type, etc.

Ultimaker Cura slicer - Version: 4.4.1

Note: Unlike its predecessor STL (*.stl) format, 3D Manufacturing Format (*.3mf) and Additive Manufacturing File (*.amf) has native support for color, materials, lattices, and constellation (groups). Therefore, it requires less post-processing to define data such as the position of your model relative to the selected 3D printer, orientation, color, materials, etc.

An OBJ (*.obj) file is a standard 3D image format that can be exported and opened by various 3D image editing programs.

Fused Filament Fabrication (FFF)

Fused filament fabrication (FFF) is a relatively new method of Additive manufacturing (also known as FDM) technology used for building three-dimensional prototypes or models layer by layer with a range of thermoplastics.

FFF technology is the most widely used form of 3D desktop printing at the consumer level, fueled by students, hobbyists, and office professionals.

The technology uses a continuous filament of a thermoplastic material. The thermoplastic filament is provided in a spool (open filament area) or in a refillable auto loading cartridge. Thermoplastic filament comes in two standard diameters: 1.75mm and 3mm (true size of 2.85mm).

🔆 Ultimaker 3 and newer machines use Near Field Communication technology with their 3mm filament. **NFC technology informs the printer of the filament type and color**. NFC spools tend to be more expensive than generic filament and you need to manually feed the filament correctly through the Bowden-tubes into the extruder (hot end). The spools are in the back of the printer. This increases the footprint of the printer.

Ultimaker 3 3D printer

🔆 Sindoh 3DWOX printers uses a replacement filament spool (1.75mm) inside their cartridge with a smart chip. The benefit is the automatic loading and unloading feature along with informing the printer of the filament type and properties. Replacement spools are available. The Sindoh 3DWOX 1 printer provides the ability to either use their filament or open source filament.

Sindoh 3DWOX replacement spool and cartridge

The thermoplastic material is heated in the nozzle (hot end) (160C - 250ºC) to form liquid, which solidifies immediately when it's deposited onto the build plate or platform. The nozzle travels at a controlled rate and moves in the X and Y direction. The build plate moves in the Z direction. This creates mechanical adhesion (not chemical).

Fused Filament Fabrication (FFF) technology can be described as the inverse process of a computer numerical cutting (CNC) machine. This technology only uses the amount of material required for the part, as opposed to a CNC machine which requires significant amounts of scrap material.

Sindoh 3DWOX DP200 printer

Fused Filament Fabrication printers are available with multiple-head extruders. The most common usage for multiple-head extruders is to print in different colors or a different material for support or increase bed adhesion (raft).

Sindoh 3DWOX 2X

A few advantages of Fused Filament Fabrication (FFF) Additive manufacturing:

- Lower cost (different entry levels) into the manufacturing environment.

- Lowers the barriers (space, power, safety, and training) to traditional subtractive manufacturing.

- Reduce part count in an assembly from traditional subtractive manufacturing (complex parts vs. assemblies).

- Numerous thermoplastic filament types and colors (open and closed source) are available and affordable.

- Remote monitoring ability with a camera and LED lighting.

- Print more than one thermoplastic filament type and color during the print cycle (multiple-head extruders).

- Change or load new filament material during a print cycle.

- Customize percent infill and infill pattern type in the print.

- Only uses the amount of material required for the part. One exception, when using a raft and or support during the print.

Ultimaker S3

- No post-curing process required.

- Reduce prototyping time.

- Faster development cycle.

- Quicker customer feedback.

- Quicker product customization and configuration.

- Parallel verticals: develop and prototype at the same time.

- Most work on either the Window or Mac OSX platform.

- Open source slicing engines. To name a few: Slic3r, Skeinforge, Netfabb, KISSkice, and Cura.

- Open source filament.

Prusa i3 MK3

A slicer takes a 3D model, most often in STereoLithography (*.stl) file format and translates the model into individual layers. It then generates the machine code that the 3D printer uses. You can also use Additive Manufacturing file (*.amf) or a 3D Manufacturing Format (*.3mf) file if supported.

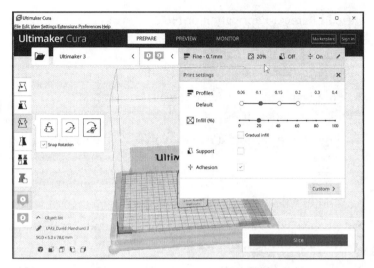

Ultimaker Cura slicer - Version: 4.4.1

Slicer software allows the user to calibrate 3D printer settings: filament type, part orientation, extruder speed, extruder temperature, bed temperature, cooling fan rate, raft type, support type, percent infill, infill pattern type, etc.

A few disadvantages of Fused Filament Fabrication (FFF) Additive manufacturing:

- Slow build rates. Many printers lay down material at a speed of one to five cubic inches per hour. Depending on the part needed, other manufacturing processes may be significantly faster.

- May require post-processing. The surface finish and dimensional accuracy may be lower quality than other manufacturing methods.

- Poor mechanical properties. Layering and multiple interfaces can cause defects in the product.

- Frequent calibration is required. Without frequent calibration, prints may not be the correct dimensions, they may not stick to the build plate, and a variety of other not-so-wanted effects can occur.

- Limited by the accuracy of the stepper motors, extruder nozzle diameter, user calibration as well as print speed.

- Print time increases linearly as part tolerances become tighter. In general, FFF print tolerances range from 0.05mm to 0.5mm.

- To print something, you require a CAD model. You either need to know how to design using CAD software (SOLIDWORKS) or download a CAD model (native format, or an (*.stl), (*.amf), or (*.3mf) file from a website (GrabCAD, Thingiverse, etc.).

StereoLithography (SLA)

Stereolithography (SLA) was the world's first 3D printing technology. It was introduced in 1988 by 3D Systems, Inc., based on work by inventor Charles Hull.

SLA is one of the most popular resin-based 3D printing technologies for professionals today.

SLA technology is a form of 3D printing in which a computer-controlled low-power, highly focused ultraviolet (UV) laser is used to create layers (layer by layer) from a liquid polymer (photopolymerization) that hardens on contact.

Markforged X7
SLS printer

When a layer is completed, a leveling blade is moved across the surface to smooth it before depositing the next layer. The platform moves by a distance equal to the layer thickness, and a subsequent layer is formed on top of the previously completed layers.

The process of tracing and smoothing is repeated until the part is complete. Once complete, the part is removed from the resin tank and drained of any excess polymer.

The part is in a "green" state. This green state differs from the completely cured state in one very important way: there are still polymerizable groups on the surface that subsequent layers can covalently bond to.

Through the application of heat and light, the strength and stability of printed parts improves beyond their original "green" state. However, each resin behaves slightly differently when post-cured, and requires different amounts of time and temperature to arrive at the material's optimum properties.

After the post-cure, you need to address the post-process. Post-processing is the removal of the supports and any needed polishing or sanding.

SLA prints are watertight and fully dense.

Resolution of SLA technology varies from 0.05mm to 0.15mm with the industry average tolerance around 0.1mm. On average, this is significantly more precise than FFF technology and is the preferred rapid prototyping solution when extremely tight tolerances are required.

 As the ultraviolet (UV) laser traces the layer, the liquid polymer solidifies and the excess areas are left as liquid in the tank.

(1micron = 1μm = 0.001mm).

One example of a popular SLA 3D desktop printer is the Formlabs Form 2. Formlabs was founded in September 2011 by three MIT Media Lab students.

The build area is 45mm × 145mm × 175mm with a layer thickness (Axis Resolution) of 25, 50, and 100 microns.

🔅 Desktop area increases significantly if you include their Form Wash and Form Cure products.

The resin (liquid polymer) is provided in a cartridge. There are different cartridges for various colors and materials. Not all resin cartridges support the 25, 50, and 100 micron resolutions.

Once the print is finished, remove it from the resin tank. Resin tanks are consumables and require replacement. Expect to replace a standard resin tank after 1,000-3,000 layers of printing (1-1.5 liters) of resin.

Wash the print with isopropyl alcohol (IPA). You should wait approximately 30 minutes for the IPA to fully evaporate after washing.

To ensure proper washing, Formlabs recommends their Form Wash. Form Wash automatically cleans uncured liquid resin printed part surfaces.

IPA dissolves uncured resin. The part is covered in uncured resin when it's removed from the resin tank. Use IPA in a well-ventilated area. Always wear protective gloves and eyewear. Find a safe way to dispose of the used IPA.

After the IPA wash and dry, perform a post-cure. At a basic level, exposure to sunlight triggers the formation of additional chemical bonds within a printed part, making the material stronger and stiffer.

To ensure proper post-curing, Formlabs recommends their post-curing unit, Form Cure. Form Cure precisely combines temperature and 405 nm light to post-cure parts for peak material performance.

🔅 You cannot use a camera or webcam inside the build area. Formlabs Dashboard feature provides the ability to keep track of what layer the printer is on, resin consumptions, and various other stats.

Formlabs Form 2
SLA 3D printer

Formlabs Form 2
resin cartridges

Formlabs
Form Wash

Formlabs
Form Cure

Formlabs uses their PreForm software. PreForm provides the ability to select the One-Click Print option, to automatically orient, support and layout the part.

Formlabs Form 2 printer works with the Window or Mac OSX platform.

💡 Clean the resin tank after a print fails. Remove any cured resin on the elastic layer, discard print failures, and filter out debris. Clean any contamination on the clear acrylic tank window.

A few advantages of Stereolithography (SLA) Additive manufacturing:

- Final parts are stronger than using FFF technology.

- Parts are watertight and fully dense. No infill is required.

- Higher resolution than FFF technology.

A few disadvantages of Stereolithography (SLA) Additive manufacturing:

- Higher cost (different entry levels) into the manufacturing environment than FFF.

- Material costs are significantly higher than FFF due to the proprietary nature and limited availability of the photopolymers.

- Significantly slower fabrication speed than FFF.

- Suitable for low volume production runs of small, precise parts.

- Print only a single material type (color) at a time.

- Requires an isopropyl alcohol (IPA) wash.

- Ability to safely dispose of the used isopropyl alcohol (IPA).

- Requires a post-cure.

- Parts are sensitive to long exposure to UV light.

- May need drain hole in the part to remove excess liquid polymer.

Selective Laser Sintering (SLS)

Selective Laser Sintering (SLS) technology is the most common additive manufacturing technology for industrial powder base applications.

3D Systems
ProX DMP 300

SLS fuses particles together layer by layer through a high energy pulse laser. Similar to SLA, this process starts with a tank full of bulk material but is in a powder form vs. liquid. As the print continues, the bed lowers itself for each new layer as done in the SLS process.

Both plastics and metals can be fused in this manner, creating much stronger and more durable prototypes.

Although the quality of the powders is dependent on the supplier's proprietary processes, the base materials used are typically more abundant than photopolymers, and therefore cheaper. However, there are additional costs in energy used for fabricating with this method which may reverse any savings realized in the material cost.

 Speed and resolution of SLS printers typically match that of SLA, with industry averages at around 0.1mm tolerances. Only suitable for low volume production runs of small, precise parts.

SLS is the preferred rapid-prototyping method of metals and exotic materials.

Select the Correct Filament Material for FFF

There are many materials that are being explored for 3D printing; however, the two most dominant plastics for FFF technology are PLA (Polylactic acid) and ABS (Acrylonitrile-Butadiene-Styrene).

Both PLA and ABS are known as thermoplastics; that is, they become soft and moldable when heated and return to a solid when cooled.

There are three key printing stages for both thermoplastics.

1. Cold to warm: The thermoplastic starts in a hard state. It stays this way until heated to its glass transition temperature.

2. Warm to hot: The thermoplastic is now in a viscous state. It stays this way until heated to its melting temperature.

3. Hot to melting: The thermoplastic is now in a liquid state.

For PLA, the glass transition temperature is approximately 60°C. The melting temperature is approximately 155°C with a printing temperature range of 190°C - 220°C.

For ABS, the glass transition temperature is approximately 100°C with a printing temperature range between 210C - 250°C. Therefore, a heated print bed is needed for ABS and is optional for PLA. The print bed should be kept near the glass transition temperature and well below the melting temperature. It also helps to keep the part from cooling too fast and warping.

PLA and ABS are hygroscopic, meaning it attracts and absorbs moisture from the air. The effects of attracting water may result in one or more of the following problems: increased brittleness, diameter augmentation (potential problems with Bowden-tube printers), or filament bubbling once reaching the extruder (hot-end).

It is recommended that you store used filament in an airtight plastic bag (container) with a few silica gel packs. Place the bag in a dry, dark, controlled environment.

Silica gel pack

PLA

PLA (Polylactic Acid) is a biodegradable thermoplastic, made from renewable resources like corn starch or sugarcane. Relevant information:

- **Strength**: High | Flexibility: Low | Durability: Medium

- **Difficulty to use**: Low

- **Print temperature**: 190°C - 220°C

- **Print bed temperature**: 20°C - 40°C (not required)

- **Shrinkage/warping**: Minimal

- **Soluble**: No

- **Hot Head**: Standard Polytetrafluoroethylene, PTFE (Teflon)

- **Filament size**: 1.75mm and 2.85mm

- **Food safety**: Refer to manufacturer guidelines

PLA has a lower printing temperature (20°C - 30°C) than ABS, and it doesn't warp as easily. PLA does not require a heated bed. Most non-heated beds are made from glass or metal.

 A removable flexible bed makes it ideal to retrieve printed parts.

PLA is normally used for its nice finish, easy and fast printing characteristics and for the large amounts of colors and varieties available.

Avoid using PLA if the print is exposed to temperatures of 60°C or higher or might be bent, twisted or dropped repeatedly.

Outside of 3D printing, PLA is typically used in medical implants, food packaging, and disposable tableware.

Darker colors and glow in the dark materials, often require higher extruder temperatures (5°C - 10°C).

PLA in general is more forgiving to temperature fluctuations and moisture than ABS and Nylon during a build cycle.

Flex/Soft PLA - Common flexible filaments are polyester-based (non-toxic). Recommended print temperature range is 220°C - 250°C. It is highly recommended to drastically lower your printing speed to around 10-20mm/s. To take advantage of the filament's properties, print it with 10% infill or less. Most flexible filament adheres well to a heated bed.

PLA - Storage

PLA is mildly hygroscopic, meaning it attracts and absorbs moisture from the air.

PLA responds somewhat differently to moisture than ABS. Over time, it can become very brittle. In addition to bubbles or spurting at the nozzle (hot end), you may see discoloration and a reduction in 3D printed part properties.

Store the filament in an airtight plastic bag (container) with a few silica gel packs. Place the bag (container) in a dry, dark, temperature controlled environment. As an extra precaution, filament manufacturers often recommend using up rolls as soon as possible.

PLA can be dried using something as simple as a food dehydrator. It is important to note that this can alter the crystallinity ratio in the PLA and will lead to changes in extrusion temperature and other extrusion characteristics.

PLA - Part Accuracy

Compared to ABS, PLA demonstrates much less part warping. PLA is less sensitive to changes in temperature than ABS.

PLA undergoes more of a phase-change when heated and becomes much more liquid. If actively cooled, sharper details can be seen on printed corners without the risk of cracking or warping. The increased flow can also lead to stronger binding between layers, improving the strength of the printed part.

In a small enclosed space, it is recommended to have your printer enclosed with a HEPA filtration filter.

ABS

ABS (Acrylonitrile-Butadiene-Styrene) is an oil-based thermoplastic, commonly found in (DWV) pipe systems, automotive trim, bike helmets, and toys (LEGO). Relevant information:

- **Strength**: High | Flexibility: Medium | Durability: High
- **Difficulty to use**: Medium
- **Print temperature**: 210°C - 250°C
- **Print bed temperature**: 80°C - 110°C (required)
- **Shrinkage/warping**: Medium
- **Soluble**: In esters, ketones, and acetone
- **Hot Head**: Standard Polytetrafluoroethylene, PTFE (Teflon)
- **Filament size**: 1.75mm and 2.85mm
- **Food safety**: Not food safe

ABS boast slightly higher strength, flexibility, and durability. ABS is more sensitive to changes in temperature than PLA, which can result in cracking and warping if the print cools too quickly.

A heated bed plate is required for ABS. Most heated beds are made from glass or metal.

HEPA FILTER

It is recommended to have a ventilated printing area when using ABS material (oil-based thermoplastic). It is also recommended to have the printer enclosed with a HEPA filtration filter.

ABS is better suited for items that are frequently handled, dropped, or heated. It can be used for mechanical parts, especially if they are subjected to stress or must interlock with other parts.

Sindoh 3DWOX 1, 2X
HEPA filter is built into
the printer.

For high temperature applications, ABS (glass transition temperature of 105°C) is more suitable than PLA (glass transition temperature of 60°C). PLA can rapidly lose its structural integrity as it approaches 60°C.

ABS - Storage

ABS is mildly hygroscopic. Diameter augmentation (potential problems with Bowden-tube printers) can be an issue.

Store the filament in an airtight plastic bag (container) with a few silica gel packs. Place the bag in a dry, dark, temperature controlled environment. As an extra precaution, filament manufacturers often recommend using up rolls as soon as possible. ABS can be easily dried using a source of hot (preferably dry) air such as a food dehydrator.

Ultimaker 3 using Bowden-tubes

ABS - Part Accuracy

For most, the single greatest hurdle for accurate parts is good bed (platform) adhesion. Start with a clean, level, heated bed. Check the bed and extruder (hot end) are set to the correct temperature.

Eliminate all build area drafts (open windows, air conditioning vents, etc.). Use dry filament. Wet filament during printing prevents good layer adhesion and greatly weakens the part.

When printing on a glass plate, a (Polyvinyl Acetate) PVA based glue stick applied to the bed helps with bed adhesion. Elmer's or Scotch permanent glue sticks are inexpensive and easily found. Remember, less is more when applying the glue stick to the build plate.

You may need to add a raft (a horizontal latticework of filament located underneath the part) to the build. Over time, a heated metal bed can warp. Check for flatness.

ABS provide a more matte appearance than PLA, but it can become very shiny after acetone vapor smoothing.

Nylon

Nylon (618, 645) is a popular family of synthetic polymers used in many industrial applications. Compared to most other filaments, it ranks as the number one contender when together considering strength, flexibility, and durability.

Relevant information for Nylon 618:

- **Strength**: High | Flexibility: High | Durability: High

- **Difficulty to use**: Medium

- **Print temperature**: 240°C - 255°C

- **Print bed temperature**: 50°C - 60°C (required)

- **Shrinkage/warping**: Medium

- **Soluble**: No

- **Hot Head**: Metal

- **Filament size**: 1.75mm and 2.85mm

- **Food safety**: Refer to manufacturer guidelines

Relevant information for Nylon 645:

- **Strength**: High | Flexibility: High | Durability: High

- **Difficulty to use**: Medium

- **Print temperature**: 255°C - 265°C

- **Print bed temperature**: 85°C - 95°C (required)

- **Shrinkage/warping**: Considerable

- **Soluble**: No

- **Hot Head**: Metal

- **Filament size**: 1.75mm and 2.85mm

- **Food safety**: Refer to manufacturer guidelines

PTFE Hot end

All metal Hot end

Nylon filament requires temperatures above 240°C to extrude. Most low-end 3D printers come standard with hot ends that use Polyether ether Ketone (PEEK) and Polytetrafluoroethylene (Teflon). Both PEEK and PTFE begin to breakdown above 240°C and will burn and emit noxious fumes. You should only use an all metal hot end.

💡 PLA and ABS are less likely to get stuck in the inner wall PTFE (Teflon) than an all metal hot end.

When should I use Nylon? Taking advantage of nylon's strength, flexibility, and durability use this 3D printer filament to create tools, functional prototypes, or mechanical parts (like hinges, buckles, or gears). Dry nylon prints buttery smooth and has a glossy finish.

Nylon - Storage

Nylon is very hygroscopic, more so than PLA or ABS. Nylon can absorb more than 10% of its weight in water in less than 24 hours. Successful 3D printing with nylon requires dry filament. When you print with nylon that isn't dry, the water in the filament explodes causing air bubbles during printing that prevents good layer adhesion and greatly weakens the part. It also ruins the surface finish. To dry nylon, place it in an oven at 50C - 60°C for 6-8 hours. After drying, store in an airtight container (vacuum bag), preferably with dry silica gel packets. Place the container in a dry, dark, temperature controlled environment.

Nylon - Part Accuracy

Compared to ABS and PLA, Nylon and ABS warp approximately the same. PLA demonstrates much less part warping. A heated plate (50°C - 90°C) is required for Nylon. When printing on a glass plate, a (Polyvinyl Acetate) PVA based glue stick applied to the bed is the best method of bed adhesion. Elmer's or Scotch permanent glue sticks are inexpensive and easily found. Remember, less is more when applying the glue stick to the plate. You will also have to clean it up after your build.

PVA (Polyvinyl Alcohol)

PVA (Polyvinyl Alcohol) filament is a water-soluble synthetic polymer. PVA filament dissolves in water. Many multi-head extruder users find PVA to be a useful support material because of its dissolvable properties. In general, PVA filament is used in conjunction with PLA not ABS. PVA adheres well to PLA and not ABS. Moreover, the extrusion temperature difference between PVA and ABS can be problematic.

PVA extrusion temperatures range between 160°C - 190°C. A heated bed is recommended. Bed temperatures range between 40°C - 50°C.

PVA is highly hygroscopic and is costly. PVA should be stored in an airtight box or container and may need to be dried before use.

Submerge the finished part in a bath of cold circulating water, until the PVA support structure is completely dissolved. This can be time consuming and messy.

💡 Do not expose PVA filament to temperatures higher than 200°C for an extended period of time. An irreversible degradation of the material will occur, known as pyrolysis. It will jam the extruder nozzle (hot end). Unlike PLA and ABS, you cannot remove the jam by increasing the temperature. Clearing the jam in the nozzle will often require it to be re-drilled or replaced altogether.

💡 A few filament companies provide breakaway material to replace the high cost of PVA. Breakaway is a support material used with multi head 3D printers. It is quick to remove and does not need further post-processing.

STereoLithography (*.stl) file

STereoLithography (*.stl) is a file format native to the Stereolithography CAD software created by 3D Systems. STL has several after-the-fact backronyms such as "Standard Triangle Language" and "Standard Tessellation Language".

An STL file describes only the surface geometry of a three-dimensional object without any representation of color, texture, or other common CAD model attributes. The STL format specifies both ASCII and Binary representations.

Binary files are more common, since they are more compact. An STL file describes a raw unstructured triangulated (point cloud) surface by the unit normal and vertices (ordered by the right-hand rule) of the triangles using a three-dimensional Cartesian coordinate system.

STL Save options allow you to control the number and size of the triangles by setting the various parameters in the CAD software.

Save an STL (*.stl) file in SOLIDWORKS

To save a SOLIDWORKS model as a STL file, click **File**, **Save As** from the Main menu or **Save A**s from the Main menu toolbar. The Save As dialog is displayed. Select **STL(*.stl)** as the Save as type.

Click the **Options** button. The dialog box is displayed. View your options and the three types of Resolution: **Coarse**, **Fine** and **Custom**. The resolution options provide the ability to control the number and size of the triangles by setting various parameters in the CAD software. The Custom setting provides the ability to control the deviation and angle for the triangles. Click **OK**. View the generate point cloud of the part. Click **Yes**. The STL file is now ready to be imported into your 3D printer software.

In SOLIDWORKS for a smoother STL file, change the Resolution to Custom. Change the deviation to 0.0005in (0.01mm). Change the angle to 5. Smaller deviations and angles produce a smoother file but increase the file size and print time.

Additive Manufacturing (*.amf) file

Additive Manufacturing (*.amf) is a file format that includes the materials that have been applied to the parts or bodies in the 3D model.

Additive Manufacturing (*.amf) is an open standard for describing objects for additive manufacturing processes such as 3D printing. The official ISO/ASTM 52915:2013 standard is an XML-based format designed to allow any computer-aided design software to describe the shape and composition of any 3D object to be fabricated on any 3D printer. Unlike its predecessor STL format, AMF has native support for color, materials, lattices, and constellation (groups). Therefore, it requires less post-processing to define data such as the position of your model relative to the selected 3D printer, orientation, color, materials, etc.

Save an Additive Manufacturing (*.amf) file in SOLIDWORKS

To save a SOLIDWORKS model as an Additive Manufacturing (*.amf) file, Click **File**, **Save As** from the Main menu or **Save As** from the Main menu toolbar. The Save As dialog is displayed. Select **Additive Manufacturing (*.amf)** as the Save as type.

Click the **Options** button. The dialog box is displayed. View your options. For most parts, utilize the default setting.

Close the dialog box.

Click **OK**. View the generate point cloud of the part.

Click **Yes**. The file is now ready to be imported into your 3D printer software or print directly from SOLIDWORKS.

3D Manufacturing Format (*.3mf) file

3D Manufacturing Format (*.3mf) is a file format developed and published by the 3MF Consortium. This format became natively supported in all Windows operating systems since Windows 8.1. 3MF has since garnered considerable support from large companies such as HP, 3D Systems, Stratasys, GE, Siemens, Autodesk and Dassault Systems although it is unknown how many actively use this file format. 3D Manufacturing Format file has similar native support for color, materials, lattices, and constellation (groups).

VRML (*.wrl)
STL (*.stl)
3D Manufacturing Format (*.3mf)
Additive Manufacturing File (*.amf)
eDrawings (*.eprt)
3D XML (*.3dxml)
Microsoft XAML (*.xaml)
CATIA Graphics (*.cgr)

Save a 3D Manufacturing Format (*.3mf) file in SOLIDWORKS

To save a SOLIDWORKS model as a 3D Manufacturing Format (*.3mf) file, Click **File**, **Save As** from the Main menu or **Save As** from the Main menu toolbar. The Save As dialog is displayed. Select **3D Manufacturing Format (*.3mf)** as the Save as type.

Click the **Options** button. The dialog box is displayed. View your options. For most parts, utilize the default setting.

Close the dialog box. Click **OK**. View the generate point cloud of the part. Click **Yes**. The file is now ready to be imported into your 3D printer software or print directly from SOLIDWORKS.

The include materials and include appearances option is not selected by default.

In SOLIDWORKS for a smoother STL file, change the Resolution to Custom. Change the deviation to 0.0005in (0.01mm). Change the angle to 5. Smaller deviations and angles produce a smoother file but increase the file size and print time.

What is a Slicer? How does a Slicer Work?

The (*.3mf), (*.amf), or (*.stl) file is imported into the printer slicer. The slicer takes the file and translates the model into individual layers. It then generates the machine code (G-code) that the printer requires.

Slicing is the process of turning the 3D model into a toolpath for the printer. Most people call it slicing because the first thing the slicing engine does is cut the 3D model into thin horizontal layers.

The G-code file contains the instructions based on settings you choose and calculates how much material the printer will need and how long it will take to print.

Slicer Parameters

Proper slicer parameters can mean the difference between a successful print, and a failed print. That's why it's important to know how slicers work and how various settings will affect the final print. There are open and close source slicers.

Layer Height

Layer height controls the resolution of the print. The setting specifies the height of each filament layer.

The higher values produce faster prints in lower resolution. Lower values produce slower prints in higher resolution.

The default value for (PLA) using the Sindoh 3DWOX 2X - Version: 1.4.2102.0 is .20mm. There are four default settings: Fast, Normal, Silent, and Slow. Settings range from .05mm - .4mm. Sindoh uses a .4mm nozzle with 1.75mm filament.

The default value for (PLA) using in the Ultimaker 3 Cura slicer - Version: 4.4.1 is .1mm. There are four default settings: Fast, Normal, Fine, and Extra Fine. Settings range from .06mm - .2mm.

Sindoh 3DWOX slicer - Version: 1.4.2102.0

Ultimaker Cura slicer - Version: 4.4.1

Shell (Wall) Thickness

Wall thickness refers to the distance between one surface of the model and the opposite sheer surface. Set the wall thickness of the outside shell in the horizontal direction. Use in combination with the nozzle size to define the number of perimeter lines and the thickness of these perimeter lines.

Sindoh 3DWOX slicer - Version: 1.4.2102.0

The default value for (PLA) using the Sindoh 3DWOX 2X - Version: 1.4.2102.0 is .80mm. The minimum is .40mm.

The default value for (PLA) using the Ultimaker 3 Cura - Version: 3.4.1 is 1mm.

Shell		
Wall Thickness	1	mm
Wall Line Count	3	
Top/Bottom Thickness	1	mm
Top Thickness	1	mm
Top Layers	10	
Bottom Thickness	1	mm
Bottom Layers	10	

Ultimaker Cura slicer - Version: 3.4.1

Infill Density/Overlap

Infill is the internal structure of your object, which can be as sparse or as substantial as you would like it to be. A higher percentage will result in a more solid object, so 100% (not recommended) infill will make your object completely solid, while 0% infill will give you something completely hollow.

The higher the infill percentage, the more material and longer the print time. It will also increase weight and material cost.

When using any infill percentage, a pattern is used to create a strong and durable structure inside the print. A few standard patterns are: Rectilinear, Honeycomb, Circular, Tri-hexagon, Cubic, Octet, and Triangular. In general, use a 10% - 15% infill with a maximum infill of 60%. 100% infill is not recommended. Part warping can be a concern.

Shells are the outer layers of a print which make the walls of an object, prior to the various infill levels being printed within. The number of shells affect stability and translucency of the model.

View the different infill and number of shells between the two models.

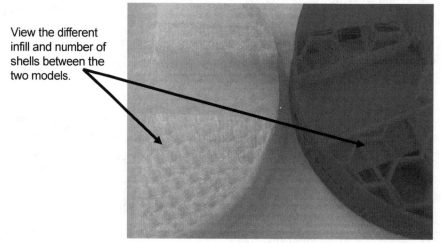

The default value for (PLA) using the Sindoh 3DWOX 2X - Version: 1.4.2102.0 is 15%.

The default value for (PLA) using the Ultimaker 3 Cura - Version: 3.4.1 is 20%.

Sindoh 3DWOX slicer - Version: 1.4.2102.0

Strength corresponds to the maximum stress the print can take before breaking.

Print Speed

Print speed refers to the speed at which the extruder travels while it lays down filament.

Print speed affects the following areas: Infill speed, Wall speed, Top/Bottom speed, Initial layer travel speed, Raft print speed, and Maximum travel resolution.

The default value for (PLA) using the Sindoh 3DWOX 2X - Version: 1.4.2102.0 is 40mm/s.

The default value for (PLA) using the Ultimaker 3 Cura - Version: 3.4.1 is 70mm/s.

Supports

Generate structures to support parts (features) of the model which have overhangs to the build plate.

Support material

Without these structures, such parts would collapse during the print process.

In order to create an overhang at any angle less than vertical, your printer offsets each successive layer. The lower the angle gets to horizontal, or 90°, the more each successive layer is offset.

General 45-degree rule. If your model has overhangs greater than 45 degrees, you need support material. If the part has numerous holes, sharp edges, long run (bridge), or thin bodies, support material may also be required.

Support Types

Touching Buildplate

The Touching Buildplate option provides supports only where the part touches the build plate. This reduces build time, clean up and support material.

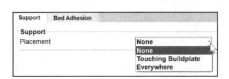

Sindoh 3DWOX slicer - Version: 1.4.2102.0

Use the touching Buildplate option when you have overhangs and tricky angles toward the bottom of a design, but do not wish to plug up holes, hollow spaces, or arches in the rest of the design.

Everywhere

The Everywhere option provides support material everywhere on the part, not only where the part touches the build plate.

Bed (Platform) Adhesion

Bed adhesion is one of the most important elements for getting a good 3D print. Set various options to ensure good bed adhesion and to prevent part warping. If needed, use blue painter's tape, hair spray or a glue stick.

Bed Adhesion Type- Raft

A Raft is a horizontal latticework of filament located underneath the part. A Raft is used to help the part stick to the build plate (heated or non-heated).

Rafts are also used to help stabilize thin tall parts with small build plate footprints.

When the print is complete, remove the part from the build plate. Peel the raft away from the part. If needed, use a scraper or spatula.

Bed Adhesion Type- Skirt

A Skirt is a layer of filament that surrounds the part with a 3mm - 4mm offset. The layer does not connect the part directly to the build plate. The Skirt primes the extruder and establishes a smooth flow of filament. In some slicers, the skirt is added automatically when you select the None option, for bed adhesion type.

Bed Adhesion Type- Brim

A Brim is basically like a Skirt for the part. A Brim has a zero offset from the part. It is a layer of filament laid down around the base of the part to increase its surface area. A Brim, however, does not extend underneath the part, which is the key difference between a brim and a raft.

Touching Build plate

Part with a Raft on the Build plate

Part with a Skirt on the Build plate

Part Orientation

Insert the file into your printer's slicer. The model is displayed in the build plate area.

Depending on your printer's slicer software, you may or may not receive a message indicating the object is too large for the current build plate.

If the object is too large, you will need to scale it down or redesign it into separate parts.

Use caution when scaling critical features if you require fasteners or a minimum wall thickness.

You should always center the part and have it lay flat on the build plate. Bed adhesion is one of the most important elements for getting a good 3D print.

If you are printing more than one part, space them evenly on the plate or position them for a single build.

In SOLIDWORKS, lay the parts out in an assembly. Save the assembly as a part. Save the part file as an STL (*.stl), or Additive Manufacturing (*.amf), or 3D Manufacturing format (*.3mf).

Consideration should be used when printing an assembly. If the print takes 20 hours, and a failure happens after 19 hours, you just wasted a lot of time versus printing each part individually.

| STL (*.stl) |
| 3D Manufacturing Format (*.3mf) |
| Additive Manufacturing File (*.amf) |
| eDrawings (*.eprt) |
| 3D XML (*.3dxml) |
| Microsoft XAML (*.xaml) |

Example 1: Part Orientation

Part orientation is very important on build strength and the amount of raft and support material required for the build. Incorrect part orientations can lead to warping, curling, and delamination.

If maximizing strength is an issue, select the part orientation on the build plate so that the "grain" of the print is oriented to maximize the strength of the part.

Example 1: First Orientation - Vertical

In the first orientation (vertical), due to the number of holes and slots, additional support material is required (with minimum raft material) to print the model.

Removing the material in these geometrics can be very time consuming.

Example 1: Second Orientation - Horizontal

In the second orientation (horizontal), additional raft material is used, and the support material is reduced.

The raft material can be easily removed with a pair of needle-nose pliers and no support material clean-up is required for the holes and slots. Note: In some cases, raft material is not needed.

Example 2: Part Orientation

The lens part is orientated in a vertical position with the large face flat on the build plate. This reduces the required support material and ensures proper contact (maximum surface area) with the build plate.

Optimize Print Direction

Some slicers (Sindoh 3DWOX desktop) provide the ability to run an optimization print direction analysis of the part under their Advanced Mode. The areas in evaluations are: **Thin Region**, **Area of Overhang Surface** and the **Amount of needed support** material. This can be very useful when you are unsure of the part orientation.

Suppress mates in an assmbly to have the model lay flat on the build plate. If the model does not lay flat, you will require a raft and additional supports. This will increase build time and material cost.

First orientation - vertical

Second orientation - horizontal

Sindoh 3DWOX slicer - Version: 1.4.2102.0

The needed support material is created mainly internal to the part to print the CBORE feature. Note: There is some outside support material on the top section of the part.

Raft material   Internal support material for the CBORE

Proper part orientation for thin parts will make the removal of the raft easier.

Remove the Model from the Build Plate

Non-heated Build Plate

Most Non-heated build plates are made from glass or metal. If needed, use blue painter's tape, glue stick or hair spray to assure good model adhesion. After the build, remove the plate from the printer.

Ultimaker 3 glass build plate

Utilize a flat edge tool (thin steel spatula). Gently work under the part, and lift the part directly away from yourself. Clean the build plate. Return the plate. Re-level the plate after every build.

The Sindoh 3DWOX DP201 printer (non-heated) has a flexible magnet removable plate that does not require blue painter's tape, glue stick or hair spray to assure good model adhesion. This eliminates the need for scrapers or any sharp tools to remove the print. After the build, remove the plate and bend.

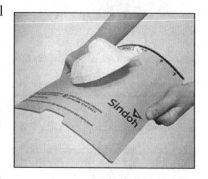

Sindoh 3DWOX DP201
flexible magnet build plate

💡 Bed adhesion is one of the most important elements for getting a good 3D print.

Heated Build Plate

Most heated build plates are made from glass or metal. If needed, use a glue stick or hair spray to assure good model adhesion. If you have a heated build plate your temperatures can range between 30°C - 90°C depending on the material, so be careful. After the build, remove the plate from the printer. Utilize a flat edge tool (thin steel spatula). Gently work under the part, and lift the part directly away from yourself. Clean the build plate. Return the plate. Re-level the plate after every build.

Prusa i3 MK3 flexible metal
magnet build plate

A few manufacturers (Sindoh 3DWOX 1 and 2X and Prusa i3 MK3) provide a flexible heated magnet metal build plate.

This eliminates the need for scrapers or any sharp tools most of the time. After the build, remove the flexible magnet plate and bend.

💡 Most heated metal bed plates are coated. Over time, the coating wears and the plate needs to be replaced.

Sindoh 3DWOX 1 and 2X
flexible metal magnet build plate

Know your Printer's Limitations

Overall part size can be an issue. Most affordable 3D printers typically are small enough to fit on your desktop. Typical build volumes range between 200 x 200 x 185 mm and 228 x 200 x 300 mm. The Sindoh 3DWOX 2X has one of the largest build volumes for an affordable desktop FFF 3D printer.

Sindoh 3DWOX 2X

There are features that are too small to be printed on a desktop 3D printer. An important, but often overlooked variable in what the printer can achieve is thread width. Thread width is determined by the diameter of the extruder nozzle. Most printers have a 0.4mm nozzle. A circle created by the printer is approximately two thread widths deep: 0.8mm thick with a 0.4mm nozzle to 1mm thick for a 0.5mm nozzle. A good rule of thumb is "The smallest feature you can create is double the thread width".

Tolerance for Interlocking Parts

For objects with multiple interlocking parts, design for a tolerance fit. Getting tolerances correct, can be difficult using FFF technology.

Ultimaker S3

In general, use the below suggested guidelines.

- Use ±0.1mm (±.004 in.) tolerance for a tight fit (press fit parts, connectors).

- Use ±0.2mm (±.008 in.) tolerance for a print in place (hinge).

- Use ±0.3mm (±.012 in.) tolerance for loose fit (pin in hole).

Test the fit yourself with the particular model to determine the right tolerance for the items you are creating and material you are using.

Tolerance may vary depending on filament type, manufacturer, color, humidity, build plate flatness, bed temperature, extruder temperature, etc.

Ultimaker Cura slicer - Version: 3.4.1

General Printing Tips

Reduce Infill (Density/Overlap)

Infill is a settable variable in most Slicers. The amount of infill can affect the top layers, bottom layers, infill line distance and infill overlap.

Reduced infill can have a negative effect on part strength. There are always trade-offs.

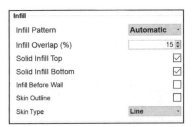

Sindoh 3DWOX slicer - Version: 1.4.2102.0

The material inside the part (infill) exerts a force on the entire printed part as it cools. More material increases cost of the part and build time.

Parts with a lower percentage of infill should have a lower internal force between layers and can reduce the chance of curling, cracking, and layer delamination along with a low build cost and time.

Sindoh 3DWOX 1 - Standard enclosed build area

Control Build Area Temperature

For a consistent quality build, control the build area and environment temperature. Eliminate all drafts and control air flow that may cause a temperature gradient within the build area.

Changes in temperature during a build cycle can cause curling, cracking and layer delamination, especially on long thin parts. From 1000s of hours in 3D printing experience, we have found that having a top cover and sides along with a consistent room temperature provides the best and repeatable builds.

Some printer companies, like Ultimaker and FlashForge, sell additional enclosure kits (doors, sides, and tops). This helps regulate the temperature and reduce drafts for improved print quality.

FlashForge 3D Printer Creator Pro

The kits also lower the sound of the printer and provide a more secure print area.

When troubleshooting issues with your printer, it is always best to know if the nozzle and heated bed are achieving the desired temperatures. A thermocouple and a thermometer come in handy. I prefer a Type-K Thermocouple connected to a multi-meter. A non-contact IR or Laser-based thermometer also works well.

Cover/Door kit for Ultimaker 3

Add Pads

Sometimes, when you are printing a large flat object, you may view warping at the corners or extremities. One way to address this is to create small pads to your part during the modeling process. Create the model for the print. Think before you print and know your printer limitations. The pads can be any size and shape, but generally, diameter 10mm cylinders that are 1-2 layers thick work well. After the part is printed, remove them.

Makerbot Image

Unique Shape or a Large Part

If you need to make parts larger than your build area or create parts that have intricate projections, here are a few suggestions:

- Fuse smaller sections together using acetone (if using ABS). Glue if using PLA.

- Design smaller parts to be attached together (without hardware).

- Design smaller parts to be screwed together (with hardware).

Safe Zone Rule

Parts may have a safe zone. The safe zone is called "self-supporting" and no support material is required to build the part.

The safe zone can range between 30° to 150°. If the part's features are below 30° or greater than 150°, it should have support material during the build cycle. This is only a rule. Are there other factors to consider? Yes. They are layer thickness, extrusion speed, material type, length of the overhang along with the general model design of features.

Design your part for your printer. Use various modeling techniques (ribs, fillets, pads, etc.) during the design process to eliminate or to minimize the need for supports and clean up.

First Layer Not Sticking

One of the toughest aspects of 3D printing is to get your prints to stick to the build surface or bed platform. Investigate the following:

- Clean the bed. Remove any residue of tape, glue, hair spray, etc.

- Apply new blue painter's tape (non-heated), hair spray or glue to the build plate if needed.

- Level the bed (build platform). Perform an automatic (Assisted Bed Leveling) or manual leveling.

- Check extruder (hot end) temperature. Different filaments require different hot end temperatures.

- Check heated build plate temperature. Different filaments require different bed temperatures.

- Control the build area temperature around the printer. Eliminate all drafts.

- Layer height. Min layer height = 1/4 nozzle diameter. Max layer height = 1/2 nozzle diameter. Layer height too low, might cause the filament to be pushed back into the nozzle (plugging). If layer height is too high, the layers won't stick to the build plate.

Level Build Platform

An unleveled build platform will cause many headaches during a print. You can quickly check the platform by performing the business card test: use a single business card to judge the height of the extruder nozzle over the build platform. Achieve a consistent slight resistance when you position the business card between the tip of the extruder and the bed platform for all leveling positions.

Most 3D printers have an automatic Assisted Bed Leveling feature as illustrated.

Sindoh 3DWOX Printers

Minimize Internal Support

Design the part for the printer. Use various modeling techniques in SOLIDWORKS (ribs, fillets, pads, etc.) during the design process to eliminate or to minimize the need for support and final part clean up.

Design a Water Tight Mesh

A water-tight mesh is achieved by having closed edges creating a solid volume. If you were to fill your geometry with water, would you see a leak? You may have to clean up any internal geometry that could have been left behind accidentally from Booleans.

Clearance

If you are creating separate or interlocking parts, make sure there is a large enough distance between tight areas. 3D printing production makes moving parts without assembly a possibility. Take advantage of this strength by creating enough clearance that the model's pieces do not fuse together or trap support material inside.

In General (FFF Printers)

- Keep your software and firmware up to date.

- Think before you print. Design the model for your printer.

- Understand the printer's limitations. Adjust one thing at a time between prints and keep notes about the settings and effect on the print. Label test prints and take photographs.

- Control the build area and environment temperature. Eliminate all drafts and control air flow that may cause a temperature gradient within the build area.

- Level and clean the build plate before a build.

- Select the correct filament (material) for the application. Materials are still an area of active exploration.

- Set the correct extruder (hot end) and bed plate temperature.

- Most low-end 3D printers use an extruder (hot end) with Polyether ether Ketone (PEEK) or Polytetrafluoroethylene (Teflon). Both PEEK and PTFE begin to breakdown above 240°C and will burn and emit noxious fumes. Use an all metal hot end above 240°C.

- Select the correct part orientation.

- Suppress mates in an assembly to have the model lay flat on the build plate.

- Most parts can be printed successfully with 15 - 20% infill.

- Select the correct settings for your Slicer. If in doubt, use the factory default settings.

- Control the filament storage environment (temperature, humidity, etc.).

- General 45 degree rule. If your model doesn't have any overhangs greater than 45 degrees, you should not need support material. If the part has numerous holes, sharp edges, long run (bridge), or thin bodies, support material may be needed.

- If in doubt, create your first build with a raft and support.

- If needed, orientate the printed part on the bed plate for maximum strength (lines perpendicular to the force being applied).

Tension load normal to layers
Part is weak

Tension load parallel to layers
Part is strong

Print Directly from SOLIDWORKS

Download and install the printer drivers.

Download and install the SOLIDWORKS slicer Add-in. In this case, the SOLIDWORKS 3DWOX.

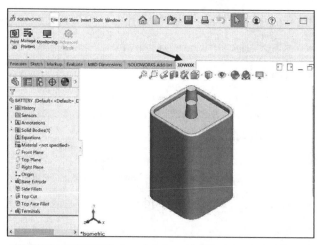

💡 Screen shots and procedure will vary depending on the 3D printer manufacture's slicer Add-in.

When the 3DWOX Add-in installation is complete, a 3DWOX tab is displayed in the CommandManager.

Open a SOLIDWORKS model.

Click the 3DWOX tab. View your options:

- **Print 3D**

- **My Printers**

- **Monitoring**

- **Advanced Mode**

The **Print 3D** button provides the ability to access the slicer within SOLIDWORKS. The SOLIDWORKS part model is automatically converted into an STL file and appears on the selected (Settings button) printer bed. The Sindoh 3DWOX slicer is displayed.

The **My Printers** button provides the ability to manage your printers (printer name, IP and availability) over a network.

The **Monitoring** button provides the ability to connect to the network and view your printing real time with an internal camera.

The **Advanced Mode** button provides the ability to access the default settings, or the Advanced settings to customize the print.

Certified SOLIDWORKS Associate Additive Manufacturing (CSWA-AM) Exam

The Certified SOLIDWORKS Associate Additive Manufacturing (CSWA-AM) exam indicates a foundation in and apprentice knowledge of today's 3D printing technology and market.

Information is provided on the login procedure at https://3dexperience.virtualtester.com/#home for students to register, examine their certifications, manage their account and take the CSWA-AM certification.

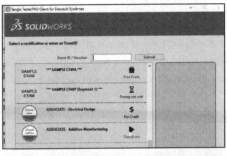

If your school is an academic certification provider, your instructor can allocate free exam credits for the CSWA (Segment 1 & 2), CSWA-SD, CSWA-S, and CSWA-AM certifications. The instructor will require your .edu email address.

The CSWA-AM exam is meant to be taken after the completion of the 10-part learning path located on MySOLIDWORKS.com.

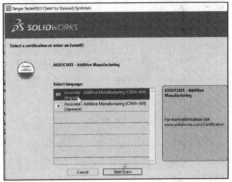

The learning content is free, but the creation of a free MySOLIDWORKS account is needed in order to access the content.

The exam is 50 questions. Each question is worth 2 points. You are allowed to answer the questions in any order. Total exam time is 60 minutes. You need a minimum passing score of 80 or higher. The exam is out of 100 points.

The Certified SOLIDWORKS Associate Additive Manufacturing (CSWA-AM) exam covers the Ultimaker 3 (FFF) and the FormLabs Form 2 (SLA) machine as examples.

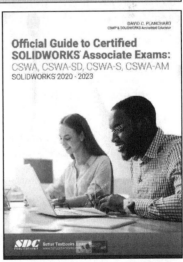

Note: The Official Guide to Certified SOLIDWORKS Associate Exams: CSWA, CSWA-SD, CSWA-S, CSWA-AM 2020 - 2023 is meant to provide all of the needed information from the online MySOLIDWORKS lessons to pass the CSWA-AM exam.

The Certified SOLIDWORKS Associate Additive Manufacturing (CSWA-AM) exam covers the Ultimaker 3 (FFF) and the FormLabs Form 2 (SLA) machine as examples.

Summary

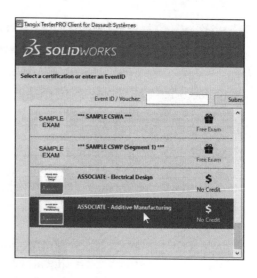

Stereolithography (SLA) was the world's first 3D printing technology. It was introduced in 1988 by 3D Systems, Inc., based on work by inventor Charles Hull.

SLA is one of the most popular resin-based 3D printing technologies for professionals today.

Selective Laser Sintering (SLS) technology is the most common additive manufacturing technology for industrial powder base applications. SLS fuses particles together layer by layer through a high energy pulse laser. Similar to SLA, this process starts with a tank full of bulk material but is in a powder form vs. liquid.

Fused filament fabrication (FFF) is an additive manufacturing technology used for building three-dimensional prototypes or models layer by layer with a range of thermoplastics.

FFF technology is the most widely used form of 3D desktop printing at the consumer level, fueled by students, hobbyists and office professionals.

The two most dominant plastics for FFF technology are PLA (Polylactic acid) and ABS (Acrylonitrile-Butadiene-Styrene).

With 1000s of hours using multiple low cost FFF 3D printers, we have found that learning about additive print technology is a great experience. Students face a multitude of obstacles with their first 3D prints. Understanding what went wrong and knowing the capabilities of the 3D printer produces positive results.

Never focus too much on one single issue. These machines are complex, and trouble often arises from multiple reasons. A slipping filament may not only be caused by a bad gear drive, but also by an obstructed nozzle, a wrong feed value, a too low (or too high) temperature or a combination of all these.

Always store filament in an airtight plastic bag (container) with a few silica gel packs. Place the bag in a dry, dark, temperature controlled environment. As an extra precaution, filament manufacturers often recommend using up rolls as soon as possible.

It's best to monitor the first 3 - 5 layers when starting a print. Never assume you will have perfect bed adhesion. If there is a bed adhesion issue, check to see if the bed is level. Check the bed and extruder (hot end) temperatures and then either apply blue painter's tape (non-heated), a glue stick or hair spray. You can also add a raft to the part.

Always clean and level the build plate before every print.

As printed parts cool (PLA, ABS and Nylon), various areas of the object cool at different rates. Depending on the model being printed and the filament material, this effect can lead to warping, curling and or layer delamination.

Acetone is often used in post processing to smooth ABS, also giving the part a glossy finish. ABS can be sanded and is often machined (for example, drilled) after printing. PLA can also be sanded and machined; however, greater care is required.

For high temperature applications, ABS (glass transition temperature of 105°C) is more suitable than PLA (glass transition temperature of 60°C). PLA can rapidly lose its structural integrity as it approaches 55°C - 60°C.

Design the part for your printer. Try to use the 45-degree rule. If your model doesn't have any overhangs greater than 45 degrees, you should not need support. There are exceptions to the 45-degree rule. The most common ones are straight overhangs, and fully suspended islands.

Use various CAD modeling techniques (ribs, fillets, pads, etc.) during the design process to eliminate or to minimize the need for supports and clean up.

From a structural situation, select the correct % infill, infill pattern and orientate the part on the build plate to print the layers perpendicular in relation to the movement of the build platform.

Notes:

Appendix

SOLIDWORKS Keyboard Shortcuts

Below are some of the pre-defined keyboard shortcuts in SOLIDWORKS:

Action:	Key Combination:
Model Views	
Rotate the model horizontally or vertically	**Arrow** keys
Rotate the model horizontally or vertically 90 degrees	**Shift** + **Arrow** keys
Rotate the model clockwise or counterclockwise	**Alt** + left of right **Arrow** keys
Pan the model	**Ctrl** + **Arrow** keys
Magnifying glass	**g**
Zoom in	**Shift + z**
Zoom out	**z**
Zoom to fit	**f**
Previous view	**Ctrl + Shift + z**
View Orientation	
View Orientation menu	**Spacebar**
Front view	**Ctrl + 1**
Back view	**Ctrl + 2**
Left view	**Ctrl + 3**
Right view	**Ctrl + 4**
Top view	**Ctrl + 5**
Bottom view	**Ctrl + 6**
Isometric view	**Ctrl + 7**
NormalTo view	**Ctrl + 8**
Selection Filters	
Filter edges	**e**
Filter vertices	**v**
Filter faces	**x**
Toggle Selection Filter toolbar	**F5**
Toggle selection filters on/off	**F6**
File menu items	
New SOLIDWORKS document	**Ctrl + n**
Open document	**Ctrl + o**
Open From Web Folder	**Ctrl + w**
Make Drawing from Part	**Ctrl + d**
Make Assembly from Part	**Ctrl + a**
Save	**Ctrl +s**
Print	**Ctrl + p**
Additional items	
Access online help inside of PropertyManager or dialog box	**F1**
Rename an item in the FeatureManager design tree	**F2**

Action:	Key Combination:
Rebuild the model	**Ctrl + b**
Force rebuild - Rebuild the model and all its features	**Ctrl + q**
Redraw the screen	**Ctrl + r**
Cycle between open SOLIDWORKS document	**Ctrl + Tab**
Line to arc/arc to line in the Sketch	**a**
Undo	**Ctrl + z**
Redo	**Ctrl + y**
Cut	**Ctrl + x**
Copy	**Ctrl + c**
Paste	**Ctrl + v**
Delete	**Delete**
Next window	**Ctrl + F6**
Close window	**Ctrl + F4**
View previous tools	**s**
Selects all text inside an Annotations text box	**Ctrl + a**

In a sketch, the **Esc** key un-selects geometry items currently selected in the Properties box and Add Relations box.

In the model, the **Esc** key closes the PropertyManager and cancels the selections.

Use the **g** key to activate the Magnifying glass tool. Use the Magnifying glass tool to inspect a model and make selections without changing the overall view.

Use the **s** key to view/access previous command tools in the Graphics window.

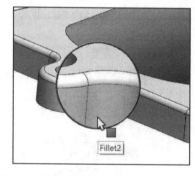

Modeling - Best Practices

Best practices are simply ways of bringing about better results in easier, more reliable ways. The Modeling - Best Practice list is a set of rules helpful for new users and users who are trying to experiment with the limits of the software.

These rules are not inflexible, but conservative starting places; they are concepts that you can default to, but that can be broken if you have good reason. The following is a list of suggested best practices:

- Create a folder structure (parts, drawings, assemblies, simulations, etc.). Organize into project or file folders.

- Construct sound document templates. The document template provides the foundation that all models are built on. This is especially important if working with other SOLIDWORKS users on the same project; it will ensure consistency across the project.

- Generate unique part filenames. SOLIDWORKS assemblies and drawings may pick up incorrect references if you use parts with identical names.

- Apply Custom Properties. Custom Properties is a great way to enter text-based information into the SOLIDWORKS parts. Users can view this information from outside the file by using applications such as Windows Explorer, SOLIDWORKS Explorer, and Product Data Management (PDM) applications.

- Understand part orientation. When you create a new part or assembly, the three default Planes (Front, Right and Top) are aligned with specific views. The plane you select for the Base sketch determines the orientation.

- Learn to sketch using automatic relations.

- Limit your usage of the Fixed constraint.

- Add geometric relations, then dimensions in a 2D sketch. This keeps the part from having too many unnecessary dimensions. This also helps to show the design intent of the model. Dimension what geometry you intend to modify or adjust.

- Fully define all sketches in the model. However, there are times when this is not practical; generally when using the Spline tool to create a freeform shape.

- When possible, make relations to sketches or stable reference geometry, such as the Origin or standard planes, instead of edges or faces. Sketches are far more stable than faces, edges, or model vertices, which change their internal ID at the slightest change and may disappear entirely with fillets, chamfers, split lines, and so on.

- Do not dimension to edges created by fillets or other cosmetic or temporary features.

- Apply names to sketches, features, dimensions, and mates that help to make their function clear.

- When possible, use feature fillets and feature patterns rather than sketch fillets and sketch patterns.

- Apply the Shell feature before the Fillet feature, and the inside corners remain perpendicular.

- Apply cosmetic fillets and chamfers last in the modeling procedure.

- Combine fillets into as few fillet features as possible. This enables you to control fillets that need to be controlled separately such as fillets to be removed and simplified configurations.

- Create a simplified configuration when building very complex parts or working with large assemblies.

- Use symmetry during the modeling process. Utilize feature patterns and mirroring when possible. Think End Conditions.

- Use global variables and equations to control commonly applied dimensions (design intent).

- Add comments to equations to document your design intent. Place a single quote (') at the end of the equation, then enter the comment. Anything after the single quote is ignored when the equation is evaluated.

- Avoid redundant mates. Although SOLIDWORKS allows some redundant mates (all except distance and angle), these mates take longer to solve and make the mating scheme harder to understand and diagnose if problems occur.

- Fix modeling errors in the part or assembly when they occur. Errors cause rebuild time to increase, and if you wait until additional errors exist, troubleshooting will be more difficult.

- Create a Library of Standardized notes and parts.

- Utilize the Rollback bar. Troubleshoot feature and sketch errors from the top of the design tree.

- Determine the static and dynamic behavior of mates in each sub-assembly before creating the top-level assembly.

- Plan the assembly and sub-assemblies in an assembly layout diagram. Group components together to form smaller sub-assemblies.

- When you create an assembly document, the base component should be fixed, fully defined or mated to an axis about the assembly origin.

- In an assembly, group fasteners into a folder at the bottom of the FeatureManager. Suppress fasteners and their assembly patterns to save rebuild time and file size.

- When comparing mass, volume and other properties with assembly visualization, utilize similar units.

- Use limit mates sparingly because they take longer to solve and whenever possible, mate all components to one or two fixed components or references. Long chains of components take longer to solve and are more prone to mate errors.

Helpful On-line Information

The SOLIDWORKS URL:
http://www.SOLIDWORKS.com
contains information on Local
Resellers, Solution Partners,
Certifications, SOLIDWORKS
users groups and more.

Use the SOLIDWORKS Resources
tab in the Task Pane to obtain
access to Customer Portals, User
Groups, Manufacturers, Solution
Partners, Labs, 3DEXPERIENCE
Marketplace, and more.

Helpful on-line SOLIDWORKS information is available from
the following URLs:

- http://www.swugn.org/

List of all SOLIDWORKS User groups.

- https://www.solidworks.com/sw/edu
  cation/certification-programs-cad-
  students.htm

The SOLIDWORKS Academic
Certification Programs.

- http://www.solidworks.com/sw/in
  dustries/education/engineering-
  education-software.htm

The SOLIDWORKS Education
Program:

- visit
  https://3dexperience.virtualtester.c
  om/#home to obtain additional
  SOLIDWORKS Certification
  exam information.

On-line tutorials are for educational
purposes only. Tutorials are
copyrighted by their respective
owners.

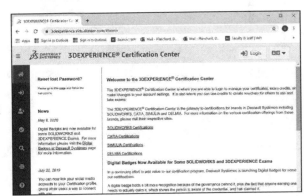

SOLIDWORKS Document Types

SOLIDWORKS has three main document file types: Part, Assembly and Drawing, but there are many additional supporting types that you may want to know. Below is a brief list of these supporting file types:

Design Documents	**Description**
.sldprt	SOLIDWORKS Part document
.slddrw	SOLIDWORKS Drawing document
.sldasm	SOLIDWORKS Assembly document

Templates and Formats	**Description**
.asmdot	Assembly Template
.asmprp	Assembly Template Custom Properties tab
.drwdot	Drawing Template
.drwprp	Drawing Template Custom Properties tab
.prtdot	Part Template
.prtprp	Part Template Custom Properties tab
.sldtbt	General Table Template
.slddrt	Drawing Sheet Template
.sldbombt	Bill of Materials Template (Table-based)
.sldholtbt	Hole Table Template
.sldrevbt	Revision Table Template
.sldwldbt	Weldment Cutlist Template
.xls	Bill of Materials Template (Excel-based)

Library Files	**Description**
.sldlfp	Library Part file
.sldblk	Blocks

Other	**Description**
.sldstd	Drafting standard
.sldmat	Material Database
.sldclr	Color Palette File
.xls	Sheet metal gauge table

CSWA Section: Answer key

Chapter 7

1. Identify the illustrated Drawing view.

- The correct answer is B: Alternative Position View.

2. Identify the illustrated Drawing view.

- The correct answer is B: Break View.

3. Identify the illustrated Drawing view.

- The correct answer is D: Aligned Section View.

4. Identify the view procedure. To create the following view, you need to insert a:

- The correct answer is B: Closed Profile: Spline.

5. Identify the view procedure. To create the following view, you need to insert a:

- The correct answer is B: Closed Spline.

6. Identify the illustrated view type.

- The correct answer is A: Crop View.

7. To create View B from Drawing View A insert which View Type?

- The correct answer is Aligned Section View.

8. To create View B it is necessary to sketch a closed spline on View A and insert which View type?

- The correct answer is Broken out Section View.

9. To create View B it is necessary to sketch a closed spline on View A and insert which View type?

- The correct answer is Horizontal Break View.

Chapter 8

1. Part 1. Calculate the overall mass of the part, volume, and locate the Center of mass with the provided illustrated information.

- Overall mass of the part = 1280.91 grams

- Volume of the part = 474411.54 cubic millimeters

- Center of Mass Location: X = 0.00 millimeters, Y = -29.17 millimeters, Z = 3.18 millimeters

2. Part 2. Calculate the overall mass of the part, volume, and locate the Center of mass with the provided information.

- Overall mass of the part = 248.04 grams

- Volume of the part = 91868.29 cubic millimeters

- Center of Mass Location: X = -51.88 millimeters, Y = 24.70 millimeters, Z = 29.47 millimeters

3. Part 3. Calculate the overall mass of the part, volume, and locate the Center of mass with the provided information.

- Overall mass of the part = 0.45 pounds, Volume of the part = 4.60 cubic inches and the Center of Mass Location: X = 0.17 inches, Y = 0.39 inches, Z = 0.00 inches

4. Part 4. Calculate the overall mass of the part, volume, and locate the Center of mass with the provided information.

- Overall mass of the part = 0.28 pounds, Volume of the part = 2.86 cubic inches and the Center of Mass Location: X = 0.70 inches, Y = 0.06 inches, Z = 0.00 inches

Mass properties of Homework 8-1
   Configuration: Default
   Coordinate system: -- default --

Density = 0.00 grams per cubic millimeter

Mass = 1280.91 grams

Volume = 474411.54 cubic millimeters

Surface area = 86851.60 square millimeters

Center of mass: (millimeters)
   X = 0.00
   Y = -29.17
   Z = 3.18

Mass properties of Homework 8-2
   Configuration: Default
   Coordinate system: -- default --

Density = 0.00 grams per cubic millimeter

Mass = 248.04 grams

Volume = 91868.29 cubic millimeters

Surface area = 29140.64 square millimeters

Center of mass: (millimeters)
   X = -51.88
   Y = 24.70
   Z = 29.47

Mass properties of Homework 8-3
   Configuration: Default
   Coordinate system: -- default --

Density = 0.10 pounds per cubic inch

Mass = 0.45 pounds

Volume = 4.60 cubic inches

Surface area = 32.66 square inches

Center of mass: (inches)
   X = 0.17
   Y = 0.39
   Z = 0.00

Mass properties of Homework 8-4
   Configuration: Default
   Coordinate system: -- default --

Density = 0.10 pounds per cubic inch

Mass = 0.28 pounds

Volume = 2.86 cubic inches

Surface area = 22.54 square inches

Center of mass: (inches)
   X = 0.70
   Y = 0.06
   Z = 0.00

5. Part 5. Calculate the overall mass and volume of the part with the provided information.

- Overall mass of the part = 888.48grams. Overall volume of the part = 111059.43 cubic millimeters.

```
Density = 0.01 grams per cubic millimeter

Mass = 888.48 grams

Volume = 111059.43 cubic millimeters

Surface area = 25814.97  square millimeters

Center of mass: ( millimeters )
   X = -15.39
   Y = 15.93
   Z = -2.65
```

6. Part 5A Modify. Calculate the Center of mass with a new Coordinate System.

- X= -80.39 millimeters, Y = 15.93 millimeters, Z = -22.65 millimeters.

```
Mass = 888.48 grams

Volume = 111059.43 cubic millimeters

Surface area = 25814.97  square millimeters

Center of mass: ( millimeters )
   X = -80.39
   Y = 15.93
   Z = -22.65
```

7. Part 5B Modify. Modify the model. Calculate the mass and volume of the model.

- Mass = 309.75 grams. Volume = 114721.22 cubic millimeters

```
Mass = 309.75 grams

Volume = 114721.22 cubic millimeters

Surface area = 27094.41  square millimeters
```

8. Part 6. Calculate the overall mass and volume of the part with the provided information.

- Overall mass of the part = 509.92 grams. Overall volume of the part = 188860.93 cubic millimeters.

```
Mass = 509.92 grams

Volume = 188860.93 cubic millimeters

Surface area = 32545.06  square millimeters

Center of mass: ( millimeters )
   X = 31.39
   Y = 16.55
   Z = 1.37
```

9. Part 6 A modify. Calculate the overall mass and volume of the part with the provided information.

- Overall mass of the part = 1465.70 grams. Overall volume of the part = 187910.60 cubic millimeters.

```
Configuration: modified
   Coordinate system: -- default --

Density = 0.01 grams per cubic millimeter

Mass = 1465.70 grams

Volume = 187910.60 cubic millimeters

Surface area = 32373.16  square millimeters

Center of mass: ( millimeters )
   X = 31.29
   Y = 16.45
   Z = 1.33
```

10. Part 6 B modify. Create a new coordinate system. Enter the Center of Mass location.

- X = 31.29 millimeters

- Y = 16.45 millimeters

- Z = -48.67 millimeters

```
Configuration: modified
   Coordinate system: Coordinate System1

Density = 0.01 grams per cubic millimeter

Mass = 1465.70 grams

Volume = 187910.60 cubic millimeters

Surface area = 32373.16  square millimeters

Center of mass: ( millimeters )
   X = 31.29
   Y = 16.45
   Z = -48.67
```

11. Part 7. Calculate the overall mass and volume of the part with the provided information.

- Mass = 4079.32 grams

- Volume = 522989.22 cubic millimeter

Mass properties of Homework 8-7
　Configuration: Default
　Coordinate system: -- default --

Density = 0.01 grams per cubic millimeter

Mass = 4079.32 grams

Volume = 522989.22 cubic millimeters

Surface area = 92824.57 square millimeters

Center of mass: (millimeters)
　X = 35.91
　Y = 0.00
　Z = 0.70

Chapter 9

1. Calculate the overall mass of the part, volume, and locate the Center of mass with the provided information.

- Overall mass of the part = 1.99 pounds

- Volume of the part = 6.47 cubic inches

- Center of Mass Location: X = 0.00 inches, Y = 0.00 inches, Z = 1.49 inches

Mass properties of Homework Problem 9-1
　Configuration: Default
　Coordinate system: -- default --

Density = 0.31 pounds per cubic inch

Mass = 1.99 pounds

Volume = 6.47 cubic inches

Surface area = 48.10 square inches

Center of mass: (inches)
　X = 0.00
　Y = 0.00
　Z = 1.49

2. Calculate the overall mass of the part, volume, and locate the Center of mass with the provided information.

- Overall mass of the part = 279.00 grams

- Volume of the part = 103333.73 cubic millimeters

- Center of Mass Location: X = 0.00 millimeters, Y = 0.00 millimeters, Z = 21.75 millimeters

Mass = 279.00 grams

Volume = 103333.73 cubic millimeters

Surface area = 29853.94 square millimeters

Center of mass: (millimeters)
　X = 0.00
　Y = 0.00
　Z = 21.75

3. Calculate the overall mass of the part, volume and locate the Center of mass with the provided information.

- Overall mass of the part = 2040.57 grams

- Volume of the part = 755765.04 cubic millimeters

- Center of Mass Location: X = -0.71 millimeters, Y = 16.66 millimeters, Z = -9.31 millimeters

Mass = 2040.57 grams

Volume = 755765.04 cubic millimeters

Surface area = 75833.18 square millimeters

Center of mass: (millimeters)
　X = -0.71
　Y = 16.66
　Z = -9.31

4. Calculate the overall mass of the part, volume and locate the Center of mass with the provided information. Create Coordinate System1 to locate the Center of mass for the model.

- Overall mass of the part = 2040.57 grams

- Volume of the part = 755765.04 cubic millimeters

- Center of Mass Location: X = 49.29 millimeters, Y = 16.66 millimeters, Z = -109.31 millimeters

```
Mass = 2040.57 grams

Volume = 755765.04 cubic millimeters

Surface area = 75833.18  square millimeters

Center of mass: ( millimeters )
    X = 49.29
    Y = 16.66
    Z = -109.31
```

5. Calculate the overall mass of the part, volume and locate the Center of mass with the provided information.

- Overall mass of the part = 37021.48 grams

- Volume of the part = 13711657.53 cubic millimeters

- Center of Mass Location: X = 0.00 millimeters, Y = 0.11 millimeters, Z = 0.00 millimeters

```
Mass = 37021.48 grams

Volume = 13711657.53 cubic millimeters

Surface area = 539553.61  square millimeters

Center of mass: ( millimeters )
    X = 0.00
    Y = 0.11
    Z = 0.00
```

6. Create a new Coordinate system with the provided information. Center of Mass location:

- X = 0.00 millimeters

- Y = 0.11 millimeters

- Z = 0.00 millimeters

```
Mass = 37021.48 grams

Volume = 13711657.53 cubic millimeters

Surface area = 539553.61  square millimeters

Center of mass: ( millimeters )
    X = 0.00
    Y = 0.11
    Z = 0.00
```

7. Build the provided model. Calculate the overall mass and volume of the part.

- Mass = 3442.00 grams

- Volume = 447012.75 cubic millimeters

```
Mass properties of Homework Part 9-7
    Configuration: Default
    Coordinate system: -- default --

Density = 0.01 grams per cubic millimeter

Mass = 3442.00 grams

Volume = 447012.75 cubic millimeters

Surface area = 67768.84  square millimeters

Center of mass: ( millimeters )
    X = 0.00
    Y = 50.00
    Z = 0.00
```

Notes:

GLOSSARY

Alphabet of Lines: Each line on a technical drawing has a definite meaning and is drawn in a certain way. The line conventions recommended by the American National Standards Institute (ANSI) are presented in this text.

Alternate Position View: A drawing view superimposed in phantom lines on the original view. Utilized to show range of motion of an assembly.

Anchor Point: The origin of the Bill of Material in a sheet format.

Annotation: An annotation is a text note or a symbol that adds specific information and design intent to a part, assembly, or drawing. Annotations in a drawing include specific note, hole callout, surface finish symbol, datum feature symbol, datum target, geometric tolerance symbol, weld symbol, balloon and stacked balloon, center mark, centerline marks, area hatch and block.

ANSI: American National Standards Institute.

Area Hatch: Apply a crosshatch pattern or solid fill to a model face, to a closed sketch profile, or to a region bounded by a combination of model edges and sketch entities. Area hatch can be applied only in drawings.

ASME: American Society of Mechanical Engineering, publisher of ASME Y14 Engineering Drawing and Documentation Practices that controls drawing, dimensioning and tolerancing.

Assembly: An assembly is a document in which parts, features and other assemblies (sub-assemblies) are put together. A part in an assembly is called a component. Adding a component to an assembly creates a link between the assembly and the component. When SOLIDWORKS opens the assembly, it finds the component file to show it in the assembly. Changes in the component are automatically reflected in the assembly. The filename extension for a SOLIDWORKS assembly file name is *.sldasm.

Attachment Point: An attachment point is the end of a leader that attaches to an edge, vertex, or face in a drawing sheet.

AutoDimension: The Autodimension tool provides the ability to insert reference dimensions into drawing views such as baseline, chain, and ordinate dimensions.

Auxiliary View: An Auxiliary View is similar to a Projected View, but it is unfolded normal to a reference edge in an existing view.

AWS: American Welding Society, publisher of AWS A2.4, Standard Location of Elements of a Welding Symbol.

Axonometric Projection: A type of parallel projection, more specifically a type of orthographic projection, used to create a pictorial drawing of an object, where the object is rotated along one or more of its axes relative to the plane of projection.

Balloon: A balloon labels the parts in the assembly and relates them to item numbers on the bill of materials (BOM) added in the drawing. The balloon item number corresponds to the order in the Feature Tree. The order controls the initial BOM Item Number.

Baseline Dimensions: Dimensions referenced from the same edge or vertex in a drawing view.

Bill of Materials: A table inserted into a drawing to keep a record of the parts and materials used in an assembly.

Block: A symbol in the drawing that combines geometry into a single entity.

BOM: Abbreviation for Bill of Materials.

Broken-out Section: A broken-out section exposes inner details of a drawing view by removing material from a closed profile. In an assembly, the Broken-out Section displays multiple components.

CAD: The use of computer technology for the design of objects, real or virtual. CAD often involves more than just shapes.

Cartesian Coordinate System: Specifies each point uniquely in a plane by a pair of numerical coordinates, which are the signed distances from the point to two fixed perpendicular directed lines, measured in the same unit of length. Each reference line is called a coordinate axis or just axis of the system, and the point where they meet is its origin.

Cell: Area to enter a value in an EXCEL spreadsheet, identified by a Row and Column.

Center Mark: A cross that marks the center of a circle or arc.

Centerline: An axis of symmetry in a sketch or drawing displayed in a phantom font.

CommandManager: The CommandManager is a Context-sensitive toolbar that dynamically updates based on the toolbar you want to access. By default, it has toolbars embedded in it based on the document type. When you click a tab below the Command Manager, it updates to display that toolbar. For example, if you click the Sketch tab, the Sketch toolbar is displayed.

Component: A part or sub-assembly within an assembly.

ConfigurationManager: The ConfigurationManager is located on the left side of the SOLIDWORKS window and provides the means to create, select and view multiple configurations of parts and assemblies in an active document. You can split the

ConfigurationManager and either display two ConfigurationManager instances, or combine the ConfigurationManager with the FeatureManager design tree, PropertyManager or third party applications that use the panel.

Configurations: Variations of a part or assembly that control dimensions, display and state of a model.

Coordinate System: SOLIDWORKS uses a coordinate system with origins. A part document contains an original origin. Whenever you select a plane or face and open a sketch, an origin is created in alignment with the plane or face. An origin can be used as an anchor for the sketch entities, and it helps orient perspective of the axes. A three-dimensional reference triad orients you to the X, Y, and Z directions in part and assembly documents.

Copy and Paste: Utilize copy/paste to copy views from one sheet to another sheet in a drawing or between different drawings.

Cosmetic Thread: An annotation that represents threads.

Crosshatch: A pattern (or fill) applied to drawing views such as section views and broken-out sections.

Cursor Feedback: The system feedback symbol indicates what you are selecting or what the system is expecting you to select. As you move the mouse pointer across your model, system feedback is provided.

Datum Feature: An annotation that represents the primary, secondary and other reference planes of a model utilized in manufacturing.

Depth: The horizontal (front to back) distance between two features in frontal planes. Depth is often identified in the shop as the thickness of a part or feature.

Design Table: An Excel spreadsheet that is used to create multiple configurations in a part or assembly document.

Detail View: A portion of a larger view, usually at a larger scale than the original view. Create a detail view in a drawing to display a portion of a view, usually at an enlarged scale. This detail may be of an orthographic view, a non-planar (isometric) view, a section view, a crop view, an exploded assembly view or another detail view.

Detailing: Detailing refers to the SOLIDWORKS module used to insert, add and modify dimensions and notes in an engineering drawing.

Dimension Line: A line that references dimension text to extension lines indicating the feature being measured.

Dimension Tolerance: Controls the dimension tolerance values and the display of non-integer dimensions. The tolerance types are *None, Basic, Bilateral, Limit, Symmetric, MIN, MAX, Fit, Fit with tolerance* or *Fit (tolerance only)*.

Dimension: A value indicating the size of the 2D sketch entity or 3D feature. Dimensions in a SOLIDWORKS drawing are associated with the model, and changes in the model are reflected in the drawing, if you DO NOT USE DimXpert.

Dimensioning Standard - Metric: - ASME standards for the use of metric dimensioning required all the dimensions to be expressed in millimeters (mm). The (mm) is not needed on each dimension, but it is used when a dimension is used in a notation. No trailing zeroes are used. The Metric or International System of Units (S.I.) unit system in drafting is also known as the Millimeter, Gram Second (MMGS) unit system.

Dimensioning Standard - U.S: - ASME standard for U.S. dimensioning uses the decimal inch value. When the decimal inch system is used, a zero is not used to the left of the decimal point for values less than one inch, and trailing zeroes are used. The U.S. unit system is also known as the Inch, Pound, Second (IPS) unit system.

DimXpert for Parts: A set of tools that applies dimensions and tolerances to parts according to the requirements of the ASME Y.14.41-2009 standard.

DimXpertManager: The DimXpertManager lists the tolerance features defined by DimXpert for a part. It also displays DimXpert tools that you use to insert dimensions and tolerances into a part. You can import these dimensions and tolerances into drawings. DimXpert is not associative.

Document: In SOLIDWORKS, each part, assembly, and drawing is referred to as a document, and each document is displayed in a separate window.

Drawing Sheet: A page in a drawing document.

Drawing Template: A document that is the foundation of a new drawing. The drawing template contains document properties and user-defined parameters such as sheet format. The extension for the drawing template filename is .DRWDOT.

Drawing: A 2D representation of a 3D part or assembly. The extension for a SOLIDWORKS drawing file name is .SLDDRW. Drawing refers to the SOLIDWORKS module used to insert, add, and modify views in an engineering drawing.

Edit Sheet Format: The drawing sheet contains two modes. Utilize the Edit Sheet Format command to add or modify notes and Title block information. Edit in the Edit Sheet Format mode.

Edit Sheet: The drawing sheet contains two modes. Utilize the Edit Sheet command to insert views and dimensions.

eDrawing: A compressed document that does not require the referenced part or assembly. eDrawings are animated to display multiple views in a drawing.

Empty View: An Empty View creates a blank view not tied to a part or assembly document.

Engineering Graphics: Translates ideas from design layouts, specifications, rough sketches, and calculations of engineers & architects into working drawings, maps, plans and illustrations which are used in making products.

Equation: Creates a mathematical relation between sketch dimensions, using dimension names as variables, or between feature parameters, such as the depth of an extruded feature or the instance count in a pattern.

Exploded view: A configuration in an assembly that displays its components separated from one another.

Export: The process to save a SOLIDWORKS document in another format for use in other CAD/CAM, rapid prototyping, web or graphics software applications.

Extension Line: The line extending from the profile line indicating the point from which a dimension is measured.

Extruded Cut Feature: Projects a sketch perpendicular to a Sketch plane to remove material from a part.

Face: A selectable area (planar or otherwise) of a model or surface with boundaries that help define the shape of the model or surface. For example, a rectangular solid has six faces.

Family Cell: A named empty cell in a Design Table that indicates the start of the evaluated parameters and configuration names. Locate Comments in a Design Table to the left or above the Family Cell.

Fasteners: Includes Bolts and nuts (threaded), Set screws (threaded), Washers, Keys, and Pins to name a few. Fasteners are not a permanent means of assembly such as welding or adhesives.

Feature: Features are geometry building blocks. Features add or remove material. Features are created from 2D or 3D sketched profiles or from edges and faces of existing geometry.

FeatureManager: The FeatureManager design tree located on the left side of the SOLIDWORKS window provides an outline view of the active part, assembly, or drawing. This makes it easy to see how the model or assembly was constructed or to examine the various sheets and views in a drawing. The FeatureManager and the Graphics window are dynamically linked. You can select features, sketches, drawing views and construction geometry in either pane.

First Angle Projection: In First Angle Projection the Top view is looking at the bottom of the part. First Angle Projection is used in Europe and most of the world. However, America and Australia use a method known as Third Angle Projection.

Fully defined: A sketch where all lines and curves in the sketch, and their positions, are described by dimensions or relations, or both, and cannot be moved. Fully defined sketch entities are shown in black.

Foreshortened radius: Helpful when the centerpoint of a radius is outside of the drawing or interferes with another drawing view: Broken Leader.

Foreshortening: The way things appear to get smaller in both height and depth as they recede into the distance.

French curve: A template made out of plastic, metal or wood composed of many different curves. It is used in manual drafting to draw smooth curves of varying radii.

Fully Defined: A sketch where all lines and curves in the sketch, and their positions, are described by dimensions or relations, or both, and cannot be moved. Fully defined sketch entities are displayed in black.

Geometric Tolerance: A set of standard symbols that specify the geometric characteristics and dimensional requirements of a feature.

Glass Box method: A traditional method of placing an object in an *imaginary glass box* to view the six principle views.

Global Coordinate System: Directional input refers by default to the Global coordinate system (X-, Y- and Z-), which is based on Plane1 with its origin located at the origin of the part or assembly.

Graphics Window: The area in the SOLIDWORKS window where the part, assembly, or drawing is displayed.

Grid: A system of fixed horizontal and vertical divisions.

Handle: An arrow, square or circle that you drag to adjust the size or position of an entity such as a view or dimension.

Heads-up View Toolbar: A transparent toolbar located at the top of the Graphic window.

Height: The vertical distance between two or more lines or surfaces (features) which are in horizontal planes.

Hidden Lines Removed (HLR): A view mode. All edges of the model that are not visible from the current view angle are removed from the display.

Hidden Lines Visible (HLV): A view mode. All edges of the model that are not visible from the current view angle are shown gray or dashed.

Hole Callouts: Hole callouts are available in drawings. If you modify a hole dimension in the model, the callout updates automatically in the drawing if you did not use DimXpert.

Hole Table: A table in a drawing document that displays the positions of selected holes from a specified origin datum. The tool labels each hole with a tag. The tag corresponds to a row in the table.

Import: The ability to open files from other software applications into a SOLIDWORKS document. The A-size sheet format was created as an AutoCAD file and imported into SOLIDWORKS.

Isometric Projection: A form of graphical projection, more specifically, a form of axonometric projection. It is a method of visually representing three-dimensional objects in two dimensions, in which the three coordinate axes appear equally foreshortened and the angles between any two of them are 120º.

Layers: Simplifies a drawing by combining dimensions, annotations, geometry and components. Properties such as display, line style and thickness are assigned to a named layer.

Leader: A solid line created from an annotation to the referenced feature.

Line Format: A series of tools that controls Line Thickness, Line Style, Color, Layer and other properties.

Local (Reference) Coordinate System: Coordinate system other than the Global coordinate system. You can specify restraints and loads in any desired direction.

Lock Sheet Focus: Adds sketch entities and annotations to the selected sheet. Double-click the sheet to activate Lock Sheet Focus. To unlock a sheet, right-click and select Unlock Sheet Focus or double click inside the sheet boundary.

Lock View Position: Secures the view at its current position in the sheet. Right-click in the drawing view to Lock View Position. To unlock a view position, right-click and select Unlock View Position.

Mass Properties: The physical properties of a model based upon geometry and material.

Menus: Menus provide access to the commands that the SOLIDWORKS software offers. Menus are Context-sensitive and can be customized through a dialog box.

Model Item: Provides the ability to insert dimensions, annotations, and reference geometry from a model document (part or assembly) into a drawing.

Model View: A specific view of a part or assembly. Standard named views are listed in the view orientation dialog box such as isometric or front. Named views can be user-defined names for a specific view.

Model: 3D solid geometry in a part or assembly document. If a part or assembly document contains multiple configurations, each configuration is a separate model.

Motion Studies: Graphical simulations of motion and visual properties with assembly models. Analogous to a configuration, they do not actually change the original assembly model or its properties. They display the model as it changes based on simulation elements you add.

Mouse Buttons: The left, middle, and right mouse buttons have distinct meanings in SOLIDWORKS. Use the middle mouse button to rotate and Zoom in/out on the part or assembly document.

Oblique Projection: A simple type of graphical projection used for producing pictorial, two-dimensional images of three-dimensional objects.

OLE (Object Linking and Embedding): A Windows file format. A company logo or EXCEL spreadsheet placed inside a SOLIDWORKS document are examples of OLE files.

Ordinate Dimensions: Chain of dimensions referenced from a zero ordinate in a drawing or sketch.

Origin: The model origin is displayed in blue and represents the (0,0,0) coordinate of the model. When a sketch is active, a sketch origin is displayed in red and represents the (0,0,0) coordinate of the sketch. Dimensions and relations can be added to the model origin but not to a sketch origin.

Orthographic Projection: A means of representing a three-dimensional object in two dimensions. It is a form of parallel projection, where the view direction is orthogonal to the projection plane, resulting in every plane of the scene appearing in affine transformation on the viewing surface.

Parametric Note: A Note annotation that links text to a feature dimension or property value.

Parent View: A Parent view is an existing view on which other views are dependent.

Part Dimension: Used in creating a part, they are sometimes called construction dimensions.

Part: A 3D object that consists of one or more features. A part inserted into an assembly is called a component. Insert part views, feature dimensions and annotations into 2D drawing. The extension for a SOLIDWORKS part filename is .SLDPRT.

Perspective Projection: The two most characteristic features of perspective are that objects are drawn: smaller as their distance from the observer increases and Foreshortened: the size of an object's dimensions along the line of sight are relatively shorter than dimensions across the line of sight.

Plane: To create a sketch, choose a plane. Planes are flat and infinite. Planes are represented on the screen with visible edges.

Precedence of Line Types: When obtaining orthographic views, it is common for one type of line to overlap another type. When this occurs, drawing conventions have established an order of precedence.

Precision: Controls the number of decimal places displayed in a dimension.

Projected View: Projected views are created for Orthogonal views using one of the following tools: Standard 3 View, Model View or the Projected View tool from the View Layout toolbar.

Properties: Variables shared between documents through linked notes.

PropertyManager: Most sketch, feature, and drawing tools in SOLIDWORKS open a PropertyManager located on the left side of the SOLIDWORKS window. The PropertyManager displays the properties of the entity or feature so you specify the properties without a dialog box covering the Graphics window.

RealView: Provides a simplified way to display models in a photo-realistic setting using a library of appearances and scenes. RealView requires graphics card support and is memory intensive.

Rebuild: A tool that updates (or regenerates) the document with any changes made since the last time the model was rebuilt. Rebuild is typically used after changing a model dimension.

Reference Dimension: Dimensions added to a drawing document are called Reference dimensions and are driven; you cannot edit the value of reference dimensions to modify the model. However, the values of reference dimensions change when the model dimensions change.

Relation: A relation is a geometric constraint between sketch entities or between a sketch entity and a plane, axis, edge or vertex.

Relative view: The Relative View defines an Orthographic view based on two orthogonal faces or places in the model.

Revision Table: The Revision Table lists the Engineering Change Orders (ECO), in a table form, issued over the life of the model and the drawing. The current Revision letter or number is placed in the Title block of the Drawing.

Right-Hand Rule: Is a common mnemonic for understanding notation conventions for vectors in 3 dimensions.

Rollback: Suppresses all items below the rollback bar.

Scale: A relative term meaning "size" in relationship to some system of measurement.

Section Line: A line or centerline sketched in a drawing view to create a section view.

Section Scope: Specifies the components to be left uncut when you create an assembly drawing section view.

Section View: You create a section view in a drawing by cutting the parent view with a cutting, or section line. The section view can be a straight cut section or an offset section defined by a stepped section line. The section line can also include concentric arcs. Create a Section View in a drawing by cutting the Parent view with a section line.

Sheet Format: A document that contains the following: page size and orientation, standard text, borders, logos, and Title block information. Customize the Sheet format to save time. The extension for the Sheet format filename is .SLDDRT.

Sheet Properties: Sheet Properties display properties of the selected sheet. Sheet Properties define the following: Name of the Sheet, Sheet Scale, Type of Projection (First angle or Third angle), Sheet Format, Sheet Size, View label, and Datum label.

Sheet: A page in a drawing document.

Silhouette Edge: A curve representing the extent of a cylindrical or curved face when viewed from the side.

Sketch: The name to describe a 2D profile is called a sketch. 2D sketches are created on flat faces and planes within the model. Typical geometry types are lines, arcs, corner rectangles, circles, polygons, and ellipses.

Spline: A sketched 2D or 3D curve defined by a set of control points.

Stacked Balloon: A group of balloons with only one leader. The balloons can be stacked vertically (up or down) or horizontally (left or right).

Standard views: The three orthographic projection views, Front, Top and Right positioned on the drawing according to First angle or Third angle projection.

Suppress: Removes an entity from the display and from any calculations in which it is involved. You can suppress features, assembly components, and so on. Suppressing an entity does not delete the entity; you can unsuppress the entity to restore it.

Surface Finish: An annotation that represents the texture of a part.

System Feedback: Feedback is provided by a symbol attached to the cursor arrow indicating your selection. As the cursor floats across the model, feedback is provided in the form of symbols riding next to the cursor.

System Options: System Options are stored in the registry of the computer. System Options are not part of the document. Changes to the System Options affect all current and future documents. There are hundreds of Systems Options.

Tangent Edge: The transition edge between rounded or filleted faces in hidden lines visible or hidden lines removed modes in drawings.

Task Pane: The Task Pane is displayed when you open the SOLIDWORKS software. It contains the following tabs: SOLIDWORKS Resources, Design Library, File Explorer, Search, View Palette, Document Recovery and RealView/PhotoWorks.

Templates: Templates are part, drawing and assembly documents that include user-defined parameters and are the basis for new documents.

Third Angle Projection: In Third angle projection the Top View is looking at the Top of the part. First Angle Projection is used in Europe and most of the world. America and Australia use the Third Angle Projection method.

Thread Class or Fit: Classes of fit are tolerance standards; they set a plus or minus figure that is applied to the pitch diameter of bolts or nuts. The classes of fit used with almost all bolts sized in inches are specified by the ANSI/ASME Unified Screw Thread standards (which differ from the previous American National standards).

Thread Lead: The distance advanced parallel to the axis when the screw is turned one revolution. For a single thread, lead is equal to the pitch; for a double thread, lead is twice the pitch.

Tolerance: The permissible range of variation in a dimension of an object. Tolerance may be specified as a factor or percentage of the nominal value, a maximum deviation from a nominal value, an explicit range of allowed values, be specified by a note or published standard with this information, or be implied by the numeric accuracy of the nominal value.

Toolbars: The toolbar menus provide shortcuts enabling you to access the most frequently used commands. Toolbars are Context-sensitive and can be customized through a dialog box.

T-Square: A technical drawing instrument, primarily a guide for drawing horizontal lines on a drafting table. It is used to guide the triangle that draws vertical lines. Its name comes from the general shape of the instrument where the horizontal member of the T slides on the side of the drafting table. Common lengths are 18", 24", 30", 36" and 42".

Under-defined: A sketch is under defined when there are not enough dimensions and relations to prevent entities from moving or changing size.

Units: Used in the measurement of physical quantities. Decimal inch dimensioning and Millimeter dimensioning are the two types of common units specified for engineering parts and drawings.

Vertex: A point at which two or more lines or edges intersect. Vertices can be selected for sketching, dimensioning, and many other operations.

View Palette: Use the View Palette, located in the Task Pane, to insert drawing views. It contains images of standard views, annotation views, section views, and flat patterns (sheet metal parts) of the selected model. You can drag views onto the drawing sheet to create a drawing view.

Weld Bead: An assembly feature that represents a weld between multiple parts.

Weld Finish: A weld symbol representing the parameters you specify.

Weld Symbol: An annotation in the part or drawing that represents the parameters of the weld.

Width: The horizontal distance between surfaces in profile planes. In the machine shop, the terms length and width are used interchangeably.

Zebra Stripes: Simulate the reflection of long strips of light on a very shiny surface. They allow you to see small changes in a surface that may be hard to see with a standard display.

Index

3 Point Arc Sketch tool, 5-17
3D Printing, 11-1
3D Sketch Placement Method, 3-83, 3-84
3D Sketch, 3-80, 3-81
3HOLE-SHAFTCOLLAR Assembly - Insert SHAFT-COLLAR, 6-10
3HOLE-SHAFTCOLLAR Assembly, 6-9
4 Bar Linkage Assembly, 2-59
5HOLE-SHAFTCOLLAR Assembly, 6-11

A

A (ANSI) Landscape, 4-7
A-ANSI-MM, 4-18, 4-19
Accelerator Keys, 1-28
Add Drawing Sheet, 4-59
Add-In Command, 1-22, 1-23, 1-24
Additive Manufacturing, 11-1
Advanced Mate, 2-33
  Angle Mate, 2-33, 3-75
  Distance Mate, 2-33
  Limit Mate, 2-33, 3-73
  Linear/Linear Coupler Mate, 2-33
  Path Mate, 2-33
  Symmetric Mate, 2-33, 3-76
  Width Mate, 2-33
Advanced Mode, 1-10
AirCylinder Assembly, 2-35
Aligned Dimension, 3-38
Alignment Mate Option, 2-32
Angle Mate, 2-32, 3-75
ANGLE-13HOLE Part, 3-15
  Boss-Extrude1 Feature, 3-19
  Fillet Feature, 3-23
  First Extruded Cut Feature, 3-20
  First Linear Pattern Feature, 3-22
  Second Extruded Cut Feature, 3-24
  Second Linear Pattern Feature, 3-25
  Third Extruded Cut Feature, 3-26
Annotation tab, 4-8
Annotation toolbar, 4-8
ANSI Drafting Standard, 2-7, 3-14, 3-47, 4-9
Appearance, 1-22, 1-29, 2-14
Appearances, Scenes and Decals, 1-32, 2-14
Apply Scene tool, 1-21
Arc Condition, 5-19
  Center, 5-19

  Max, 5-19
  Min, 5-19
Arrow Leader Direction, 5-7
Arrow Size, 4-10
ASME Y14.3M, 3-6, 3-8
Assembly CommandManager, 1-24
  Assembly tab, 1-24
  Evaluate tab, 1-24
  Layout tab, 1-24
  Sketch tab, 1-24
  SOLIDWORKS Add-in tab, 1-24
Assembly References, 2-36
Assembly tool - Insert Component, 2-37
Assembly tool - Mate, 2-39
Auto Balloons Drawing tool, 4-50
Axis PropertyManager, 5-37
Axis tool, 5-37
AXLE Part, 2-10
  Appearance, 2-14, 2-15
  Base Sketch, 2-10
  Edit Color, 2-14
  Extruded Base Feature, 2-12
  Save As, 2-13
  View Modes, 2-15
AXLE-3000 Part, 5-45

B

Base Feature, 2-12
Base Sketch, 2-10
Bill of Materials PropertyManager, 4-48
Bill of Materials tool, 4-48
Binder Clip Assembly, 3-72
Blind End Condition, 3-13, 3-36, 3-80
BOM, 4-48
Boss-Extrude PropertyManager, 2-12
Boss-Extrude1 Feature, 3-11, 3-13, 3-19

C

Cam Mate, 2-34, 3-88
Center Rectangle Sketch tool, 5-9, 5-10
Center of Mass Point, 4-66
Centerline Sketch tool, 3-26, 3-34
Centerpoint Arc Sketch tool, 5-16
Centerpoint Straight Slot Sketch tool, 2-26
Chamfer Feature, 3-54

Circle PropertyManager, 2-10
Circle Sketch tool, 2-10, 3-11, 3-37
Circular Pattern Feature, 3-44, 3-53, 5-37
Circular Pattern PropertyManager, 5-36
Coincident Mate, 2-32, 2-38
Coincident Relation, 3-14, 3-40
Coincident SmartMate, 3-64
Collinear Relation, 5-29
Collision Detection, 3-92
CommandManager, 1-22
CommandManager - Assembly, 1-24
  Assembly tab, 1-24
  Evaluate tab, 1-24
  Layout tab, 1-24
  Sketch tab, 1-24
  SOLIDWORKS Add-in tab, 1-24
CommandManager - Dock/Float, 1-25
CommandManager - Drawing, 1-23
  Annotation tab, 1-23
  Evaluate tab, 1-23
  Sketch tab, 1-23
  SOLIDWORKS Add-in tab, 1-23
  View Layout tab, 1-23
CommandManager - Float, 1-25
CommandManager - Part, 1-22
  DimXpert tab, 1-22
  Evaluate tab, 1-22
  Features tab, 1-22
  Sketch tab, 1-22
  SOLIDWORKS Add-in, 1-22
Company Logo, 4-15
Component Patterns in the Assembly, 6-44
Component Properties, 6-54
Concentric/Coincident SmartMate, 3-64
Concentric Mate, 2-32, 2-37
Concentric Relation, 3-28
ConfigurationManager, 4-37, 4-44
ConfigurationManager tab, 1-27, 1-28
Confirmation Corner, 1-19
Consolidated Slot toolbar, 3-27
Constant Radius Fillet Type, 3-23
Construction Line, 5-3, 5-33
Convert Entities Sketch tool, 3-51
Copy a Drawing View, 4-59
Corner Rectangle Sketch tool, 3-39
Cosmetic Thread, 3-14, 3-80
Counter Weight Assembly, 3-86
CSWA Exam, 7-1
  Advanced Part and Modification, 9-1
  Assembly and Modification, 10-1
  Basic Part and Intermediate Part, 8-1
  Introduction, 7-1
Custom Properties, 4-11, 4-37
Customize CommandManager, 1-23
Cut-Revolve PropertyManager, 5-30

D
Default Datum Planes, 3-5
Default Reference Planes, 1-12
  Front Plane, 1-12
  Right Plane, 1-12
  Top Plane, 1-12
Define Drawing Material Property, 4-36
Define Drawing Property, 4-36
Design Library, 1-30
  Toolbox, 1-30
Design Table, 4-55, 4-64
Detail Drawing, 4-28
Detail View Drawing tool, 4-54
Diameter Dimension, 3-48, 3-49
Dimension Cursor icon, 1-18
Dimension Feedback icon, 1-18
Dimension Font, 4-10
Dimension Precision, 4-32
DimXpertManager tab, 1-26, 1-28
Display Style, 1-20, 2-16
  Hidden Lines Removed, 1-20, 2-16
  Hidden Lines Visible, 1-20, 2-16
  Shaded With Edges, 1-20, 2-16
  Shaded, 1-20, 2-16
  Wireframe, 1-20, 2-16
DisplayManager tab, 1-26
Distance/Limit Mate, 2-33
Distance Mate, 2-33
Dock CommandManager, 1-25
Document Dialog box, 2-7
Document Properties, 2-9, 2-19, 3-10
Document Properties - ANSI, 2-9, 2-19
Document Properties - Decimal Place, 2-19
Document Properties - IPS, 2-9, 2-19
Document Properties - Units, 2-9, 2-19
Dome Feature, 5-20
Dome PropertyManager, 5-20
Double-click, 1-35
Draw1, 4-8
Drawing - Leader arrow direction, 5-7
Drawing CommandManager, 1-23
  Annotation tab, 1-23
  Evaluate tab, 1-23
  Sketch tab, 1-23

SOLIDWORKS Add-in tab, 1-23
  View Layout tab, 1-23
Drawing Custom Property, 4-12
Drawing Dimension Text, 4-31
Drawing Document Properties, 4-9
Drawing Radius Text, 4-35
Drawing Template, 4-5
Drawing View Component Properties, 4-26
Drawing View Edge Properties, 4-26
Drawing View Properties, 4-26
Drop-down Menu, 1-9
Dual Dimensions display, 2-9
Dynamic Mirror Sketch tool, 3-27, 3-34

E

Edge Cursor icon, 1-18
Edge Feedback icon, 1-18
Edit Appearance tool, 2-14
Edit Command, 3-13, 3-23, 5-40
Edit Design Table, 4-57
Edit Feature, 3-23, 5-40, 6-10
Edit Sheet Format Mode, 4-5
Edit Sheet Mode, 4-5
Edit Sketch, 3-13, 3-50
End Chain, 3-34
End Condition - Blind, 2-12, 3-36
End Condition - Mid Plane, 2-12, 3-52
End Condition - Through All, 2-23, 3-21
Equal Relation, 3-18, 3-28, 3-38, 3-40
Exit Sketch tool, 5-11
Explicit Equation Driven tool, 5-59
Exploded View tool, 4-44
Extruded Base Feature, 2-10, 2-20
Extruded Boss/Base Feature, 2-10, 2-20
Extruded Cut Feature, 2-22, 3-20
Extruded Cut PropertyManager, 2-23
Extruded Thin Feature, 3-16, 3-18

F

F Key, 2-11
Face Feedback icon, 1-18
FeatureManager, 1-26, 1-27
  Customize, 1-24
  Rollback bar, 1-13
  Split, 1-27
FeatureManager Design Tree, 1-26
  ConfigurationManager tab, 1-26
  DimXpertManager tab, 1-26
  DisplayManager tab, 1-26
  PropertyManager tab, 1-26

FeatureManager tab, 1-26
File Command, 1-8
File Explorer, 1-30
Fillet Feature, 3-23
Fillet PropertyManager, 3-23
  FilletXpert tab, 3-23
  Manual tab, 3-23
FilletXpert tab, 3-23
Filter icon, 5-12
Filter, 1-27
First Angle Projection, 3-7, 3-8
Fixed Component, 3-57
FLATBAR Drawing, 4-21
  Front, Top, Right View, 4-22
  Insert Annotations, 4-30
  Insert Dimension, 4-30
  Isometric View, 4-23
  Linked Note, 4-38
  Position View, 4-27
  Sheet2, 4-59
FLATBAR Part, 2-25
  Base Sketch, 2-26
  Design Table, 4-55
  Extruded Base Feature, 2-28
  Extruded Cut Feature, 2-29
  Linear Pattern Feature, 2-30
  Save, 2-31
FLATBAR Sub-Assembly, 6-9
FLATBAR-SHAFTCOLLAR Assembly, 4-61
  Insert First SHAFTCOLLAR, 4-62
  Insert FLATBAR, 4-65
  Insert Second SHAFTCOLLAR, 4-63
Flexible State, 6-61, 6-62
Float CommandManager, 1-25
Float Component, 3-57
Fly-out FeatureManager, 1-28
For Construction, 5-31, 5-34
Formatting Dialog box, 4-12
FRONT-SUPPORT Assembly, 3-56
  Insert ANGLE-13HOLE, 3-56
  Insert First HEX-STANDOFF, 3-58
  Insert First SCREW, 3-64
  Insert Fourth SCREW, 3-65
  Insert Second HEX-STANDOFF, 3-60
  Insert Second SCREW, 3-65
  Insert Third SCREW, 3-65
  Insert TRIANGLE, 3-61
Front View, 1-16

G

Gear Mate, 2-34, 3-77, 3-87
Gem Binder Clip Assembly, 3-72
Geometric Relation - Coincident, 3-14, 3-40
Geometric Relation - Collinear, 5-31
Geometric Relation - Concentric, 3-28
Geometric Relation - Equal, 3-18, 3-28, 3-38
Geometric Relation - Horizontal, 3-12, 3-28
Geometric Relation - Make Equal, 3-18
Geometric Relation - Make Horizontal, 3-12
Geometric Relation - Parallel, 3-59
Geometric Relation - Tangent, 3-29, 3-40
Geometric Relation - Vertical, 5-17
Geometry Pattern, 3-41
Geometry Pattern Option, 2-30
Graphics Window, 1-11

H

Heads-up View toolbar, 1-15, 1-19
  3D Drawing View, 1-21, 1-34
  Apply Scene, 1-21
  Display Style, 1-20
  Edit Appearance, 1-21
  Hide/Show Items, 1-21, 1-27
  Previous View, 1-19
  Rotate, 1-15, 1-21
  Section View, 1-20
  View Orientation, 1-20
  View Settings, 1-21
  Zoom to Area, 1-15, 1-19
  Zoom to Fit, 1-15, 1-19
Helix/Spiral Feature, 5-57
Help Command, 1-16
HEX-STANDOFF Part, 3-10
  Base Feature, 3-13
  Hole Wizard Feature, 3-14
Hidden Lines Removed, 1-20
Hide Components, 6-48
Hide Drawing Dimensions, 4-33
Hide Drawing View, 4-33
Hide Origin, 4-24
Hide Sketch, 3-44
Hide/Show Items tool, 1-21
Hinge Mate, 2-34
Hole Specification PropertyManager, 3-14
Hole Wizard Feature, 3-14, 3-83, 3-84
HOOK Part, 5-14
  Dome Feature, 5-20
  Swept Base Feature, 5-20
  Thread Feature, 5-21

Horizontal Relation, 3-12, 3-28, 5-18

I - J - K

Insert a Plane, 5-8
Insert Axis, 5-37
Insert Component Assembly tool, 2-37
Insert Component, 3-61
Instant3D Feature, 5-13
IPS - (inch, pound, second), 2-9, 3-17
Isometric View, 1-20

L

Leader Arrow Direction, 5-7
Lightweight State, 2-58
Limit Mate, 2-33, 3-73
Line Sketch tool, 3-18, 5-17
Linear Component Pattern, 6-46
Linear Pattern Feature, 2-30, 3-22, 3-25
Linear Pattern Feature, Instances, 3-22, 3-26
Linear Pattern Feature, Spacing, 3-22, 3-26
Linear Pattern PropertyManager, 2-30
Linear/Linear Coupler Mate, 2-33
Link Notes, 4-36
Link to Property tool, 4-39
LINKAGE Assembly, 2-31, 2-36, 2-37
  Basic Motion, 2-45
  Insert AirCylinder, 2-35
  Insert AXLE, 2-37
  Insert First FLATBAR, 2-39
  Insert First HEX-STANDOFF, 6-32
  Insert First SHAFT-COLLAR, 2-42
  Insert Second AXLE, 6-35
  Insert Second SHAFT-COLLAR, 2-43
  Insert Second FLATBAR, 2-40
  Insert Second HEX-STANDOFF, 6-34
LINKAGE Assembly Drawing, 4-41
  Animation, 4-47
  Automatic Balloons, 4-50
  Bill of Materials, 4-48
  Sheet1, 4-41
  Sheet2, 4-51
  Sheet2 - Detail View, 4-53
  Sheet2 - Section View, 4-53
Linked Note, 4-38
Lock Mate, 2-32
Lofted Boss/Base Feature, 5-12
Lofted Feature, 5-12
Logo, 4-15
LPattern Feature, 3-22, 3-24

M

Make Assembly from Part/Assembly, 4-25
Make Drawing from Part/Assembly, 4-25
Make Equal Relation, 3-18
Make Horizontal Relation, 3-12
Mate Error, 6-12
Mate tool, 3-58
Mate Types, 2-32
Max Arc Condition, 5-19
Measure tool, 6-21
Mechanical Mates, 2-34
   Cam Mate, 2-34, 3-88
   Gear Mate, 2-34, 3-77
   Hinge Mate, 2-34, 3-91
   Rack Pinion Mate, 2-34, 3-90
   Screw Mate, 2-34, 3-74
   Slot Mate, 2-34, 3-78
   Universal Joint Mate, 2-34
Menu Bar Menu, 1-8
   Edit, 1-8
   File, 1-8
   Help, 1-8
   Insert, 1-8
   Pin, 1-8
   Tools, 1-8
   View, 1-8
Menu Bar Toolbar, 1-8
   File Properties, 1-8
   New Document, 1-8
   Open Document, 1-8
   Options, 1-8
   Print Document, 1-8
   Rebuild Document, 1-8
   Save Document, 1-8
   Select Sketch Entities, 1-8
   Undo Last Action, 1-8
Mid Plane - End Condition, 2-12, 3-52
Mirror Components Assembly tool, 6-51
Mirror Components PropertyManager, 6-51
Mirror Feature, 3-41
MMGS - (millimeter, gram, second), 2-9
Model Items tool, 4-30
Model View PropertyManager, 4-7, 4-22
Model View tool, 4-52
Modify Feature, 4-34
Modify Part, 5-39
Modify Arc Condition, 5-19
Modify Dimensions, 3-20, 5-40
Modify Drawing Dimension Text, 4-32
Modify Drawing Radius Text, 4-35

Motion Study tab, 1-32
MotionManager, 1-32
   Animation tab, 1-32
   Basic Motion tab, 1-32
Mouse Wheel, 1-35
Mouse Pointer, 1-18
Move Drawing View, 4-25
Move with Triad, 6-17, 6-19
Multiple Drawing View, 4-22

N

New Assembly Document, 3-56, 3-57, 6-9
New Document, 2-7, 3-10, 3-17
New Drawing Document, 4-7, 4-19, 4-22
New Drawing, 4-7, 4-19, 4-22
New Part Document, 1-6, 2-7, 3-10, 3-17
New Part, 2-7, 3-10, 3-17, 3-33, 5-7
New Sketch, 3-11
Note tool, 4-32
Novice Mode, 2-7

O

Offset Entities Sketch tool, 3-35
Offset Sketch, 3-35
OnSurface Sketch Relation, 3-85
Open Document, 1-9
Options, 1-8, 1-9
Origin, 1-11
Orthographic Projection, 3-5, 3-6, 3-7

P - Q

Pack and Go tool, 2-36
Parallel Mate, 3-59, 3-61, 3-63
Parallel Relation, 3-28
Parallelogram Sketch tool, 3-42
Parametric Notes, 4-32
Part CommandManager, 1-22
   DimXpert tab, 1-26
   Evaluate tab, 1-22
   Features tab, 1-22
   Sketch tab, 1-22
   SOLIDWORKS Add-in tab, 1-22
Part Template, 5-7
PART-ANSI-IN, 5-7
Paste a Drawing View, 4-59
Pattern Direction Box, 3-25
Pattern Driven Component Pattern tool, 6-45
Perpendicular Mate, 2-32
Pin Option, 1-8
Plane Feature, 5-9

Plane PropertyManager, 5-9
Pneumatic Test Module Layout, 6-7
PNEUMATIC-TEST-MODULE, 6-7, 6-30
  Component Pattern Assembly tool, 6-44
  Insert AIR-RESERVOIR-SUPPORT, 6-42
  Insert FLAT-PLATE, 6-31
  Insert FRONT-SUPPORT Assembly, 6-48
  Insert HEX-STANDOFF Part, 6-32
  Insert LINKAGE Assembly, 6-40
  Insert WHEEL-AND-AXLE Assembly, 6-54
  Linear Component Pattern tool, 6-46
  Mirror Components tool, 6-51, 6-52
  MirrorFRONT-SUPPORT, 6-53
Point Sketch tool, 3-14
Polygon Sketch tool, 3-11, 5-26
Primary Datum Plane, 3-5
Primary Units, 2-9
Print Document, 1-8
Properties of a Drawing Sheet, 4-25
PropertyManager tab, 1-26

R

Rack Pinion Mate, 2-34, 3-90
Rebuild Command, 2-23, 4-50
Recover from a Mate Error, 6-11, 6-12
Reference Axis, 5-37
Reference Planes, 3-5
Rename Feature, 2-47
Rename Sketch, 2-47
Reorder Features, 5-44
Replace Components tool, 6-11
Replace PropertyManager, 6-11
Revolved Base Feature tool, 3-49
Revolved Cut Feature, 5-30
Rigid State, 6-56
Right-click, 1-18, 1-35
Right-click End Chain, 3-34
Right-click Select, 3-14, 3-18
Right View, 1-16
ROBOT Assembly, 6-62
  Insert basic_integration, 6-62, 6-65
  Insert HEX-STANDOFF, 6-32, 6-64
  Insert PNEUMATIC-TEST-MODULE, 6-63
  Insert Robot-Platform, 6-63
Rollback Bar, 1-13
Rotate Model, 1-15

S

Save, 1-8, 2-16
Save an Assembly, 2-44

Save As, 2-19, 3-11, 5-39
Save as Copy and Continue, 5-40, 6-11
Save As Drawing Template, 4-18
Save Command, 1-8, 2-16, 5-39
Save Document, 1-8, 2-16
Save Drawing Template, 4-17, 4-18
Save Part Template, 5-7
Save Sheet Format, 4-17, 4-18
Screw Mate, 2-34, 3-74
SCREW Part, 3-45
  Chamfer Feature, 3-54
  Circular Pattern Feature, 3-53
  Extruded Cut Feature, 3-51
  Fillet Feature, 3-53
  Revolved Base Feature, 3-47
Secondary Datum Plane, 3-5
Section View Drawing tool, 4-53
Section View PropertyManager, 4-53
Section View, 5-44
Selection Filter toolbar, 5-12
SHAFT-COLLAR Part, 2-18
  Appearance, 2-25
  Base Sketch, 2-21
  Boss-Extrude Feature, 2-21
  Edit Color, 2-23
  Extruded Base Feature, 2-21
  Extruded Cut Feature, 2-23
  Modify Dimension, 2-23
  Save As, 2-19
SHAFTCOLLAR-500 Part, 5-46
Sheet Format, 4-5
Sheet Format/Size Dialog box, 4-7
Sheet Properties, 4-9, 4-24, 4-26
Sheet Scale, 4-9
Sheet1, 4-8
Show, 3-37, 5-8, 5-31
Sketch Fillet Sketch tool, 3-36
Sketch Plane, 1-12, 2-10, 3-5, 3-11
Sketch Point tool, 3-84
Sketch Status, 2-14
  Fully Defined, 2-14
  Over Defined, 2-14
  Under Defined, 2-15
Sketch tools, 2-10
  3 Point Arc, 5-18
  Center Rectangle, 5-9, 5-10
  Centerline, 3-26, 3-34, 3-37, 3-47, 5-32
  Centerpoint Arc, 5-16
  Centerpoint Straight Slot, 2-26
  Circle, 2-22, 3-20, 3-37

Convert Entities, 3-51
Corner Rectangle, 3-39, 5-12
Line, 3-18, 3-51
Offset Entities, 3-35
Parallelogram, 3-42
Point, 3-14
Polygon, 3-11, 5-26
Sketch Fillet, 3-36
Smart Dimension, 2-11, 3-35
Straight Slot, 3-27, 3-42
Tangent Arc, 3-40, 3-50
Trim Entities, 3-39, 3-50
Slot Mate, 2-34, 3-78
Slot PropertyManager, 3-27
Smart Dimension Sketch tool, 2-11
SmartMate, 3-63, 6-10
SolidWorks Graphics Window, 1-11
  CommandManager, 1-11
  FeatureManager, 1-11
  Heads-up View Toolbar, 1-11
  Origin, 1-11
  Task Pane, 1-11
SolidWorks Help, 1-16
SolidWorks Resources, 1-29
SolidWorks Tutorials, 1-17
SolidWorks User Interface, 1-7, 1-11, 1-15
  CommandManager, 1-22
  Confirmation Corner, 1-19
  Drop-down menu, 1-9
  FeatureManager Design Tree, 1-26
  Fly-out FeatureManager, 1-28
  Heads-up View toolbar, 1,15, 1-19
  Menu Bar Menu, 1-8
  Menu Bar Toolbar, 1-8
  Motion Study tab, 2-45
  Right-click Pop-up menu, 1-18
  System feedback icons, 1-18
  Tags, 1-27
  Task Pane, 1-29
Split ConfigurationManager, 1-27
Split FeatureManager, 1-27
Split Line Feature, 5-62
Standard Mates, 2-32
  Alignment, 2-32
  Angle Mate, 2-32
  Coincident Mate, 2-32
  Concentric Mate, 2-32
  Distance Mate, 2-32
  Lock Mate, 2-32
  Parallel Mate, 2-34, 6-28, 6-33

  Perpendicular Mate, 2-32
  Tangent Mate, 2-32
Standard Sheet Size, 4-7
Standard Views Toolbar, 1-19
Start a SolidWorks Session, 1-4
Start Screen, 1-4
Straight Slot Sketch tool, 2-26, 3-27
Suppress Feature, 3-54
Suppress, 3-54
Swept Base Feature, 5-14, 5-15, 5-20
Swept Boss/Base Feature, 5-20
Swept Path, 5-14
Swept Profile, 5-14
Symbol Dialog Box, 4-14
Symmetric Mate, 2-33, 3-76
Symmetry, 3-22, 8-10
System Feedback icon, 1-18
  Dimension, 1-18
  Edge, 1-18
  Face, 1-18
  Vertex, 1-18
System Options - File Locations, 4-18

T
Tangent Arc Sketch tool, 3-32, 3-40
Tangent Edges Removed, 4-43
Tangent Edges Visible, 4-43
Tangent Edges with Font, 4-43
Tangent Mate, 2-32
Tangent Relation, 3-29, 3-40
Task Pane, 1-29
  3DEXPERIENCE MARKETPLACE, 1-29
  Appearances, Scenes and Decals, 1-32
  Custom Properties, 1-32
  Design Library, 1-31
  File Explorer, 1-31
  SolidWorks Resources, 1-30
  View Palette, 1-31
Templates - Default, 2-7
Templates tab, 2-7
Temporary Axis, 3-43, 3-53, 5-36
Tertiary Datum Plane, 3-5
Text Format Box, 4-14, 4-39
Third Angle Projection, 3-7, 3-8, 4-6
Thread Feature, 5-21
Thread Profile, 5-23
Through All End Condition, 2-23, 3-25
Tile Horizontally, 3-64
Title Block, 4-10
Tolerance Block, 4-13
Tolerance/Precision Box, 4-13

Toolbars, 1-8
Top View, 1-16
Trailing Zeroes, 4-31
TRIANGLE Part, 3-31
  Boss-Extrude Feature, 3-36
  Circular Pattern Feature, 3-44
  First Extruded Cut Feature, 3-37
  Mirror Feature, 3-41
  Second Extruded Cut Feature, 3-39
  Third Extruded Cut Feature, 3-42
Trim Entities Sketch tool, 3-39
Trim Lines, 3-39
Trim to Closest, 3-39, 3-50
Trimetric View, 1-16
Tutorial tab, 2-7
Type of Projection, 4-9

U
Under Defined Sketch, 2-15
Undo Document, 2-23
Undo Last Action, 2-23
Undo tool, 2-23
Units, 2-9, 2-11, 2-19, 3-10
Units - Dual Dimensions Display, 2-9
Units - Primary, 2-9
Units - Secondary, 2-9
Universal Joint Mate, 2-34
User Specified Name, 4-37, 4-49

V
Vertex Cursor icon, 1-18
Vertex Feedback icon, 1-18
Vertical Relation, 5-17, 5-29
View Axis, 6-24
View Layout tab, 4-8
View Layout toolbar, 4-8
View Modes, 1-16, 1-20
  Front, 1-16, 1-20
  Isometric, 1-20
  Right, 1-16, 1-20
  Top, 1-16, 1-20

View Orientation tool, 1-21
View Origin, 2-8, 2-15
View Palette tab, 1-31, 4-23
View Reference Axis, 5-37
View Sketch Relations, 2-27
View Temporary Axes, 3-44, 3-53
View Toolbars, 1-4, 1-15, 1-16

W
Wake up, 2-28
WEIGHT Part, 5-6
  Extruded Cut Feature, 5-13
  Instant3D Feature, 5-13
  Lofted Feature, 5-12
WEIGHT-HOOK Assembly, 3-70
WEIGHT-LINK Assembly, 3-71
What's Wrong Dialog box, 6-12
WHEEL Part, 5-24
  Boss-Extrude Feature, 5-27
  Circular Pattern Feature, 5-36
  First Extruded Cut Feature, 5-31
  Revolved Cut Feature, 5-30
  Second Extruded Cut Feature, 5-33
WHEEL-AND-AXLE Assembly, 6-23
  Insert AXLE-3000 Part, 6-23
  Insert HEX-ADAPTER Part, 6-26
  Insert SHAFTCOLLAR-500 Part, 6-28
WHEEL-FLATBAR Assembly, 6-14
  Insert 3HOLE-SHAFTCOLLAR, 6-17
  Insert 5HOLE-SHAFTCOLLAR, 6-19
  Insert WHEEL, 6-15
Width Mate, 2-33
Windows Command, 1-8
Window-Select Command, 1-15
Wireframe, 1-20, 3-27

X - Y - Z
Zoom In, 1-15
Zoom Out, 1-15
Zoom to Area, 1-15, 1-19
Zoom to Fit, 1-15, 1-19